Direct Methods in the Calculus of Variations

Direct Methods in the Calculus of Variations

Enrico Giusti
Università di Firenze, Italy

World Scientific
New Jersey • London • Singapore • Hong Kong

Published by

World Scientific Publishing Co. Pte. Ltd.
5 Toh Tuck Link, Singapore 596224
USA office: 27 Warren Street, Suite 401-402, Hackensack, NJ 07601
UK office: 57 Shelton Street, Covent Garden, London WC2H 9HE

British Library Cataloguing-in-Publication Data
A catalogue record for this book is available from the British Library.

DIRECT METHODS IN THE CALCULUS OF VARIATIONS

ISBN-13 978-981-238-043-2
ISBN-10 981-238-043-4

Contents

Introduction

Pro minimis adhiberi possunt quasi minima.

Leibniz

The fundamental problem of the calculus of variations consists of the research of a function $u(x)$ minimizing the integral functional

$$\mathcal{F}(u, \Omega) = \int_{\Omega} F(x, u(x), Du(x))dx \tag{0.1}$$

among all the functions u satisfying suitable conditions, the most usual of which consists of taking prescribed values $U(x)$ on the boundary of Ω:

$$u = U \quad \text{on } \partial\Omega\,. \tag{0.2}$$

This problem, that bears the name of DIRICHLET, is beyond doubt the most studied, and probably the most important; in conforming to that tradition, we too shall limit our discussion to it.

In (0.1) Ω is an open set in $\mathbf{R}^n$, whose generic point we shall denote by $x = (x_1, x_2, \ldots, x_n)$, and the boundary datum $U(x)$, and by consequence the unknown function $u(x)$, are functions with values in $\mathbf{R}^N$, with components respectively U^α and u^α, $\alpha = 1, 2, \ldots, N$.

The first examples of problems in the calculus of variations, of course in the simplest case $n = N = 1$, date from the beginning of the infinitesimal calculus, and are all founded on the EULER *equation* of the functional $\mathcal{F}$. To derive it, we assume that $F(x, u, z)$ is of class C^1, and that $u(x)$ is a minimum of the corresponding functional $\mathcal{F}$. Let φ be a function equal to zero on $\partial\Omega$, so that $u + t\varphi$ takes the same values as u on the boundary for every real number t, and set

$$g(t) = \mathcal{F}(u + t\varphi, \Omega).$$

The function $g(t)$ has a minimum for $t = 0$, and hence we must have $g'(0) = 0$. We then compute g' by writing (0.1) for $u+t\varphi$ and differentiating under the integral sign. We get[1]

$$\int_\Omega \left(\frac{\partial F}{\partial z_i^\alpha}(x, u(x), Du(x))\; D_i\varphi^\alpha + \frac{\partial F}{\partial u^\alpha}(x, u(x), Du(x))\; \varphi^\alpha \right) dx = 0.$$

If F is of class C^2, integrating by parts and remembering that $\varphi = 0$ on $\partial\Omega$:

$$\int_\Omega \left(\frac{\partial}{\partial x_i}\frac{\partial F}{\partial z_i^\alpha}(x, u(x), Du(x)) - \frac{\partial F}{\partial u^\alpha}(x, u(x), Du(x)) \right) \varphi^\alpha \, dx = 0. \qquad (0.3)$$

The preceding equation must be satisfied for every φ equal to zero on $\partial\Omega$, so that we must have

$$\frac{\partial}{\partial x_i}\left(\frac{\partial F}{\partial z_i^\alpha}(x, u(x), Du(x)) \right) - \frac{\partial F}{\partial u^\alpha}(x, u(x), Du(x)) = 0 \qquad (0.4)$$

for every $\alpha = 1, 2, \ldots, N$.

A necessary condition for $u(x)$ to minimize the integral (0.1) is therefore that u be a solution of the partial differential Eq. (0.4),[2] which carries the name EULER, or sometimes EULER–LAGRANGE equation.

If, as is often the case, the functional $\mathcal{F}$ is convex, that is if it satisfies the relation

$$\mathcal{F}(tu + (1-t)v) \leq t\mathcal{F}(u) + (1-t)\mathcal{F}(v)$$

for every couple of functions u, v and for every $t \in [0, 1]$, then $g(t)$ is convex itself, and since $g'(0) = 0$, we conclude that g has a minimum in 0 for every function φ, and therefore that u minimizes $\mathcal{F}$.

In conclusion, in the case of convex functionals, the Eq. (0.4) is a necessary and sufficient condition for $u(x)$ to minimize the functional $\mathcal{F}$.

---

[1] We shall always sum over repeated indices; the latin indices running from 1 to n, and the greeks ones from 1 to N.

[2] Actually, when $N > 1$, it is a system of partial differential equations, that will reduce to ordinary differential equations when $n = 1$. Needless-to-say, we shall be interested only to the case $n > 1$; even if in principle the result of this book would hold for ordinary differential equations, the problems for those are of a different kind. We shall use the term "equation" to denote both the equations ($N = 1$) and the systems of differential equations. The context will make clear what we are talking about.

The EULER equation is particularly useful when one wants to find an explicit solution (possibly in the form of a series) of the minimum problem; in particular when we arrive at an ordinary differential equation that sometimes can be integrated explicitely, or else reduce to quadratures. The first book completely dedicated to the calculus of variations, due to L. EULER [1], contains a large number of problems solved in this way.

On the other hand, when we pass to higher dimensional integrals, that lead to partial differential equations, the method above shows evident limitations, due to the difficulty of resolving explicitly such equations. Already in the simplest case of the DIRICHLET integral

$$\int_{\Omega} |Du|^2 \, dx \,, \tag{0.5}$$

which leads to the LAPLACE equation

$$\Delta u =: \sum_{i=1}^{n} \frac{\partial^2 u}{\partial x_i^2} = 0 \,,$$

the way to an explicit solution is restricted to very few cases, and can be carried out successfully only when Ω is a domain with spherical ($n = 2$ or 3) or cylindrical symmetry ($n = 3$).

It was exactly the need for harmonic functions (that is of solutions of the equation $\Delta u = 0$) taking prescribed values at the boundary of arbitrary domains, that induced RIEMANN [1] to reverse the usula point of view, and to reduce this problem to that of minimizing the integral (0.5) among all the functions taking on $\partial\Omega$ the given values. For that purpose, he introduced the DIRICHLET *principle*, which consists essentially of considering the functional (0.5) as a map from the manifold V of the functions taking on $\partial\Omega$ the given values into $\mathbf{R}$, and of the assumption that it is possible to apply to that mapping a generalization of the WEIERSTRASS theorem, assuring the existence of the minimum (and of the maximum) of any continuous function. Once the existence of the minimum has been assumed, it will automatically be a harmonic function.

In this way RIEMANN gave birth to the so-called *direct methods* in the calculus of variations, which consist of proving the existence of the minimum of an integral functional $\mathcal{F}$ (and more generally in discovering its properties, in the first place its regularity) without recourse to the EULER equation, but deducing it directly from the properties of the functional $\mathcal{F}$, considered as a map from V into $\mathbf{R}$.

After some unsuccessful attempts by ARZELÀ [1], the proof of the DIRICHLET principle for the functional (0.5) was given by HILBERT [1, 2].

On the other hand, a complete treatment of the minimum problem for the general functional (0.1) could not be carried our without two conditions. Firstly place, one had to recognize that the semicontinuity, and not the continuity, was the main assumption for a successful application of the WEIERSTRASS theorem to the functional $\mathcal{F}$; and secondly, it was necessary to introduce new function spaces (beyond those already studied for continuous functions and the like) and to prove for them compactness results analogous to those of ASCOLI and ARZELÀ.

The first step was carried out by TONELLI, who introduced the lower semicontinuity, and obtained a series of existence results, mostly in the case of one independent variable, for functionals of the type

$$\Phi(u) =: \int_a^b F(t, u(t), u'(t))\, dt \tag{0.6}$$

in the framework of *absolutely continuous* functions, and of the functions *with bounded variation.*

The introduction of the semicontinuity is the key to dealing with the general functionals (0.1). Actually, in order to apply the WEIERSTRASS theorem, it is necessary that the functional $\mathcal{F}$ be lower semicontinuous, and that the set V in which one looks for the minimum be compact. These two properties are in some sense in competition; in order to have the semicontinuity it is preferable to endow V with a relatively strong topology: the fewer convergent sequences exist, the easier the functional is semicontinuous. On the contrary, for the compactness it is better to have the opposite: the weaker the topology, the easier for a sequence to converge.[3]

The absence of suitable function spaces prevented TONELLI from going beyond the functionals (0.6). The extension to the case of many independent variables will take place thanks to the progress of the functional analysis, and to the introduction by SOBOLEV [1] (and independently by CALKIN [1] and MORREY [1]) of the spaces carrying his name. The setting of the minimum problem in these spaces makes it possible to prove general existence theorems, covering a large class of functionals.[4] In particular we have semicontinuity theorems in the weak topology of $W^{1,p}$ (the space of the functions whose derivatives are p-summable) under the

[3] The same can be said reasoning in terms of coverings. On the other hand the two notions of compactness coincide for metrizable spaces.

[4] We recall however that some important problems, such as that of minimal surfaces, fall outside this general setting, and require the introduction of more general spaces.

assumption of convexity of $F(x, u, z)$ with respect to z in the scalar case, and of quasi-convexity in the vector case. The last condition, introduced by MORREY [2], is essentially equivalent to the assumption that the linear functions $u = a + \langle \lambda, x \rangle$ minimize the "frozen" functional

$$\mathcal{F}^0(v, Q) = \int_Q F(x_0, u_0, Dv(x))\, dx\,. \tag{0.7}$$

On the other hand, the solution of the existence problem in the SOBOLEV spaces opens up another series of questions. Actually, the functions of these spaces have derivatives only in a weak sense, and in general are not even continuous. Once a minimum has been found, the problem arises of proving that it is continuous, or differentiable, and so on; briefly the problem of the regularity of the solutions. This problem remained unsolved for long time; its solution, limited to the *scalar* case ($N = 1$), begun with the fundamental paper by DE GIORGI [1].[5]

In it, DE GIORGI proved the Hölder continuity of the solutions to differential equations in divergence form:

$$D_i(a_{ij}(x) D_j v) = 0 \tag{0.8}$$

where the coefficients a_{ij} are assumed only to be measurable and bounded (therefore possibly discontinuous), and to satisfy the ellipticity condition

$$a_{ij}(x)\xi_i\xi_j \geq \nu|\xi|^2\,, \quad \nu > 0\,. \tag{0.9}$$

Of course, Eq. (0.8) has to be interpreted as holding in a weak sense; more precisely we shall assume that the integral equation

$$\int_\Omega a_{ij} D_i v D_j \varphi\, dx = 0 \tag{0.10}$$

is satisfied for every function φ of class $C_0^\infty(\Omega)$.

The application to minima of functionals is immediate. Actually, any function $u(x)$ minimizing the functional

$$\mathcal{F}(u, \Omega) = \int_\Omega F(Du)\, dx\,,$$

satisfies the EULER equation in weak form:

$$\int_\Omega F_{z_i}(Du(x)) D_i \zeta(x)\, dx = 0$$

---

[5] A similar result was obtained independently about at the same time by NASH [1].

for every test function ζ.[6] Taking $\zeta = D_s\varphi$ and integrating by parts, we find then

$$\int_\Omega F_{z_i z_j}(Du(x))D_j(D_s u)D_i\varphi\, dx = 0 \tag{0.11}$$

and hence the derivatives $D_s u$ are weak solutions of an equation of type (0.10), with

$$a_{ij}(x) = F_{z_i z_j}(Du(x)).$$

Assuming now that the second derivatives of F are bounded, and that F is strictly convex, that is that Eq. (0.11) is elliptic, we can apply DE GIORGI's theorem, and thus conclude that derivatives Du are Hölder-continuous functions.

DE GIORGI's theorem was widely studied and extended in variuos directions, so as to apply it to the EULER equation (0.3) of the general functional (0.1). We shall quote here only the volume by LADYŽENSKAYA and URAL'CEVA [1], where one can find the relevant results, obtained mostly by the two authors. A new and particularly elegant proof was given by MOSER [1, 2], who extended to general elliptic equations the classical HARNACK inequality.

The regularity theorem does not extend to the vector case ($N > 1$). This was shown by DE GIORGI [5] himself, who constructed a linear elliptic system with bounded measurable coefficients having discontinuous solutions. Shortly after this example was extended by GIUSTI and MIRANDA [1] to nonlinear systems with regular coefficients, and to minima of functionals.[7] In the meantime, adapting some ideas introduced by DE GIORGI [3] and developed by ALMGREN [1] in the study of minimal surfaces, MORREY [4] was able to prove the *partial regularity* of the weak solutions of nonlinear elliptic systems, or more precisely their regularity in an open set $\Omega_0 \subset \Omega$, and to show that the measure of the singular set $\Omega - \Omega_0$ was zero.[8] MORREY's result has been extended by GIUSTI and MIRANDA [2], and later by GIUSTI [2], GIAQUINTA and GIUSTI [1], and others.

[6] We have denoted by F_{z_i} the derivative of F with respect to z_i.

[7] Other examples were found independently by MAZ'YA [1]; more general examples were later found by NEČAS [2].

[8] Unlike DE GIORGI's theorem, the proof does not make use of a similar theorem for a linear system with discontinuous coefficients, but it is based in an essential way on the nonlinearity and on the regularity of the coefficients. It should be noted that the result does not hold for a linear system with discontinuous coefficients, as a has been shown by SOUČEK [1], who constructed a linear system whose solution is discontinuous in a dense set.

All these regularity theorems apply to the minima of functionals by the intermediate of the EULER equation, and therefore make only a marginal use of the most characteristic properties of the minima, those properties that distinguish true minima from simple extremals, let alone from solutions of general elliptic equations and systems.

Now, whereas there is a total coincidence between minima and extremals when the functional $\mathcal{F}$ is convex, this identity is no longer true in the general case, where one can have extremals that are not even local minima of the functional in question. As a consequence, one can expect that the assumptions necessary to prove the regularity of extremals are substantially heavier than those conducting to the regularity of the minima; first of all the differentiability of F with respect to u, and even more so the growth of the derivative F_u, indispensable already in the deduction of the EULER equation.

A first step towards the use of direct methods in the regularity problem (that is the proof of the regularity directly from the minimum property, without passing through the EULER equation), was taken by GIAQUINTA and GIUSTI [2], who have shown, in the case of scalar functionals, that the minima of the functional (0.1), independently of any assumption of differentiability of F, satisfy the assumptions of DE GIORGI's regularity theorem, and therefore are Hölder-continuous functions. The same result holds in general for *quasi-minima*, that is for functions u for which

$$\mathcal{F}(u, K) \leq Q\mathcal{F}(u + \varphi, K) \tag{0.12}$$

for every φ with compact support $K \subset \Omega$.

The notion of quasi-minimum includes of course that of minimum, to which it reduces when $Q = 1$. Actually, it is substantially more general, since it includes solutions of linear and nonlinear elliptic equations and systems (and in the vector case, quasi-regular mappings). We have thus, under the general notion of quasi-minimum, a unified treatment of the regularity of the minima of functionals in the calculus of variations, and of the solutions of elliptic equations and systems in divergence form.

Of course in the vector case it is not a question of getting global regularity (i.e., Hölder continuity); one can only prove that generally speaking a quasi-minimum has derivatives summables with an exponent larger than that of the SOBOLEV space to which it belongs *a priori*. This result, that was proved originally by BOJARSKI [1] and by MEYERS [1] in the case of solutions of linear elliptic equations, has become an important tool in the study of the partial regularity in the vector case.

The further regularity results need more stringent assumptions, and the notion of quasi-minimum, that has accomplished its duty in the proof of the higher summability of the derivatives in the general case, and of the Hölder continuity in the scalar case, must be abandoned in favor of more particular assumptions. In the case of the regularity of the first derivatives, it can be replaced by the notion of ω-minimum, introduced by the ANZELLOTTI [1], and analogous to ALMGREN [1] approach to minimal surfaces. Given a continuous increasing function $\omega(R)$, defined for $R \geq 0$, and such that $\omega(0) = 0$, we call the ω-minimum of $\mathcal{F}$ a function $u(x)$ such that for every cube $Q_R = Q(x_0, R)$, of side $2R$ and center x_0, and for every function φ with support contained in Q_R we have

$$\mathcal{F}(u, Q_R) \leq (1 + \omega(R))\mathcal{F}(u + \varphi, Q_R). \tag{0.13}$$

For scalar ω-minima with $\omega(R) = cR^\alpha$, one proves the Hölder continuity of the first derivatives under suitable assumptions of continuity for the function F and of differentiability with respect to z, without assuming the existence of the derivatives with respect to u, a result that can be extended to the solutions of quasi-linear elliptic equations in divergence form.

In the vector case, as we have already said, the best we can expect is partial regularity, that is regularity outside a singular set Σ, generally non-empty, with in addition an estimate of the dimension of Σ. After a certain number of results relative to functions $F(x, u, z)$ convex in z, a first regularity theorem for the minima of strictly quasi-convex functionals was proved by EVANS [1], adapting methods introduced in the study of minimal surfaces. According to that result, and to its extensions by various authors, the ω-minima of quasi-convex functionals are of class $C^{1,\alpha}$ in an open set $\Omega_0 \subset \Omega$, and the singular set $\Sigma = \Omega - \Omega_0$ has zero measure.

Better estimates for the singular set Σ can be found in the case of quadratic functionals

$$\mathcal{Q}(u, \Omega) = \int_\Omega A^{ij}_{\alpha\beta}(x, u) D_i u^\alpha D_j u^\beta \, dx \tag{0.14}$$

for which one proves that the dimension of Σ is less than $n-2$ (GIAQUINTA and GIUSTI [2]). This result can be further ameliorated for separated coefficients:

$$A^{ij}_{\alpha\beta}(x, u) = g^{ij}(x) G_{\alpha\beta}(u),$$

a situation that arises for instance in the theory of harmonic mappings between Riemannian manifolds. In this last case, the dimension of the singular set does not exceeds $n - 3$ (SCHOEN and UHLENBECK [1, 2],

GIAQUINTA and GIUSTI [6]), and decreases to $n-7$ if the target manifold is the n-dimensional sphere (SCHOEN and UHLENBECK [3], GIAQUINTA and SOUČEK [1]).

The further regularity concerns only the minima, and merges with the theory of elliptic differential equations. Here the results are nowadays classical, and the minima u are as regular as the function F permits. In particular, if F is of class $C^\infty(\Omega)$, every minimum u of the relative functional will be of class $C^\infty(\Omega_0)$, where $\Omega_0 = \Omega$ in the scalar case, whereas in the vector case we have meas $(\Omega - \Omega_0) = 0$.

The structure of this volume follows essentially the above lines. In the first introductory chapter we study scalar functionals, dependent only on the gradient:

$$\mathcal{F}(u,\Omega) = \int_\Omega F(Du(x))\,dx\,, \tag{0.15}$$

and we prove the existence of minima for the DIRICHLET problem in the space of Lipschitz-continuous functions. The results are obtained by means of elementary techniques, that make constant use of the maximum principle. In particular, this chapter does not require the knowledge of SOBOLEV spaces, and the notions of functional analysis do not go beyond the theorem of ASCOLI–ARZELÀ. We obtain nevertheless some significant theorems, in particular for the area functional, a functional that, as we have remarked, is not included within those discussed later.

In the second chapter we gather most of the results relative to spaces of summable functions. In particular, we introduce the spaces L^p and L^p-weak, as well as MORREY and CAMPANATO spaces, and we prove the Hölder continuity of the functions belonging to the last ones. In this chapter one can find the CALDERON–ZYGMUND covering theorem, the lemmas of JOHN–NIRENBERG concerning BMO functions, and the interpolation theorems of MARCINKIEWICZ and STAMPACCHIA. The last section is devoted to the HAUSDORFF measure.

In the third chapter we introduce the SOBOLEV spaces, that we shall use throughout the book. We note that on some occasions, as for instance in the characterization of traces, we have preferred simplicity to generality, for which we refer to books explicitly devoted to that subject. Even in this case we have never left without proof a result essential in the sequel.

The next two chapters deal with semicontinuity theorems, both under assumptions of convexity (Chapter 4) and of quasi-convexity. In the last situation, the most precise result is due to ACERBI and FUSCO [1]. We have followed the proof given by MARCELLINI [1], that besides requiring slightly less restrictive assumptions, represents an excellent example of the interrelations between the theory of semicontinuity and that of regularity of the quasi-minima, which we develop in the subsequent chapter.

Once the semicontinuity has been proved, the existence of minima depends on the coerciveness of the functional under discussion. The question is treated briefly at the end of the chapter.

The remaining Chapters 6–10 concern the regularity. In the sixth we introduce the notion of quasi-minimum, we examine its relationships with solutions of elliptic equations and systems, and we prove the L^p summability of the derivatives of cubical quasi-minima. Here as in the following chapters the main role is played by an inequality, that permits one to estimate the derivatives by the function, to which is associated the name of CACCIOPPOLI, who was the first, as far as I know, to prove it for solutions of linear elliptic equations [3].

The seventh chapter is all devoted to the Hölder regularity of the scalar quasi-minima. We introduce some function classes, which we have named after DE GIORGI who introduced them in his paper [1] already quoted, and we prove that the functions in these classes are Hölder-continuous. Moreover, following DI BENEDETTO and TRUDINGER [1] (see also DI BENEDETTO [1]), we prove the HARNACK inequality for these functions. The regularity of the scalar quasi-minima is a consequence of the fact that they belong to suitable DE GIORGI classes. We prove in addition the boundary regularity of solutions of the DIRICHLET problem.

In the following chapter we continue the study of the regularity in the scalar case, proving the Hölder continuity of the derivatives of the ω-minima of functionals, as well as of the solutions of nonlinear elliptic equations in divergence form, under assumptions in many ways more general than usual.

The core of the proof consists of integral estimates for solutions of elliptic equations of the type

$$D_i A^i(Du) = 0\,,$$

with coefficients A^i dependent only on the gradient, which we have obtained following the techniques introduced by LEWIS [1]. From these estimates we obtain the regularity in the general case, considering the dependence on x and u as a perturbation. An important ingredient of the proofs are the spaces of MORREY and CAMPANATO [1] and [2], the last ones expressing the Hölder continuity of the functions in terms of integral estimates.

The ninth chapter deals with the partial regularity of the ω-minima of quasi-convex functionals. We consider first quadratic functionals (0.14), proving the regularity up to a closed set K of dimension less than $n-2$.

We pass then to more general functionals (0.1), and we show that every ω-minimum is of class $C^{1,\alpha}$, except possibly on a singular set Σ, closed in Ω and of zero measure.

Finally, the tenth and last chapter is concerned with the regularity of the higher derivatives of the solutions of elliptic equations, and as a consequence of minima of functionals. The treatment does not distinguish any more between the scalar and the vector case, and it is made both in the SOBOLEV spaces (HILBERT regularity) and in CAMPANATO spaces (HÖLDER regularity). Except for some reference to results previously proved, and in any case it is not difficult to obtain directly, this chapter is independent of the others, and can be considered as a brief introduction to the regularity of the solutions of linear elliptic systems with regular coefficients.

The volume, as far as it is possible, is self-sufficient, and does not use unproved results, except for what concerns the LEBESGUE integral, the first properties of L^p spaces, and some elementary notions of functional analysis. The notations are the standard ones, and do not need special explainations. It is only appropriate to remark that, following again a well-established rule, we have denoted by c a generic constant, in general dependent on the data of the problem, and that may change within the same formula, as for instance

$$ab \leq c(a^2 + b^2) \leq c\left(a^4 + b^6 + \frac{1}{4}\right).$$

Only when we want to stress the dependence of such constants on one or more parameters λ, μ, etc., we shall write $c(\lambda)$, $c(\lambda, \mu)$ and the like.

This book has three different origins. The first are some notes of a course given at the Nankai Institute of Mathematical in Tianjin in 1985 that for some reason were never completed. The second is a small volume *Equations ellittiche del secondo ordine* [5], published as one of the "Quaderni dell'U. M. I." and that has been out of print for a long time (it was reprinted in 2001). Finally, the research on direct methods in the regularity of the minima of variational integrals, initiated by GIAQUINTA and GIUSTI [2–4] and involving several authors, that have permitted us to give a unified treatment of the regularity of the minima of functionals and of the solutions to partial differential elliptic equations systems.

This work could never be finished without the help of several friends, in particular M. GIAQUINTA, P. MARCELLINI and G. MODICA, with whom I have frequently discussed these subjects.

Finally, I feel obliged to apologize for the title rather abused these days. Unfortunately, I could not find any other which would adequately describe the content.

Chapter 1

Semi-Classical Theory

In this chapter we shall illustrate a simple but meaningful use of direct methods in the calculus of variations.

The method is limited to functionals of the type:

$$\mathcal{F}(u, \Omega) = \int_\Omega F(Du)\, dx\,, \tag{1.1}$$

with the function $F(z)$ depending only on the gradient of a *scalar* function, namely of a function $u : \Omega \to \mathbf{R}$. On the other hand, it is independent of the growth of the function F, and therefore it covers situations that cannot be treated with the more refined methods of the subsequent chapters. An example of some importance is that of the area functional, that we shall examine in detail.

The setting in which we shall treat our minimum problem is the space $\mathrm{Lip}\,(\Omega) = C^{0,1}(\bar{\Omega})$ of Lipschitz-continuous functions, namely of the functions $u(x)$ continuous in $\bar{\Omega}$ and such that

$$[u]_{0,1} = \sup_{\substack{x,y\in\Omega \\ x\neq y}} \frac{|u(x) - u(y)|}{|x-y|} < +\infty\,. \tag{1.2}$$

It is not difficult to see that $\mathrm{Lip}\,(\Omega)$ is a BANACH space, with the norm

$$\|u\|_{0,1} = \sup_\Omega |u| + [u]_{0,1}\,.$$

Lipschitz-continuous functions are almost everywhere differentiable (see for instance SAKS [1], p. 311), and their derivatives are bounded functions, which coincide with the distributive derivatives.

We shall consider functionals of type (1.1), with F convex in $z \in \mathbf{R}^n$, and we shall prove that under suitable hypotheses it is possible to solve the DIRICHLET problem, that is to show that the functional (1.1) takes its minimum in the class of Lipschitz-continuous functions with prescribed values on the boundary of Ω.

1.1 The Maximum Principle

We shall denote by $\mathrm{Lip}_k(\Omega)$ the set of Lipschitz-continuous functions in Ω, whose Lipschitz constant is less than or equal to k:

$$\mathrm{Lip}_k(\Omega) = \{u \in \mathrm{Lip}\,(\Omega) : [u]_{0,1} \le k\}. \tag{1.3}$$

Moreover, if U is a Lipschitz-continuous function defined on $\partial\Omega$, we shall set:

$$\mathrm{Lip}\,(\Omega, U) = \{u \in \mathrm{Lip}\,(\Omega) : u = U \text{ on } \partial\Omega\} \tag{1.4}$$

and

$$\mathrm{Lip}_k(\Omega, U) = \{u \in \mathrm{Lip}_k(\Omega) : u = U \text{ on } \partial\Omega\}. \tag{1.5}$$

Proposition 1.1 *Let $F(z)$ be a convex function, and let Ω be a bounded open subset of $\mathbf{R}^n$. Let U be a Lipschitz-continuous function in $\bar{\Omega}$, and let $k \ge [U]_{0,1}$. Then, the functional $\mathcal{F}(u, \Omega)$ takes its minimum in* $\mathrm{Lip}_k(\Omega, U)$.

Proof. Let $\{u_j\}$ be a minimizing sequence, that is a sequence of functions in $\mathrm{Lip}_k(\Omega, U)$ such that

$$\lim_{j\to\infty} \mathcal{F}(u_j, \Omega) = \inf\{\mathcal{F}(u, \Omega);\ u \in \mathrm{Lip}_k(\Omega, U)\} =: \mu\,.$$

We have $[u_j]_{0,1} \le k$, and moreover

$$\sup_\Omega |u_j| \le \sup_\Omega |U| + [u_j]_{0,1}\,\mathrm{diam}\,(\Omega) \le \sup_\Omega |U| + k\,\mathrm{diam}\,(\Omega)\,,$$

so that the sequence $\{u_j\}$ is bounded in $\mathrm{Lip}\,(\Omega)$. By ASCOLI–ARZELÀ's theorem it is possible to extract a subsequence, which we shall denote again by $\{u_j\}$, uniformly convergent to a function $u \in \mathrm{Lip}_k(\Omega, U)$. Since F is a convex function, there exists a Borel function $\lambda : \mathbf{R}^n \to \mathbf{R}^n$, bounded on compact sets, and such that[1]

$$F(w) \geq F(z) + \langle \lambda(z), w - z \rangle$$

for every $z, w \in \mathbf{R}^n$. We have therefore:

$$\int_\Omega F(Du_j)\, dx \geq \int_\Omega F(Du)\, dx + \int_\Omega \langle \lambda(Du), Du_j - Du \rangle\, dx\,.$$

We can evaluate the last integral observing that the function $\lambda(Du(x))$ is bounded and measurable, and whence[2] that for every $\epsilon > 0$ there exists a function $g : \Omega \to \mathbf{R}^n$ of class $C^1(\Omega)$ such that

$$\int_\Omega |\lambda(Du) - g|\, dx < \epsilon\,.$$

We have

$$\int_\Omega \langle \lambda(Du), Du_j - Du \rangle\, dx \geq \int_\Omega \langle g(x), Du_j - Du \rangle\, dx \\ - \int_\Omega |\lambda(Du) - g||Du_j - Du|\, dx\,.$$

Since both Du_j and Du belong to $\mathrm{Lip}_k(\Omega)$, the last integral can be estimated by $2k\epsilon$, whereas for the preceding one, taking into account the fact that $u = u_j = U$ on $\partial\Omega$, we have

$$\int_\Omega \langle g(x), Du_j - Du \rangle\, dx = -\int_\Omega (u_j - u)\, \mathrm{div}\, g\ dx\,,$$

so that it tends to zero as $j \to \infty$. We have in conclusion:

$$\mu = \lim_{j \to \infty} \mathcal{F}(u_j, \Omega) \geq \mathcal{F}(u, \Omega) - 2k\epsilon\,.$$

Since ϵ is arbitrary, we have $\mathcal{F}(u, \Omega) = \mu$, so that u minimizes the functional $\mathcal{F}$ in $\mathrm{Lip}_k(\Omega, U)$. □

We shall call u^k a minimizing function in $\mathrm{Lip}_k(\Omega, U)$. Generally speaking, when k increases, the minimum value of $\mathcal{F}$ in Lip_k decreases, whereas the Lipschitz constant $[u^k]_{0,1}$ of its minimizer increases. We have

---

[1] We shall denote by $\langle\, , \rangle$ the scalar product in $\mathbf{R}^n$. If $F \in C^1$, we have $\lambda = F_z$.

[2] See next chapter, in particular Corollary 2.1. As usual, we denote by C^k the space of functions having continuous derivatives up to the order k.

Proposition 1.2 *Let u^k a minimizing function in* $\mathrm{Lip}_k(\Omega, U)$. *If* $[u^k]_{0,1} < k$, *the function* u^k *minimizes* $\mathcal{F}$ *in* $\mathrm{Lip}\,(\Omega, U)$.

Proof. Let $v \in \mathrm{Lip}\,(\Omega, U)$, and for $t \in [0,1]$ define

$$v_t = u^k + t(v - u^k)\,.$$

We have $v_t = U$ on $\partial\Omega$, and moreover $[v_t]_{0,1} < k$ for t small enough. Since u^k minimizes in Lip_k we obtain, taking into account the convexity of $\mathcal{F}$:

$$\mathcal{F}(u^k, \Omega) \le \mathcal{F}(v_t, \Omega) \le (1-t)\mathcal{F}(u^k, \Omega) + t\mathcal{F}(v, \Omega)\,,$$

and hence $t\mathcal{F}(u^k, \Omega) \le t\mathcal{F}(v, \Omega)$, so that u^k minimizes $\mathcal{F}$ in $\mathrm{Lip}\,(\Omega, U)$.

□

The above proposition will be useful in the proof of the existence of a minimum in $\mathrm{Lip}\,(\Omega, U)$. For that, we shall look for an estimate of the Lipschitz constant of u^k.

In order to simplify the notation, we shall omit from now on the index k; moreover we shall write "u minimizes $\mathcal{F}$ in $\mathrm{Lip}_k(\Omega)$", understanding "among all the functions of Lip_k that coincide with u on $\partial\Omega$". It is evident that if a function u minimizes $\mathcal{F}$ in $\mathrm{Lip}_k(\Omega)$, and if $u \in \mathrm{Lip}_h(\Lambda)$, with $h \le k$ and $\Lambda \subset \Omega$, then u minimizes $\mathcal{F}$ in $\mathrm{Lip}_h(\Lambda)$.

Our principal tool will be the maximum principle. To establish it in its suitable generality, we shall introduce the notions of *sub-minimum* and *super-minimum.*

Definition 1.1 *A function* $w \in \mathrm{Lip}_k(\Omega)$ *is a* super-minimum *(resp. a* sub-minimum*) for the functional* $\mathcal{F}$ *if for every* $\vartheta \in \mathrm{Lip}_k(\Omega, w)$, *with* $\vartheta \ge w$ *(resp.* $\vartheta \le w$*), we have*

$$\mathcal{F}(w, \Omega) \le \mathcal{F}(\vartheta, \Omega)\,.$$

In particular, a minimum in $\mathrm{Lip}_k(\Omega)$ is both a sub-minimum and a super-minimum. It is not difficult to show that the converse is also true: if a function is at the same time a sub-minimum and a super-minimum, then it is a minimum. However, this result is not relevant for our purposes, and its proof is left to the reader.

Lemma 1.1 (Maximum principle I) *Let* $F(z)$ *be a strictly convex function, and let* $v(x)$ *and* $w(x)$ *be respectively a super-minimum and a sub-minimum in* $\mathrm{Lip}_k(\Omega)$ *for the functional* $\mathcal{F}$. *Suppose, moreover, that* $w \le v$ *on* $\partial\Omega$. *Then,* $w \le v$ *in* Ω.

Proof. Suppose, on the contrary, that the open set

$$A = \{x \in \Omega : v(x) < w(x)\}$$

is non-empty, and define $\vartheta(x) = \max\{v(x), w(x)\}$. The function ϑ belongs to $\mathrm{Lip}_k(\Omega, v)$, and $\vartheta \geq v$ in Ω. Since v is a super-minimum, we shall have $\mathcal{F}(v, \Omega) \leq \mathcal{F}(\vartheta, \Omega)$, or otherwise, what is the same:

$$\mathcal{F}(v, A) \leq \mathcal{F}(w, A).$$

In a similar way, comparing w with $y = \min\{v, w\}$ we obtain

$$\mathcal{F}(v, A) \geq \mathcal{F}(w, A),$$

and in conclusion

$$\mathcal{F}(v, A) = \mathcal{F}(w, A).$$

Since $w = v$ on ∂A and $w > v$ in A, we have $Dw \neq Dv$ on a set of positive measure, and therefore for the strict convexity of $\mathcal{F}$:

$$\mathcal{F}\left(\frac{v+w}{2}, A\right) < \frac{1}{2}\mathcal{F}(v, A) + \frac{1}{2}\mathcal{F}(w, A) = \mathcal{F}(v, A).$$

On the other hand, the function $u = \frac{v+w}{2}$ verifies $u = v$ on ∂A and $u \geq v$ in A, and hence

$$\mathcal{F}\left(\frac{v+w}{2}, A\right) \geq \mathcal{F}(v, A),$$

contradicting the preceding inequality. □

As a simple consequence of the maximum principle we have the following

Lemma 1.2 *Let $F(z)$ be a strictly convex function, and let $v(x)$ and $w(x)$ be respectively a super-minimum and a sub-minimum in* $\mathrm{Lip}_k(\Omega)$ *for the functional $\mathcal{F}$. Then:*

$$\sup_{\Omega}(w - v) = \sup_{\partial\Omega}(w - v). \tag{1.6}$$

Proof. It will be sufficient to remark that for every $\alpha \in \mathbf{R}$, the function $v + \alpha$ is a super-minimum in $\mathrm{Lip}_k(\Omega)$, and that for every $x \in \partial\Omega$ we have

$$w(x) \leq v(x) + \sup_{\partial\Omega}(w - v).$$

By the preceding lemma the same inequality holds in the whole Ω. From it, the relation (1.6) follows immediately, since the opposite inequality is trivial. □

In particular, if u and v minimize $\mathcal{F}$ in $\mathrm{Lip}_k(\Omega)$, the relation (1.6) holds both for $u - v$ and for per $v - u$, and therefore in his case we have

$$\sup_{\Omega} |u - v| = \sup_{\partial\Omega} |u - v| \,. \tag{1.7}$$

From that, it follows immediately the uniqueness of the minimum for the functional $\mathcal{F}$ in $\mathrm{Lip}_k(\Omega, U)$, and also in $\mathrm{Lip}\,(\Omega, U)$.

It is clear that none of the preceding results holds if we suppose that the functional $\mathcal{F}$ is convex but not strictly convex. For instance, if we define

$$F(z) = \max\{0, |z| - M\}\,,$$

every function $u \in \mathrm{Lip}_M(\Omega)$ minimizes the corresponding functional $\mathcal{F}$. Nevertheless, in many cases it is possible to prove the existence of minima for convex functionals by considering first the strictly convex function $F(z) + \epsilon|z|^2$ and then taking the limit as $\epsilon \to 0$.

1.2 The Bounded Slope Condition

Lemma 1.3 (Reduction to the boundary) *Let $F(z)$ be a strictly convex function, and let $u(x)$ be a minimum for $\mathcal{F}$ in $\mathrm{Lip}_k(\Omega)$. Then:*

$$[u]_{0,1} = \sup_{\substack{x\in\Omega \\ y\in\partial\Omega}} \frac{|u(x) - u(y)|}{|x - y|} \,. \tag{1.8}$$

Proof. Let $x_1 \neq x_2$ be two points in Ω, and let $\tau = x_2 - x_1$. The function

$$u_\tau(x) = u(x + \tau)$$

minimizes $\mathcal{F}$ in $\mathrm{Lip}_k(\Omega_\tau)$, with

$$\Omega_\tau = \{x \in \mathbf{R}^n : x + \tau \in \Omega\}\,.$$

The open set $\Omega \cap \Omega_\tau$ is non-empty, since it contains x_1; and the functions u and u_τ both minimize $\mathcal{F}$ in $\mathrm{Lip}_k(\Omega \cap \Omega_\tau)$. From (1.7) we conclude that there exists a point $x_0 \in \partial(\Omega \cap \Omega_\tau)$ such that

$$\begin{aligned} |u(x_1) - u(x_2)| &= |u(x_1) - u_\tau(x_1)| \\ &\leq |u(x_0) - u_\tau(x_0)| \\ &= |u(x_0) - u(x_0 + \tau)|\,. \end{aligned} \tag{1.9}$$

On the other hand, at least one of the points x_0, $x_0 + \tau$ belongs to $\partial\Omega$, and therefore, indicating by M the right-hand side of (1.8), we get from (1.9):

$$|u(x_1) - u(x_2)| \leq M|x_1 - x_2|,$$

and the lemma is proved. □

In view of an application of Proposition 1.2, we are led to look for a bound for the difference $|u(x) - u(y)|$ when one of these points, for instance y, belongs to $\partial\Omega$. For that, we shall make recourse again to the maximum principle, comparing the function u with suitable sub-minima and super-minima of the functional $\mathcal{F}$.

An important class of comparison functions is that of affine functions:

$$w(x) = a + \langle z_0, x\rangle$$

with $a \in \mathbf{R}$ and $z_0 \in \mathbf{R}^n$. It is easy to see that $w \in \mathrm{Lip}_{|z_0|}(\Omega)$ and that it minimizes $\mathcal{F}$ in $\mathrm{Lip}\,(\Omega)$, whence in $\mathrm{Lip}_{|z_0|}(\Omega)$. Actually, let $\eta(x)$ be a function in $\mathrm{Lip}\,(\Omega)$, with $\eta = 0$ on $\partial\Omega$. By the convexity of $F(z)$ there exists a vector $\lambda \in \mathbf{R}^n$ such that

$$F(z_0 + \xi) \geq F(z_0) + \langle\lambda, \xi\rangle$$

for every $\xi \in \mathbf{R}^n$. Taking $\xi = D\eta(x)$ and integrating on Ω we get

$$\mathcal{F}(w + \eta) = \int_\Omega F(z_0 + D\eta)\,dx \geq F(z_0)|\Omega| + \int_\Omega \langle\lambda, D\eta\rangle\,dx\,,$$

where $|\Omega| = \mathrm{meas}\,(\Omega)$. The last integral is zero, since $\eta = 0$ on $\partial\Omega$; it follows that

$$\mathcal{F}(w + \eta, \Omega) \geq F(z_0)|\Omega| = \mathcal{F}(w, \Omega)$$

and therefore w minimizes $\mathcal{F}$ in $\mathrm{Lip}\,(\Omega)$.

Definition 1.2 *We say that the function $U : \partial\Omega \to \mathbf{R}$ satisfies a* bounded slope condition *(briefly, a* B.S.C.*) with constant $Q > 0$ if for every $x_0 \in \partial\Omega$ there exist us two affine functions $w^+ = w^+_{x_0}$ and $w^- = w^-_{x_0}$ such that*

$$w^-(x) \leq U(x) \leq w^+(x) \text{ in } \partial\Omega\,; \tag{1.10}$$

$$w^-(x_0) = U(x_0) = w^+(x_0)\,; \tag{1.11}$$

$$[w^-]_{0,1} \leq Q\,; \quad [w^+]_{0,1} \leq Q\,. \tag{1.12}$$

Note that, unless the function U is itself an affine function, the B.S.C. cannot be satisfied if Ω is not a convex set. In fact, if U is not an affine function, we have $w^+ \neq w^-$, and therefore the set

$$V = \{x \in \mathbf{R}^n : w^-(x) \leq w^+(x)\}$$

is a closed half-space, and $x_0 \in \partial V$. On the other hand, we must have $\overline{\Omega} \subset V$, because otherwise (1.10) could not be satisfied, so that in conclusion every point of the boundary of Ω has a supporting hyperplane, and Ω is convex. Of course, in general, the convexity alone is not sufficient for the B.S.C., as one can easily see if $\partial\Omega$ has a flat portion Σ, and if U is not constant on Σ. Actually, even the strict convexity of Ω is not sufficient for the B.S.C., as the reader will verify by taking $\Omega = \{(x,y) \in \mathbf{R}^2 : x^4 < y < 1\}$ and $U(x,y) = x^2$.

On the other hand, we have

Theorem 1.1 (Miranda [1]) *Let Ω be an open bounded subset of $\mathbf{R}^n$, and suppose that there exists a positive constant $c = c(\Omega)$, and for every $x_0 \in \partial\Omega$ a hyperplane Π_{x_0} through x_0, such that for every $x \in \partial\Omega$ it holds that*

$$|x - x_0|^2 \leq c \operatorname{dist}(x, \Pi_{x_0}). \tag{1.13}$$

Then, every function U of class $C^2(\mathbf{R}^n)$ satisfies the B.S.C. on $\partial\Omega$.

Proof. We remark that the assumptions imply that Ω is convex.

For every $x_0 \in \partial\Omega$ we must find two affine functions $w^\pm(x)$ satisfying conditions (1.10), (1.11) and (1.12). Since Ω is bounded, we can assume that U has compact support.[3] Always without loss of generality we can assume that $x_0 = 0$ and that Π_{x_0} is the hyperplane $x_n = 0$.

Let μ be a real number, and set

$$w(x) = U(0) + \langle DU(0), x\rangle + \mu x_n .$$

We have $w(0) = U(0)$. Suppose now that for some $\bar{x} \in \partial\Omega$ we have $w(\bar{x}) = U(\bar{x})$. Then:

$$\mu = \frac{1}{\bar{x}_n}[U(\bar{x}) - U(0) - \langle DU(0), \bar{x}\rangle] = \frac{1}{2\bar{x}_n}\langle D^2U(\xi)\bar{x}, \bar{x}\rangle$$

[3] We remember that the *support* of a measurable function f, defined in $\mathbf{R}^n$, is the set $\mathbf{R}^n - A$, where A is the largest open set in which $f(x) = 0$ almost everywhere. The support of f is denoted by $\operatorname{supp}(f)$.

for some $\xi \in \mathbf{R}^n$. Therefore, if $w(\bar{x}) = U(\bar{x})$, we will have

$$|\mu| \leq \frac{|\bar{x}|^2}{|\bar{x}_n|} \sup |D^2U| \leq c \sup |D^2U| .$$

As a consequence, taking $|\mu| = c \sup |D^2U| + 1$, we have either $w > U$ or $w < U$ in $\partial\Omega - \{x_0\}$, depending on the sign of μ. For these affine functions $w^{\pm}$ we have

$$[w^{\pm}]_{0,1} \leq |DU(0)| + |\mu| \leq 1 + \sup |DU| + c \sup |D^2U| ,$$

and the theorem follows. □

From a geometric point of view, condition (1.13) says that the boundary of Ω cannot lie too close to its tangent hyperplane. If $\partial\Omega$ is of class C^2, it is equivalent to the assumption that all the principal curvatures of $\partial\Omega$ are strictly positive (and hence, by continuity, that they are bounded below by a positive constant). The proof is left to the reader.

Theorem 1.2 *Let $F(z)$ be a convex function in $\mathbf{R}^n$, and let $U : \partial\Omega \to \mathbf{R}$ be a function satisfying the B.S.C. with constant Q. Then, the functional*

$$\mathcal{F}(u, \Omega) = \int_\Omega F(Du)\, dx \tag{1.14}$$

has a minimum in Lip(Ω, U). *Moreover, at least one of the minimizing functions satisfies the estimate*

$$[u]_{0,1} \leq Q . \tag{1.15}$$

Proof. Let us assume first that F is strictly convex. The class $\mathrm{Lip}_Q(\Omega, U)$ is non-empty, since it contains at least the function

$$\psi^+(x) = \inf_{x_0 \in \partial\Omega} w^+_{x_0}(x) .$$

Let $k > Q$, and let u be the function minimizing $\mathcal{F}$ in $\mathrm{Lip}_k(\Omega, U)$, a function necessarily unique by the strict convexity of the functional. By the maximum principle we have

$$w^-_{x_0}(x) \leq u(x) \leq w^+_{x_0}(x)$$

for every $x \in \Omega$, and hence, since $w^{\pm}_{x_0}(x_0) = u(x_0)$,

$$w^-_{x_0}(x) - w^-_{x_0}(x_0) \leq u(x) - u(x_0) \leq w^+_{x_0}(x) - w^+_{x_0}(x_0) .$$

From Lemma 1.3 and from (1.12) we get then

$$[u]_{0,1} \le Q\,,$$

and the conclusion follows from Proposition 1.2.

Let now $F(z)$ be simply convex, and set $F_\epsilon(z) = F(z) + \epsilon|z|^2$. If u_ϵ is the function minimizing the functional $\mathcal{F}_\epsilon$ in $\mathrm{Lip}\,(\Omega, U)$, we shall have $[u_\epsilon]_{0,1} \le Q$, and by the maximum principle $\sup_\Omega |u_\epsilon| = \sup_{\partial\Omega} |U|$. We can therefore find a sequence $\{u_k\} = \{u_{\epsilon_k}\}$, with $\epsilon_k \to 0$, converging uniformly to a function $u \in \mathrm{Lip}_Q(\Omega, U)$. Arguing as in the proof of Proposition 1.1 we conclude that u minimizes $\mathcal{F}$ in $\mathrm{Lip}\,(\Omega, U)$. □

1.3 Barriers

In the preceding section we have proved the existence of minima for convex functionals depending only on the gradient, but otherwise arbitrary, under special assumptions of convexity on the domain Ω. If we want to weaken these conditions, so as to treat more general domains Ω, on the one hand, we must restrict the class of functionals under examination, and on the other we shall use new comparison functions, more general than the affine functions of the B.S.C.

In this section we shall treat some simple but meaningful situations, following the methods introduced by BERNSTEIN [1] and SERRIN [3].

For $x \in \Omega$, we denote by $d(x)$ the distance between x and $\partial\Omega$. Moreover, for $t > 0$, we set

$$\Sigma_t = \{x \in \Omega : d(x) < t\}\,, \tag{1.16}$$

$$\Gamma_t = \{x \in \Omega : d(x) = t\}\,. \tag{1.17}$$

Definition 1.3 *Let Ω be a bounded open set in $\mathbf{R}^n$, and let U be a Lipschitz-continuous function in $\partial\Omega$. An* upper barrier *(relative to the functional $\mathcal{F}$) is a function v^+, Lipschitz-continuous in some Σ_t, $t > 0$, and such that*

$$v^+ = U \text{ on } \partial\Omega\,; \tag{1.18}$$

$$v^+ \textit{ is a super-minimum in } \Sigma_t\,; \tag{1.19}$$

and

$$v^+ \ge \sup_{\partial\Omega} U \textit{ on } \Gamma_t\,. \tag{1.20}$$

In the same way we can define a *lower barrier* v^-, simply substituting (1.19) and (1.20), respectively, with

$$v^- \textit{ is a sub-minimum in } \Sigma_t\,, \tag{1.21}$$

$$v^- \le \inf_{\partial\Omega} U \textit{ on } \Gamma_t\,. \tag{1.22}$$

It is easily seen that if v^+ is an upper barrier relative to the function $-U$, $-v^+$ is a lower barrier relative to U, so that the existence of upper barriers for a class of functions U, closed with respect to the change of sign, implies that of lower barriers, and *vice versa*. We can therefore treat only the case of upper barriers.

The following result is in some respect a generalization of Theorem 1.2.

Theorem 1.3 *Let $F(z)$ be a strictly convex function, let U be a Lipschitz-continuous function on $\partial\Omega$, and suppose that there exists an upper barrier v^+ and a lower barrier v^-. Then, the functional $\mathcal{F}$ has a minimum in* $\mathrm{Lip}\,(\Omega, U)$.

Proof. For $x \in \Omega$ set

$$w^+(x) = \begin{cases} \min\left\{v^+(x),\ \sup_{\partial\Omega} U\right\} & \text{if } x \in \Sigma_t\,, \\ \sup_{\partial\Omega} U & \text{if } x \in \Omega - \Sigma_t\,, \end{cases}$$

$$w^-(x) = \begin{cases} \max\left\{v^-(x),\ \inf_{\partial\Omega} U\right\} & \text{if } x \in \Sigma_t\,, \\ \inf_{\partial\Omega} U & \text{if } x \in \Omega - \Sigma_t\,. \end{cases}$$

□

The functions w^+ and w^- are Lipschitz-continuous in $\bar{\Omega}$; let Q be the largest of their Lipschitz constants.

If $k > Q$, the class $\mathrm{Lip}_k(\Omega, U)$ is non-empty, and the functional $\mathcal{F}$ has a minimum in it. Let u be the minimizing function. We have obviously:

$$\inf_{\partial\Omega} U \le u(x) \le \sup_{\partial\Omega} U\,;$$

moreover, since $v^- \le u \le v^+$ in $\partial\Sigma_t$, we get from Lemma 1.1:

$$w^-(x) \le u(x) \le w^+(x) \text{ in } \Omega\,.$$

On the other hand, we have

$$w^-(x) = u(x) = w^+(x) \text{ on } \partial\Omega$$

and therefore, arguing as in Theorem 1.2, we can infer that

$$[u]_{0,1} \leq Q\,.$$

The conclusion follows from Proposition 1.3.

Remark 1.1 We remark that the functions w^+ and w^- defined above are, respectively, a super-minimum and a sub-minimum for the functional $\mathcal{F}$ in Ω. It follows that a necessary and sufficient condition for a strictly convex functional $\mathcal{F}$ to have a minimum in Lip(Ω, U) is that there exists a super-minimum and a sub-minimum taking at the boundary of Ω the value U.

Moreover, the preceding theorem is really a generalization of Theorem 1.2, at least as far as strictly convex functionals are concerned, since one can show that the functions

$$\psi^+(x) = \inf_{x_0 \in \partial\Omega} w^+_{x_0}(x) \text{ and } \psi^-(x) = \sup_{x_0 \in \partial\Omega} w^-_{x_0}(x)$$

are respectively a super-minimum and sub-minimum.

The extension of the above result to general convex functionals depends on the existence of barriers for the approximating functionals

$$\mathcal{F}_\epsilon(u) = \int_\Omega (F(Du) + \epsilon|Du|^2)\,dx\,.$$

If these functionals admit barriers (at least for ϵ small enough) and if the Lipschitz constants of these barriers remain bounded as $\epsilon \to 0$, one can argue as in Theorem 1.2 and prove the existence of a solution in this case. □

It remains to discuss the construction of barriers. We shall investigate the case of upper barriers; the lower barriers can be treated similarly.

The classical argument leading to EULER's equation gives a differential inequality for super-minima.

Let F be of class C^2, and let $v(x)$ be a super-minimum of class C^2 in some open set Σ. If η is a non-negative function with compact support in Σ, the function

$$g(t) = \mathcal{F}(v + t\eta, \Sigma)\,, \quad t \geq 0\,,$$

has a minimum for $t = 0$, and therefore we must have $g'(0) \geq 0$. Differentiating under the integral sign we easily get the inequality[4]:

$$g'(0) = \int_\Sigma F_{z_j}(Dv) D_j \eta \, dx \geq 0$$

for every $\eta \in C_0^\infty(\Sigma)$, $\eta \geq 0$, and hence, integrating by parts and taking into account the arbitrariness of $\eta \geq 0$:

$$D_j F_{z_j}(Dv) \leq 0 \, .$$

Developing the derivatives, we get in conclusion:

$$\mathcal{L}(v) =: F_{z_i z_j}(Dv) \frac{\partial^2 v}{\partial x_i \partial x_j} \leq 0 \quad \text{in } \Sigma \, . \tag{1.23}$$

Reciprocally, if a function $v(x)$ of class $C^2(\Sigma)$ verifies the inequality (1.23), we will have $g'(0) \geq 0$ and therefore, by the convexity of g, $g(0) \leq g(1)$, so that $v(x)$ is a super-minimum. We note that a function satisfying the inequality (1.23) is a *supersolution* of the corresponding differential equation, so that in a certain sense the relation between super-minimum and supersolution is similar to that between minimum and solution of the EULER equation.

Our problem is then reduced to the search for a solution to the differential inequality (1.23) in a neighborhood Σ_t of the boundary of Ω, taking on $\partial\Omega$ the prescribed value $U(x)$, and greater than $\sup_{\partial\Omega} U$ on Γ_t.

For the sake of simplicity, we shall write $A^{ij}(z)$ instead of $F_{z_i z_j}(z)$. Since F is a convex function, the matrix A^{ij} is positive semi-definite; moreover if we assume, as we shall always do, that F is strictly convex, we will have for every $\xi \in \mathbf{R}^n$:

$$\lambda(z)|\xi|^2 \leq A^{ij}(z)\xi_i\xi_j \leq \Lambda(z)|\xi|^2 \, , \tag{1.24}$$

where $\Lambda(z)$ and $\lambda(z) > 0$ are, respectively, the largest and the smallest eigenvalues of the matrix $A(z)$.

We shall also introduce the BERNSTEIN function:

$$\mathcal{E}(z) = A^{ij}(z) z_i z_j \, . \tag{1.25}$$

---

[4]We denote by F_{z_i} the derivative of F with respect to z_i, and with F_z the gradient of the function F. Remember that we always sum over repeated indices.

We have, trivially:

$$\lambda(z)|z|^2 \leq \mathcal{E}(z) \leq \Lambda(z)|z|^2 . \tag{1.26}$$

For what concerns the open set Ω, we shall suppose that it is bounded and that its boundary $\partial\Omega$ is of class C^2. For every point $y \in \partial\Omega$ there will then exist a ball $B \subset \Omega$ tangent to $\partial\Omega$ in y, that is such that $\bar{B} \cap \partial\Omega = \{y\}$. Let $r(y)$ be the supremum of the radii of the balls B with that property; since $\partial\Omega$ is compact, the continuous function $r(y)$ is bounded below by a positive constant τ. It is easily seen that τ^{-1} is an upper bound for the principal curvatures of $\partial\Omega$, oriented in such a way that a convex set has positive curvatures.

For $t < \tau$, every point $x \in \Sigma_t$ has a unique point $y = y(x)$ of least distance on $\partial\Omega$, that is such that $d(x) = |x - y(x)|$. The points x and y are connected by the relation:

$$x = y + \nu(y)d(x) ,$$

where $\nu(y)$ is the interior normal to $\partial\Omega$ at y.

Lemma 1.4 *Let Ω be a bounded open set in $\mathbf{R}^n$, with boundary of class C^k, $k \geq 2$. Then, the distance function $d(x)$ is of class C^k in Σ_τ, and for every $x \in \Sigma_\tau$ we have*

$$\Delta d(x) = -\sum_{i=1}^{n-1} \frac{\kappa_i(y)}{1 - \kappa_i(y)d(x)} , \tag{1.27}$$

where $\kappa_i(y)$ are the principal curvatures of $\partial\Omega$ at y and Δ is the LAPLACE *operator*:

$$\Delta =: \sum_{i=1}^{n} D_i D_i .$$

Proof. Let $x_0 \in \Sigma_\tau$ and let y_0 be the point corresponding to x_0 on $\partial\Omega$. We can assume that $y_0 = 0$ and that the tangent plane to $\partial\Omega$ in 0 is the horizontal plane $x_n = 0$. In a neighborhood W of 0 the boundary $\partial\Omega$ is the graph of a function $\vartheta(\bar{y})$, $\bar{y} = (y_1, y_2, \ldots, y_{n-1})$, with $D\vartheta(0) = 0$; we can assume that Ω lies locally below the graph of ϑ.

Modulo a possible rotation around the vertical axis, we can also assume that the matrix $\{D^2\vartheta(0)\}$ is diagonal, so that its elements are the opposites of the principal curvatures of $\partial\Omega$ in 0:

$$D^2\vartheta(0) = \operatorname{diag}[-\kappa_1, -\kappa_2, \ldots, -\kappa_{n-1}] . \tag{1.28}$$

Let V be the intersection of W with the plane $x_n = 0$; in $V \times \mathbf{R}$ we define

$$\gamma(\bar{y}, d) = y + \nu(y)d\,; \quad y = (\bar{y}, \vartheta(\bar{y}))\,.$$

We have $\gamma \in C^{k-1}$; recalling that

$$\nu_n(y) = -\frac{1}{\sqrt{1 + |D\vartheta(\bar{y})|^2}}\,,$$

$$\nu_i(y) = -\nu_n D_i \vartheta(\bar{y}) \quad (i = 1, 2, \ldots, n-1)\,;$$

we obtain from (1.28):

$$D\gamma(0, d) = \operatorname{diag}\left[1 - \kappa_1 d, 1 - \kappa_2 d, \ldots, 1 - \kappa_{n-1} d, 1\right]. \tag{1.29}$$

If $d < \tau$, the Jacobian determinant $D\gamma$ is positive, and therefore it is possible to express y and d as functions of x, both of class C^{k-1} in a neighborhood of $x_d =: (0, d)$. Moreover, $Dd(x)$ is the normal to $\Gamma_{d(x)}$ at x, and has the same direction as the normal to $\partial\Omega$ at the corresponding point $y(x)$:

$$Dd(x) = \nu_{\partial\Omega}(y(x)) = \nu_{\Gamma_d}(x)\,. \tag{1.30}$$

From the preceding relation it follows at once that $d(x)$ is of class C^k, because the right-hand side is of class C^{k-1}. Moreover, the mean curvature of $\Gamma_{d(x_0)}$ at x_0 is given by

$$H(x_0) =: \frac{1}{n-1}\sum_{h=1}^{n-1} \kappa_i(x_0) =: -\frac{1}{n-1}\sum_{i=1}^{n} D_i \nu_i(x_0) = -\frac{1}{n-1}\Delta d(x_0)\,.$$

We can evaluate the last quantity by means of (1.30):

$$\Delta d(x_0) = \sum_{i=1}^{n} D_i(\nu_i \circ y)(x_0) = \sum_{i=1}^{n}\sum_{h=1}^{n-1} D_h \nu_i(0) D_i y_h(x_0)\,.$$

On the other hand, $D_n y_h = 0$, and taking into account (1.28):

$$D_y \nu = \operatorname{diag}\left[\kappa_1, \kappa_2, \ldots, \kappa_{n-1}, 0\right].$$

Finally, $\frac{\partial(\bar{y}, d)}{\partial x}$ is the inverse matrix of $D\gamma$, so that from (1.29) we have

$$Dy(x_0) = \operatorname{diag}\left[\frac{1}{1 - \kappa_1 d}, \frac{1}{1 - \kappa_2 d}, \ldots, \frac{1}{1 - \kappa_{n-1} d}\right].$$

This concludes the proof. □

From (1.27) we immediately get

$$-\Delta d(x) \geq \sum_{i=1}^{n-1} \kappa_i(y(x)) = (n-1)H(y)\,, \tag{1.31}$$

where $H(y)$ is the mean curvature of $\partial\Omega$ at y. We note that the inequality (1.31) reduces to an equality for $x \in \partial\Omega$.

For what concerns the assumptions on the boundary datum U, we shall suppose that this function is the restriction to $\partial\Omega$ of a function of class $C^2(\mathbf{R}^n)$, that we shall again denote by U. Since Ω is bounded, it is not restrictive to suppose that U has compact support.

We shall look for upper barriers of the form:

$$v(x) = U(x) + \psi(d(x))\,, \tag{1.32}$$

where $\psi(t)$ is a regular function, satisfying the relations:

$$\psi(0) = 0; \quad \psi'(t) > 0; \quad \psi''(t) < 0\,. \tag{1.33}$$

We compute easily:

$$\mathcal{L}(v) = A^{ij}U_{ij} + \psi' A^{ij} d_{ij} + \frac{\psi''}{(\psi')^2}(\mathcal{E} + A^{ij}U_iU_j - 2A^{ij}v_iU_j)\,, \tag{1.34}$$

where as usual we have set $v_i = \frac{\partial v}{\partial x_i}$, etc.

We have

$$A^{ij}U_{ij} \leq c\Lambda$$

$$0 < \alpha =: A^{ij}U_iU_j \leq c\Lambda$$

and from the SCHWARTZ inequality:

$$2|A^{ij}v_iU_j| \leq 2\sqrt{\mathcal{E}\alpha} \leq \frac{1}{2}\mathcal{E} + 2\alpha\,.$$

In conclusion (remember that $\psi'' < 0$) we have

$$\mathcal{L}(v) \leq c\Lambda + \psi' A^{ij} d_{ij} + \frac{\psi''}{(\psi')^2}\left(\frac{1}{2}\mathcal{E} - c\Lambda\right). \tag{1.35}$$

From this inequality we shall discuss two different situations, in a sense opposite to each other. In the first case, we shall assume that the domain Ω is convex, and we shall make the fewest possible assumptions on the functional $\mathcal{F}$; in the second we shall consider general domains and we shall impose conditions on $\mathcal{E}$ and Λ (and hence on the function $F(z)$) that permit the construction of barriers for every boundary datum U.

Case 1. Convex domains. In this case the matrix $\{d_{ij}\}$ is negative semidefinite, and therefore we have $A^{ij}d_{ij} \leq 0$. It follows that

$$\mathcal{L}(v) \leq c\Lambda + \frac{\psi''}{(\psi')^2}\left(\frac{1}{2}\mathcal{E} - c\Lambda\right). \tag{1.36}$$

Assume now that

$$\lim_{|z|\to\infty} \frac{\Lambda(z)}{\mathcal{E}(z)} = 0. \tag{1.37}$$

Since $|Dv| \geq \psi' - \sup|DU|$ we can conclude that if ψ' is greater than some constant L, there will result $c\Lambda < \frac{1}{4}\mathcal{E}$ and therefore, since $\psi'' < 0$:

$$\mathcal{L}(v) \leq \frac{1}{4}\mathcal{E}\left\{\frac{\psi''}{(\psi')^2} + 1\right\}. \tag{1.38}$$

Let now t_0 and σ be two positive numbers, that we shall fix later, and let

$$\psi(d) = \log(1 + \sigma d).$$

For $x \in \Sigma_{t_0}$ we have

$$\psi'(d(x)) = \frac{\sigma}{1 + \sigma d(x)} \geq \frac{\sigma}{1 + \sigma t_0}. \tag{1.39}$$

If the last quantity is greater than L we can use (1.38); observing that with our choice of ψ we have $\psi'' = -(\psi')^2$, we can conclude that the function $v = U(x) + \psi(d(x))$ is a supersolution in Σ_{t_0}. It will be an upper barrier if it satisfies the relation:

$$\psi(t_0) = \log(1 + \sigma t_0) \geq \sup U - \inf U = L_1.$$

It is now easy to see that these two inequalities can be satisfied taking $t_0\sqrt{\sigma} = 1$ and choosing σ sufficiently large. In conclusion, we have proved the following theorem:

Theorem 1.4 *Let Ω be convex, and let* (1.37) *hold. Then, the functional $\mathcal{F}$ has a minimum in* Lip(Ω, U).

Case 2. General domains. If we eliminate the assumption of the convexity of Ω, it is necessary to take into account the second term in (1.35):

$$\psi' A^{ij} d_{ij} \leq c(1 + |Dv|)\Lambda,$$

and therefore instead of (1.36) we arrive at the weaker inequality:

$$\mathcal{L}(v) \leq c(1 + |Dv|)\Lambda + \frac{\psi''}{(\psi')^2}\left(\frac{1}{2}\mathcal{E} - c\Lambda\right).$$

This inequality will be sufficient if we assume that

$$\limsup_{|z|\to\infty} \frac{|z|\Lambda(z)}{\mathcal{E}(z)} < +\infty\,. \tag{1.40}$$

In this case, if ψ' is large enough, we have

$$c\Lambda \le \frac{1}{4}\mathcal{E}\,, \quad (1+|Dv|)\Lambda \le \frac{1}{4}c\mathcal{E}\,,$$

and therefore

$$\mathcal{L}(v) \le \frac{1}{4}\mathcal{E}\left\{\frac{\psi''}{(\psi')^2} + c\right\}.$$

Choosing $\psi(d) = c^{-1}\log(1+\sigma d)$ and arguing as above, we obtain the required barrier, and hence the following:

Theorem 1.5 *If the relation* (1.40) *is satisfied, the functional $\mathcal{F}$ has a minimum in the class* Lip (Ω, U), *for every bounded open set Ω with regular boundary, and for every function U of class C^2.*

We note that (1.40) is satisfied if the differential operator $\mathcal{L}(v)$ is uniformly elliptic; in other words if $\lambda(z) \ge \nu\Lambda(z)$, for some $\nu > 0$. This happens for instance when $F(z) = |z|^p$, or when $F(z) = (1+|z|^2)^{p/2}$, with $p > 1$, in particular in the case of the DIRICHLET integral:

$$\mathcal{D}(v) = \int_\Omega |Dv|^2\,dx\,.$$

1.4 The Area Functional

When $F(z) = \sqrt{1+|z|^2}$, the corresponding functional:

$$\mathcal{A}(u,\Omega) = \int_\Omega \sqrt{1+|Du|^2}\,dx \tag{1.41}$$

gives the area of the graph S of the function u. In this case, we have

$$A^{ij}(z) = (1+|z|^2)^{-3/2}\{\delta_{ij}(1+|z|^2) - z_i z_j\} \tag{1.42}$$

and therefore

$$\Lambda = (1+|z|^2)^{-1/2}\,; \quad \lambda = (1+|z|^2)^{-3/2}\,; \quad \mathcal{E} = |z|^2\lambda\,. \tag{1.43}$$

It is easily seen that (1.37), let alone the stronger assumption (1.40), is not satisfied, and hence it is not possible to apply to this case the results of the preceding section. Nevertheless, due to the special form of the matrix

$\{A^{ij}\}$, it is possible to construct barriers, and hence to prove the existence of non-parametric minimal surfaces, that is of minima of the functional (1.41) assuming prescribed values on the boundary, for a large class of domains Ω, containing all the convex sets.

Once again, we shall look for barriers of the form:

$$v(x) = U(x) + \psi(d(x)) \tag{1.44}$$

with the function ψ satisfying conditions (1.33). For such functions v we have

$$\mathcal{L}(v) = A^{ij}(U_{ij} + \psi' d_{ij}) + \psi'' A^{ij} d_i d_j \,. \tag{1.45}$$

Assume now that there exists a constant c such that

$$A^{ij} d_{ij} \leq c(|Dv| + 1)\lambda \,. \tag{1.46}$$

In this case, observing that $A^{ij} d_i d_j \geq \lambda |d_x|^2 = \lambda$, we get

$$\mathcal{L}(v) \leq c\Lambda + \lambda\{c\psi'(|Dv| + 1) + \psi''\} \,,$$

and since $|Dv| \leq c + \psi'$:

$$\mathcal{L}(v) \leq \lambda \left\{ \psi'' + c\psi'(1 + \psi') + c\frac{\Lambda}{\lambda} \right\} .$$

Assuming now that $\psi' \geq 1$, and remembering that

$$\frac{\Lambda}{\lambda} = (1 + |Dv|^2) \leq c + \psi'^2 \leq c\psi'^2 \,,$$

we obtain in conclusion:

$$\mathcal{L}(v) \leq \lambda\{\psi'' + c\psi'^2\} \,. \tag{1.47}$$

Arguing as in the preceding section, it is now simple to verify that the function:

$$\psi(d) = c^{-1} \log(1 + \sigma d)$$

is an upper barrier in Σ_{t_0}, provided we choose $t_0 = \frac{1}{\sqrt{\sigma}}$ and σ large enough.

It remains to discuss the geometric meaning of condition (1.46). Taking into account the fact that $d_{ij} d_j = \frac{1}{2} \frac{\partial}{\partial x_i} |d_x|^2 = 0$, we have

$$\begin{aligned} A^{ij}(Dv) d_{ij} &= \lambda\{(1 + |Dv|^2)\Delta d - v_i v_j d_{ij}\} \\ &= \lambda\{(1 + |Dv|^2)\Delta d - U_i U_j d_{ij}\} \,. \end{aligned}$$

Since the function $d(x)$ is of class C^2 in a neighborhood of $\partial\Omega$, the last term can be estimated by a constant, and hence we can conclude that (1.46) holds if

$$\Delta d \leq 0 \tag{1.48}$$

in a neighborhood of $\partial\Omega$, or otherwise, in virtue of (1.31), if the mean curvature of $\partial\Omega$ is non-negative. Thus we have:

Theorem 1.6 *Let Ω be a bounded open set in $\mathbf{R}^n$, whose boundary is a C^2 manifold with non-negative mean curvature. Then, for every function U of class C^2 the area functional* (1.41) *has a unique minimum in* $\mathrm{Lip}(\Omega, U)$.

In particular, the DIRICHLET problem has a solution for every C^2 boundary datum if Ω is convex. However, except in the two-dimensional case ($n = 2$), the condition of non-negative mean curvature is obviously more general than that of convexity.

Remark 1.2 Since the function $v(x)$ given by (1.44) satisfies the estimate $|Dv| \leq c + |DU|$, observing that in the proof of the preceding theorem we only need the first and second derivatives of U, we can conclude that the inequality (1.46) is satisfied only if we assume that $\Delta d \leq \epsilon_0$, with $\epsilon_0 > 0$ depending only on the C^2 norm of the function U. □

1.5 Non-Existence of Minimal Surfaces

We shall now examine in more detail the condition of non-negative mean curvature, and we shall prove that in a certain sense it is necessary for the general solvability of the DIRICHLET problem for the area functional. More precisely, we shall show that if the mean curvature $H(x_0)$ of $\partial\Omega$ is negative at some point x_0 of the boundary, then there exists a regular function U for which the area functional has no minimum in $\mathrm{Lip}(\Omega, U)$.

For that, we need a second version of the maximum principle. Assume that Ω is connected[5] and that its boundary $\partial\Omega$ is the union of two disjoints sets:

$$\partial\Omega = \partial^0\Omega \cup \partial^1\Omega$$

with $\partial^1\Omega$ open in $\partial\Omega$ (that is $\partial^1\Omega = \partial\Omega \cap A$ for some open set A) and $\partial^0\Omega$ non-empty.

[5] This assumption is not restrictive, since otherwise it is possible to consider separately its connected components.

Lemma 1.5 (Maximum principle II) *Let Ω as above, and let u be a function minimizing the area functional in* $\mathrm{Lip}(\Omega)$.

Let $v \in C^1(\Omega) \cap C(\bar{\Omega})$ be a super-minimum such that

$$u \leq v \ in \ \partial^0\Omega, \tag{1.49}$$

$$\liminf_{t\to 0^+} \inf_{A\cap\Gamma_t} \frac{\partial v}{\partial \nu} > [u]_{0,1}, \tag{1.50}$$

where $\frac{\partial v}{\partial \nu}$ is the derivative in the direction of the exterior normal to Γ_t.

Then, $u \leq v$ in Ω.

Proof. Assume first that $u < v$ on $\partial^0\Omega$. By continuity, there exists a $t_0 > 0$ such that for every $t < t_0$:

$$\frac{\partial v}{\partial \nu} > [u]_{0,1} \text{ in } \Gamma_t \cap A, \tag{1.51}$$

$$v \geq u \text{ in } \Gamma_t - A. \tag{1.52}$$

Suppose now that $u(x_0) - v(x_0) > 0$ for some $x_0 \in \Omega$, and let $t < t_0$ be such that $x_0 \in \Omega_t =: \Omega - \overline{\Sigma_t}$. By Lemma 1.1 the restriction to Ω_t of the function $w = u - v$ will take its (positive) maximum at some point of Γ_t, or better of $\Gamma_t \cap A$, since $v \geq u$ in $\Gamma_t - A$. Let x_1 be such a point; if we indicate by ν the exterior normal to Γ_t at x_1, we get from (1.51):

$$\liminf_{h\to 0^+} \frac{w(x_1 - h\nu) - w(x_1)}{h} \geq -[u]_{0,1} + \frac{\partial v}{\partial \nu} > 0.$$

This inequality contradicts the assumption that w takes its maximum in x_1, and the lemma is proved if $u < v$ on $\partial^0\Omega$. The general case follows easily by writing $v + \epsilon$ ($\epsilon > 0$) instead of v, and letting ϵ tend to zero. □

We remark that the above lemma holds for every strictly convex functional. Inequality (1.50) is trivially satisfied if

$$\frac{\partial v}{\partial \nu} = +\infty \text{ on } \partial^1\Omega$$

as will happen in the following.

We can now prove the non-existence of non-parametric minimal surfaces.

Theorem 1.7 *Let Ω be a connected bounded open set in $\mathbf{R}^n$ with C^2 boundary $\partial\Omega$, and let $H(x_0) < 0$ at some point $x_0 \in \partial\Omega$.*

Then, there exists a regular function U such that the area functional $\mathcal{A}(u, \Omega)$ does not have a minimum in $\mathrm{Lip}(\Omega, U)$.

Proof. Consider a function u minimizing the functional $\mathcal{A}$ in $\mathrm{Lip}(\Omega, U)$, and let us begin by estimating u in $\Omega - \overline{B_R(x_0)}$, $R > 0$.

For x outside $\overline{B_R}$ let $\delta(x) = \mathrm{dist}\,(x, B_R) = |x - x_0| - R$, and

$$v(x) = K + \psi(\delta(x)), \quad K > 0.$$

Recalling that $|D\delta|^2 = 1$ and therefore $\delta_{ij}\delta_j = 0$, we get from (1.45):

$$\mathcal{L}(v) = \lambda\{[\psi' + (\psi')^3]\Delta\delta + \psi''\} \tag{1.53}$$

so that, choosing $\psi(\delta) = -B\sqrt{\delta}$:

$$\mathcal{L}(v) \leq \lambda\{(\psi')^3\Delta\delta + \psi''\} \leq \lambda\frac{B\delta^{-3/2}}{4}\left\{\frac{1-n}{2\,\mathrm{diam}\,(\Omega)}B^2 + 1\right\} \tag{1.54}$$

since

$$\Delta\delta = \frac{n-1}{|x - x_0|} \geq \frac{n-1}{\mathrm{diam}\,(\Omega)}.$$

Taking $B^2 = 2\frac{\mathrm{diam}\,(\Omega)}{n-1}$, we get $\mathcal{L}(v) \leq 0$, so that v is a supersolution. We have obviously $\frac{\partial v}{\partial n} = +\infty$ on ∂B_R, and hence, choosing

$$K = \sup_{\partial\Omega - \overline{B_r}} U + B\sqrt{\mathrm{diam}\,(\Omega)},$$

we obtain from the preceding lemma the estimate

$$\sup_{\Omega - \overline{B_R}} u \leq \sup_{\partial\Omega - \overline{B_R}} U + B\sqrt{\mathrm{diam}\,(\Omega)},$$

from which in particular:

$$\sup_{\partial B_R \cap \Omega} u \leq \sup_{\partial\Omega - \overline{B_R}} U + B\sqrt{\mathrm{diam}\,(\Omega)}. \tag{1.55}$$

We shall now estimate the supremum of u in $\Omega \cap B_R$.

Since the mean curvature of $\partial\Omega$ at x_0 is negative, and $\partial\Omega$ is of class C^2, taking into account (1.31) we can assume that there exists two positive numbers ϵ_0 and R such that

$$\Delta d \geq \epsilon_0 \quad \text{in } \Omega \cap B_R(x_0)$$

in which as usual we have set $d(x) = \mathrm{dist}\,(x, \partial\Omega)$.

Taking again $v = \psi(d) = \alpha - \beta\sqrt{d}$ we get as above

$$\mathcal{L}(v) \leq \lambda\{(\psi')^3\Delta d + \psi''\} \leq \frac{\lambda\beta}{4d^{3/2}}(1 - \epsilon_0\beta^2) < 0$$

provided $\epsilon_0\beta^2 > 1$. Moreover, setting

$$\alpha = \sup_{\partial B_R \cap \Omega} u + \beta\sqrt{\operatorname{diam}(\Omega)}$$

it is possible to use Lemma 1.5 once again to obtain, taking into account (1.55), the estimate

$$\sup_{\Omega \cap B_R} u \leq \sup_{\partial\Omega - B_R} U + (\beta + B)\sqrt{\operatorname{diam}(\Omega)}$$

and hence in particular:

$$\sup_{\partial\Omega \cap B_R} U \leq \sup_{\partial\Omega - B_R} U + (\beta + B)\sqrt{\operatorname{diam}(\Omega)}\,. \tag{1.56}$$

The above inequality gives a necessary condition for the solvability of the DIRICHLET problem for the area functional.

If this condition does not hold for the boundary datum U, as it is the case if we take

$$U = 0 \quad \text{on } \partial\Omega - B_R$$

and

$$U(x_0) > (\beta + B)\sqrt{\operatorname{diam}(\Omega)}\,,$$

the area functional cannot have minimum in $\operatorname{Lip}(\Omega, U)$. □

The example that follows show what we can expect in that situation.

Example 1.1 Let $n = 2$ and let A_ϱ^R be the annulus

$$A_\varrho^R = \{x \in R : \varrho < |x| < R\}\,.$$

Consider the function:

$$U = \begin{cases} 0 & \text{on } \partial B_R\,, \\ M & \text{on } \partial B_\varrho\,, \end{cases}$$

where M is a positive constant. By the strict convexity of the area functional and the symmetry of the domain A_ϱ^R and of the datum U, the only minimum in $\operatorname{Lip}(A_\varrho^R, U)$, if it exists at all, must be a function $u(r)$ depending only on $r = |x|$.

The EULER equation in this case becomes

$$u'' + \frac{1}{r}u'[1 + (u')^2] = 0$$

and its solution, taking into account the condition $u(R) = 0$, is

$$u(r) = c \log \frac{R + \sqrt{R^2 - c^2}}{r + \sqrt{r^2 - c^2}}\,.$$

The constant c, $0 \leq c \leq \varrho$ must be determined from the condition $u(\varrho) = M$. We have

$$u(\varrho) = c \log \frac{R + \sqrt{R^2 - c^2}}{\varrho + \sqrt{\varrho^2 - c^2}} \leq \varrho \log \frac{R + \sqrt{R^2 - \varrho^2}}{\varrho} = M_0(R, \varrho)$$

and hence the DIRICHLET problem can be solved only if $M \leq M_0$.

In the limit case $M = M_0$ we have $c = \varrho$ and the normal derivative

$$u_r = -\frac{\varrho}{\sqrt{r^2 - \varrho^2}}$$

becomes infinite on the internal circumference. If $M > M_0$ there are no solutions in A_ϱ^R assuming the given values at the boundary. In this case the minimal surface is given by the graph of the solution $u(r)$ corresponding to the limit value M_0, plus the portion of the vertical cylinder having for base the internal circumference of radius ϱ, that lies between the levels M_0 and M.

1.6 Notes and Comments

The methods of this chapter take their origin from the ideas of S. BERNSTEIN [1]–[5], as generalized by SERRIN [3]. In his paper, SERRIN obtained *a priori* estimates for solutions of general elliptic equations:

$$A^{ij}(x, u, Du) D_{ij} u = B(x, u, Du)\,,$$

from which, using the fixed point theorem of SCHAUDER, he could prove the existence of solutions of the DIRICHLET problem. Our point of view is slightly different, and it is inspired by the papers of HARTMAN and STAMPACCHIA [1], and even more so by that of M. MIRANDA [1] on non-parametric minimal surfaces. These papers in turn echo the methods introduced by HILBERT [1], [2] in his proof of the existence of harmonic functions in two-dimensional domains (DIRICHLET principle), later adapted by HAAR [1], RADO [1], [2] and others to the minimal surface equation in dimension two. In particular, HILBERT [1] introduced the so-called *three points condition*, which in dimension two is equivalent to the B.S.C.

The conclusion of Theorem 1.6 remains valid only if we assume that the boundary datum U is of class $C^{1,\alpha}$, with $\alpha > 0$, but not in general if U is only Lipschitz-continuous (GIUSTI [4]).

On the other hand it can be proved, always under the hypothesis of non-negative mean curvature of the boundary $\partial\Omega$, that for every continuous boundary datum U the DIRICHLET problem has a solution u of class $C^2(\Omega) \cap C^0(\bar{\Omega})$.

The main ingredient in the proof of the last result is the *a priori* inequality for the gradient:

$$|Du(x_0)| \le c_1 \exp\left\{ c_2 \frac{\operatorname{osc}(u, B_R(x_0))}{R} \right\} \tag{1.57}$$

in which $\operatorname{osc}(u, B_R(x_0))$ indicates the oscillation of u in the ball of radius R centered at x_0 (BOMBIERI, DE GIORGI and MIRANDA [1]).

Consider now a sequence U_k of regular functions, uniformly convergent to U on $\partial\Omega$. Denoting by u_k the solution of the problem with datum U_k, whose existence follows from Theorem 1.6, we have by the maximum principle:

$$\sup_{\Omega} |u_k - u_h| \le \sup_{\partial\Omega} |U_k - U_h|,$$

and therefore the sequence u_k converges uniformly in Ω to a function u, with $u = U$ on $\partial\Omega$. By inequality (1.57) the sequence u_k has first derivatives locally equibounded in Ω, and hence u it is Lipschitz-continuous.

To conclude the proof, we observe that the minima of the functionals of this chapter, under assumptions of regularity for the function $F(z)$ and of uniform convexity:

$$F_{z_i z_j}(z)\xi_i\xi_j \ge \nu(z)|\xi|^2$$

are regular functions (see later; Chapters 8 and 10). In particular, this is true for the function u, which is therefore a classical solution of the DIRICHLET problem for the area functional.[6]

The proof of the non-solvability of the DIRICHLET problem for the minimal surface equation in open sets whose boundary has negative mean curvature at some point is due to FINN [1] and to JENKINS and SERRIN [2].

---

[6]For further information on minimal surfaces, one can see the books by GIUSTI [6] and by MASSARI and MIRANDA [1].

This case can be treated by considering the "relaxed" functional:

$$\mathcal{F}(u,\Omega) = \int_\Omega \sqrt{1+|Du|^2} + \int_{\partial\Omega} |u-U| dH_{n-1}$$

in the class $BV(\Omega)$ of functions whose derivatives are Radon measures in Ω. It can be proved that the functional $\mathcal{F}$ has minimum in $BV(\Omega)$ for arbitrary open set Ω and boundary datum $U \in L^1(\partial\Omega)$, and that the minimizing function $u(x)$ is regular in the interior of Ω (once again a major role is played by the *a priori* inequality (1.57) for the gradient). Moreover, if $\partial\Omega$ has non-negative mean curvature in a neighborhood of a point $x_0 \in \partial\Omega$ and if U is continuous at x_0, then $u(x_0) = U(x_0)$ (MIRANDA [2]). Therefore if $\partial\Omega$ has non-negative mean curvature at every point, and if U is a continuous function, the minimum of $\mathcal{F}$ is the solution of the DIRICHLET problem for the minimal surface equation.

Chapter 2

Measurable Functions

2.1 L^p Spaces

The purpose of this section is to list some definitions and properties of L^p spaces that will be useful later. For the proofs, the reader is referred to one of the many books on Lebesgue integrals, such as ROYDEN [1] or SAKS [1].

Of course, when we speak of measurable sets and functions, we shall always refer to Lebesgue measure; as usual, we shall not distinguish between functions that differ only on a set of zero measure, so that for instance the statement "the function $f(x)$ is continuous in A" means strictly speaking "the function f coincides almost everywhere with a function $\tilde{f}$ continuous in A." If E is a measurable set, we shall indicate its measure with meas (E), or briefly with $|E|$.

We begin by recalling the definitions of some well-known function spaces.

We shall denote by $C^k(\Omega)$ $(k = 0, 1, \ldots)$ the space of the functions having continuous derivatives up to and including the order k (if $k = 0$, it will be the space of continuous functions); and with $C^\infty(\Omega)$ the space of infinitely differentiable functions in Ω, that is the intersection of all the spaces $C^k(\Omega)$. With $C^k(\bar{\Omega})$ we will indicate the space of functions in $C^k(\Omega)$, whose derivatives up to the order k can be extended to continuous functions up to the boundary $\partial\Omega$, and with $C_0^k(\Omega)$ $(k = 0, 1, \ldots)$ the subspace of $C^k(\bar{\Omega})$ of the functions with compact support contained in Ω.

The spaces $C^k(\bar{\Omega})$ are Banach spaces, with the norm[1]:

$$\|u\|_{C^k} = \sum_{|\beta| \le k} \sup_{x \in \Omega} |D^\beta u(x)| .$$

If $0 < \alpha \le 1$, and D is a domain in $\mathbf{R}^n$, (namely, the closure of a bounded open set), we shall denote by $C^{0,\alpha}(D)$ the space of *Hölder-continuous* functions in D; that is continuous functions for which[2]

$$[u]_{0,\alpha} =: \sup_{\substack{x,y \in D \\ x \neq y}} \frac{|u(x) - u(y)|}{|x - y|^\alpha} < +\infty . \tag{2.1}$$

More generally, we shall denote by $C^{k,\alpha}(D)$ the space of functions whose derivatives up to order k are Hölder-continuous[3] in D. The spaces $C^{k,\alpha}(D)$ are Banach spaces, with the norm

$$\|u\|_{k,\alpha} = \|u\|_{C^k} + \sum_{|\beta| = k} [D^\beta u]_{0,\alpha} .$$

Finally, if Ω is an open set in $\mathbf{R}^n$, we shall indicate with $C^{k,\alpha}(\Omega)$ the space of the functions belonging to $C^{k,\alpha}(D)$ for every domain $D \subset \Omega$.

We recall the following results relative to the Lebesgue integral:

Theorem 2.1 (LUSIN) *Let $f(x)$ be a measurable function in $\mathbf{R}^n$, with $f = 0$ outside an open set A of finite measure. For every $\epsilon > 0$ there exists a function $g_\epsilon \in C_0^0(A)$ such that*

$$\operatorname{meas}\{x \in \mathbf{R}^n : f(x) \neq g_\epsilon(x)\} < \epsilon$$

and

$$\sup_{\mathbf{R}^n} |g_\epsilon| \le \sup_{\mathbf{R}^n} |f| .$$

Theorem 2.2 (EGOROV) *Let E be a measurable set with $|E| < +\infty$, and let $\{f_j\}$ be a sequence of measurable functions in E, converging almost everywhere to a function f. Then, for every $\epsilon > 0$ there exists a measurable set N with $|N| < \epsilon$ and such that $f_j \to f$ uniformly in $E - N$.*

---

[1] For a full explanation of the notation for the derivatives, see next chapter.

[2] Hölder-continuous functions with $\alpha = 1$ coincide with the Lipschitz-continuous functions introduced in the preceding chapter.

[3] More precisely, functions k times differentiable in the interior of D, whose derivatives extend to Hölder-continuous functions in D.

Theorem 2.3 (Absolute continuity of the integral) *If $f(x)$ is a summable function in $\mathbf{R}^n$, then for every $\epsilon > 0$ there exists a $\delta > 0$ such that if E is measurable and $|E| < \delta$, then*

$$\int_E |f| dx < \epsilon .$$

We recall the following well-known definitions:

Definition 2.1 *Let Ω be an open set in $\mathbf{R}^n$, and let $1 \leq p < +\infty$. We denote by $L^p(\Omega, \mathbf{R}^N)$ the space of measurable functions $f : \Omega \to \mathbf{R}^N$ such that*

$$\|f\|_{p,\Omega} = \left\{ \int_\Omega |f|^p \, dx \right\}^{\frac{1}{p}} < +\infty . \tag{2.2}$$

Moreover, by $L^\infty(\Omega, \mathbf{R}^N)$ we indicate the space of bounded measurable functions in Ω.

When no possible confusion might arise, we shall write simply $L^p(\Omega)$ or even L^p, without explicit mention of the codomain $\mathbf{R}^N$.

The spaces $L^p(\Omega)$, $(1 \leq p < +\infty)$ and $L^\infty(\Omega)$ are Banach spaces, respectively, with the norm (2.2), and

$$\|f\|_{\infty,\Omega} = \sup_{x \in \Omega} |f(x)| , \tag{2.3}$$

where

$$\sup_{x \in \Omega} |f(x)| = \inf\{\lambda \in \mathbf{R} : |f| \leq \lambda \text{ a.e. in } \Omega\} .$$

Theorem 2.4 *From every sequence f_k in $L^1_{\text{loc}}(\Omega)$,*[4] *converging to a function f in $L^1_{\text{loc}}(\Omega)$ (i.e. such that $f_k \to f$ in $L^1(K)$ for every compact set $K \subset \Omega$) we can extract a subsequence converging to f almost everywhere in Ω.*

If f is in $L^p(\Omega)$, the set

$$F_t = \{x \in \Omega : |f(x)| > t\}$$

is measurable, and since

$$\int_\Omega |f|^p \, dx \geq \int_{F_t} |f|^p \, dx \geq t^p |F_t| ,$$

[4] We recall that if $V(\Omega)$ is a space of functions in Ω, $V_{\text{loc}}(\Omega)$ is the space of functions belonging to $V(\Lambda)$ for every open set $\Lambda \subset\subset \Omega$ (that is such that $\bar{\Lambda}$ is a compact set contained in Ω).

we have

$$|F_t| \le t^{-p} \|f\|_{p,\Omega}^p . \tag{2.4}$$

Finally, we recall the well-known formula:

$$\int_\Omega |f|^p \, dx = p \int_0^{+\infty} t^{p-1} |F_t| dt . \tag{2.5}$$

An immediate consequence of (2.4) and of Theorems 2.1 and 2.3 is the following:

Theorem 2.5 *Let Ω be an open set in $\mathbf{R}^n$, and let $p < +\infty$. Then $C_0^0(\Omega)$ is dense in $L^p(\Omega)$.*

Proof. Let $f \in L^p(\Omega)$, and let $\tau > 0$. There will exist a ball $B_R \subset \mathbf{R}^n$ such that $\int_{\Omega - B_R} |f|^p \, dx < \tau$. The function

$$g(x) = \begin{cases} \max(-T, \min(f, T)) & \text{in } \Omega \cap B_R , \\ 0 & \text{otherwise,} \end{cases}$$

is measurable and has compact support; by Lusin's theorem, for every $\epsilon > 0$ there exists a function $g_\epsilon \in C_0^0(\Omega \cap B_R)$ coinciding with g outside a set Σ with measure less than ϵ, and such that $|g_\epsilon| \le T$. We have then:

$$\begin{aligned} \int_\Omega |f - g_\epsilon|^p \, dx &\le c \int_{\Omega \cap B_R} |f - g|^p \, dx + c \int_{\Omega \cap B_R} |g - g_\epsilon|^p \, dx \\ &\quad + \int_{\Omega - B_R} |f|^p \, dx \le c \int_{F_T} |f|^p \, dx + cT\epsilon + \tau . \end{aligned}$$

Taking T large enough, and using (2.4) and Theorem 2.3, the last integral can be made smaller than τ. Choosing ϵ so that $cT\epsilon < \tau$ we get

$$\int_\Omega |f - g_\epsilon|^p \, dx < 3\tau$$

and the theorem is proved. □

Concerning the functional structure of L^p-spaces, we have the following results:

Theorem 2.6 *For $1 < p < +\infty$, L^p is a reflexive Banach space, whose dual is isomorphic to the space L^q, with $\frac{1}{p} + \frac{1}{q} = 1$. The space L^1 has L^∞*

as its dual, but is not reflexive. Finally, L^2 is a Hilbert space, with scalar product given by

$$(f, g) = \int_\Omega f(x)g(x)dx\,. \tag{2.6}$$

We have in addition:

Proposition 2.1 (HÖLDER's inequality) *Let $p > 1$, $f \in L^p$ and $g \in L^q$, with $\frac{1}{p} + \frac{1}{q} = 1$. Then $fg \in L^1$, and*

$$\int_\Omega |f(x)g(x)|dx \leq \|f\|_{p,\Omega}\|g\|_{q,\Omega}\,. \tag{2.7}$$

It is easily seen that inequality (2.7), that for $p = q = 2$ bears also the name of Schwartz inequality, remains valid for $p = 1$ and $q = +\infty$.

In the particular case when Ω has finite measure, if $1 \leq s < r \leq +\infty$, we may take $f = |u|^s$, $g = 1$ and $p = r/s$ in (2.7), obtaining

$$\|u\|_{s,\Omega} \leq |\Omega|^{\frac{1}{s}-\frac{1}{r}}\|u\|_{r,\Omega} \tag{2.8}$$

so that, if $|\Omega|$ is finite and $r > s$, we have $L^r(\Omega) \subset L^s(\Omega)$ algebraically and topologically (that is, the topology of L^r is stronger than that of L^s).

Inequality (2.8) can be stated in a more suggestive way by saying that for $u \in L^r$ the function

$$\gamma(p) = \left\{ ⨍_\Omega |u|^p\, dx \right\}^{\frac{1}{p}} =: \left\{ \frac{1}{|\Omega|} \int_\Omega |u|^p\, dx \right\}^{\frac{1}{p}}$$

increases in the interval $[1, r]$. It is not difficult to show that $\gamma(p)$ is a continuous function, and that for $u \in L^\infty$ we have

$$\lim_{p \to +\infty} \gamma(p) = \|u\|_{\infty,\Omega}\,.$$

We conclude this section with the following:

Theorem 2.7 (LEBESGUE) *Let $f(x)$ be a function in $L^1(\Omega)$. For almost every $x \in \Omega$ we have*

$$\lim_{R \to 0^+} ⨍_{Q(x,R)} |f(y) - f(x)|dy = 0 \tag{2.9}$$

and hence

$$\lim_{R \to 0^+} ⨍_{Q(x,R)} f(y)dy = f(x)\,. \tag{2.10}$$

2.2 Test Functions and Mollifiers

The functions of class C_0^∞ are also called *test functions.* A typical test function is

$$\eta(x) = \begin{cases} \exp\left(\dfrac{1}{|x|^2 - 1}\right) & \text{if } |x| < 1\,, \\ 0 & \text{if } |x| \geq 1 \end{cases}$$

($\exp(t) = e^t$), whose support is the closure of the unit ball B.

Definition 2.2 *A function* $\varphi \in C_0^\infty(\mathbf{R}^n)$ *such that*

(i) $\varphi(x) \geq 0$,
(ii) $\operatorname{supp}\varphi \subset \bar{B}$,
(iii) $\int \varphi(x)dx = 1$

is called a mollifier.

For instance, the function $\eta(x)$ defined above becomes a mollifier when multiplied by a suitable constant so that its integral becomes equal to 1. For our purposes it will be sufficient to consider only spherically symmetric mollifiers such as $\eta(x)$, depending only on $|x|$.

Given a function $f(x) \in L^1_{\text{loc}}(\mathbf{R}^n))$, we call ϵ*-regularized* (or simply *regularized*) of f the function

$$\begin{aligned} f_\epsilon(x) &= \int f(y)\varphi_\epsilon(x-y)dy = \int f(x-z)\varphi_\epsilon(z)dz \\ &= \int f(x-\epsilon y)\varphi(y)dy\,, \end{aligned} \tag{2.11}$$

where

$$\varphi_\epsilon(x) = \epsilon^{-n}\varphi\left(\frac{x}{\epsilon}\right). \tag{2.12}$$

By differentiating under the integral sign, it follows immediately from (2.11) that for every $\epsilon > 0$, f_ϵ is an infinitely differentiable function in $\mathbf{R}^n$. Moreover, if the support of f is contained in K, we will have

$$\operatorname{supp} f_\epsilon \subset K^\epsilon = \{x \in \mathbf{R}^n : \operatorname{dist}(x, K) \leq \epsilon\}\,.$$

Definition 2.2 can be reformulated in terms of the convolution product, recalling that the convolution of two functions u and v in $\mathbf{R}^n$, at least one of which has compact support, is defined by

$$u * v(x) = \int u(x-y)v(y)dy .$$

We have then

$$f_\epsilon = f * \varphi_\epsilon = \varphi_\epsilon * f .$$

Lemma 2.1 *If $u \in L^p$ then $u_\epsilon \in L^p$, and*

$$\|u_\epsilon\|_p \leq \|u\|_p . \tag{2.13}$$

Proof. We have

$$|u_\epsilon(x)|^p \leq \left\{ \int |u(y)| \{\varphi_\epsilon(x-y)\}^{\frac{1}{p}} \{\varphi_\epsilon(x-y)\}^{1-\frac{1}{p}} dy \right\}^p .$$

Using Hölder's inequality and remarking that

$$\int \varphi_\epsilon(x-y)dy = \int \varphi_\epsilon(x-y)dx = 1 ,$$

we easily get

$$|u_\epsilon(x)|^p \leq \int |u(y)|^p \varphi_\epsilon(x-y)dy$$

from which (2.13) follows by integrating with respect to x. □

Theorem 2.8 *Let $u(x)$ be a function in $\mathbf{R}^n$. When ϵ tends to zero,*

(i) *If u is continuous, u_ϵ converges to u uniformly on every compact set $K \subset \mathbf{R}^n$;*
(ii) *if $u \in L^p(\mathbf{R}^n)$, $1 \leq p < +\infty$, u_ϵ converges to u in $L^p(\mathbf{R}^n)$.*

Proof. We have

$$u_\epsilon(x) - u(x) = \int [u(x-z) - u(x)]\varphi_\epsilon(z)dz ,$$

where the integral is made on the ball $|z| < \epsilon$.

Let us prove (i) first. Let K be a compact set in $\mathbf{R}^n$, and let $\tau > 0$. Since $u(x)$ is uniformly continuous on compact sets, there will exist $\sigma > 0$ such that for $|z| < \sigma$ we have $|u(x-z) - u(x)| < \tau$ for every $x \in K$. Taking $\epsilon < \sigma$ we conclude that

$$\sup_K |u_\epsilon(x) - u(x)| < \tau \int \varphi_\epsilon(z)dz = \tau .$$

In order to show (ii) we recall that C_0^0 is dense in L^p; and hence for every $\tau > 0$ there exists a function $w \in C_0^0$ with $\|w - u\|_p < \tau$. On the other hand:

$$\|u_\epsilon - u\|_p \leq \|u_\epsilon - w_\epsilon\|_p + \|w_\epsilon - w\|_p + \|w - u\|_p$$

and hence by Lemma 2.1

$$\|u_\epsilon - u\|_p < 2\tau + \|w_\epsilon - w\|_p$$

for every $\epsilon > 0$.

Passing to the limit as $\epsilon \to 0$, and taking into account (i) (remember that w has compact support), we have

$$\limsup_{\epsilon \to 0} \|u_\epsilon - u\|_p \leq 2\tau$$

from which (ii) follows at once. □

In particular, we have proved that $C_0^\infty(\mathbf{R}^n)$ is dense in $L^p(\mathbf{R}^n)$. By a similar argument we can show the following:

Corollary 2.1 *For any open set $\Omega \subset \mathbf{R}^n$ and for any p, $1 \leq p < +\infty$, $C_0^\infty(\Omega)$ is dense in $L^p(\Omega)$.*

Proof. For every $u \in L^p(\Omega)$ and every $\tau > 0$ there will exist by Theorem 2.5 a function $w \in C_0^0(\Omega)$ such that $\|w - u\|_{p,\Omega} < \tau$. On the other hand, if ϵ is small enough, the support of w_ϵ will be contained in Ω, and $\|w - w_\epsilon\|_p < \tau$; whence in conclusion $\|w_\epsilon - u\|_p < 2\tau$. □

2.3 Morrey's and Campanato's Spaces

These spaces of integrable functions, introduced and studied by Morrey and Campanato, have proved particularly useful in the study of elliptic partial differential equations.

By $Q(x, R)$ we indicate the cube of $\mathbf{R}^n$, with sides parallel to the coordinate axes, having a center at x and side $2R$:

$$Q(x, R) = \{y \in \mathbf{R}^n : \max_{1 \leq i \leq n} |y_i - x_i| < R\}.$$

When no confusion may arise, we shall write simply Q_R, without indication of the center.

Definition 2.3 (MORREY spaces) *Let Ω be a bounded open set in $\mathbf{R}^n$, and let $1 \leq p < +\infty$ and $\lambda \geq 0$. By $L^{p,\lambda}(\Omega, \mathbf{R}^N)$ we denote the space of* *functions* $u \in L^p(\Omega, \mathbf{R}^N)$ *such that*

$$\|u\|_{p,\lambda}^p =: \sup_{\substack{x_0\in\Omega \\ \varrho>0}} \varrho^{-\lambda} \int_{\Omega(x_0,\varrho)} |u|^p \, dx < +\infty \,, \tag{2.14}$$

where $\Omega(x_0, \varrho) = \Omega \cap Q(x_0, \varrho)$.

It is clear that condition (2.14) only depends on the behavior for small radii, since for $\varrho > \epsilon$ we have

$$\varrho^{-\lambda} \int_{\Omega(x_0,\varrho)} |u|^p \, dx \leq \epsilon^{-\lambda} \int_{\Omega} |u|^p \, dx \,.$$

As in the case of L^p spaces, we shall often write $L^{p,\lambda}(\Omega)$ or simply $L^{p,\lambda}$, and we shall omit the indication of the codomain $\mathbf{R}^N$ whenever no misunderstanding is possible. It is easily seen that $\|u\|_{p,\lambda}$ is a norm, and that the space $L^{p,\lambda}$ is complete. It is also evident that if u is a function defined in $\Omega = \Omega_1 \cup \Omega_2$ and if the restrictions of u to Ω_1 and to Ω_2 belong, respectively, to $L^{p,\lambda}(\Omega_1)$ and $L^{p,\lambda}(\Omega_2)$, then u is in $L^{p,\lambda}(\Omega)$.

Finally, $\|u\|_{p,0} = \|u\|_p$, so that $L^{p,0} = L^p$. More generally, using Hölder's inequality, one proves easily that if $s \geq p$ and $\frac{n-\lambda}{p} \geq \frac{n-\mu}{s}$ the following holds:

$$\|u\|_{p,\lambda} \leq \operatorname{diam}(\Omega)^{\frac{n-\lambda}{p} - \frac{n-\mu}{s}} \|u\|_{s,\mu} \,,$$

and therefore the immersion

$$L^{s,\mu} \hookrightarrow L^{p,\lambda} \tag{2.15}$$

is continuous.

Proposition 2.2 *The space* $L^{p,n}$ *is isomorphic to* L^∞, *and*

$$\|u\|_{p,n} = 2^{\frac{n}{p}} \|u\|_\infty \,.$$

Proof. If $u \in L^\infty$ we have

$$\varrho^{-n} \int_{\Omega_\varrho} |u|^p \, dx \leq 2^n \sup_{\Omega} |u|^p$$

and hence

$$\|u\|_{p,n} \leq 2^{\frac{n}{p}} \|u\|_\infty \,.$$

Conversely, let $u \in L^{p,n}$. For almost every $x_0 \in \Omega$ we have

$$|u(x_0)| = \lim_{\varrho \to 0} \fint_{Q(x_0,\varrho)} |u| dx \,.$$

On the other hand

$$⨍_{Q(x_0,\varrho)} |u| dx \le \left(⨍_{Q(x_0,\varrho)} |u|^p \, dx \right)^{\frac{1}{p}} \le 2^{-\frac{n}{p}} \|u\|_{p,n} \,,$$

and therefore

$$|u(x_0)| \le 2^{-\frac{n}{p}} \|u\|_{p,n} \,,$$

proving the opposite inclusion. □

In particular, the spaces $L^{p,n}$ $(p \ge 1)$ are all isomorphic.

Definition 2.4 (Campanato [1]) *We denote by $\mathcal{L}^{p,\lambda}(\Omega, \mathbf{R}^N)$ the space of functions $u \in L^p(\Omega, \mathbf{R}^N)$ such that*

$$[u]^p_{p,\lambda} =: \sup_{\substack{x_0 \in \Omega \\ \varrho > 0}} \varrho^{-\lambda} \int_{\Omega(x_0,\varrho)} |u - u_{x_0,\varrho}|^p \, dx < +\infty \,, \tag{2.16}$$

where

$$u_{x_0,\varrho} =: ⨍_{\Omega(x_0,\varrho)} u \, dx$$

is the average of u in $\Omega(x_0, \varrho)$.

Remark 2.1 Instead of the cubes $Q(x, \varrho)$ we could define Campanato's spaces by means of balls $B(x, \varrho)$, or generally speaking by means of any family of neighborhoods $I(x, \varrho)$. All these spaces are isomorphic, provided there exist two constants α and β such that for every $\varrho < \operatorname{diam}(\Omega)$ one has[5]

$$I(x, \alpha\varrho) \subset Q(x, \varrho) \subset I(x, \beta\varrho) \,. \tag{2.17}$$

Actually, assuming that these relations are satisfied, we have for every $\xi \in \mathbf{R}^N$:

$$|u_{x,\varrho} - \xi|^p = \left| ⨍_{\Omega(x,\varrho)} (u(y) - \xi) dy \right|^p \le ⨍_{\Omega(x,\varrho)} |u - \xi|^p \, dy$$

and therefore

$$\int_{\Omega(x,\varrho)} |u - u_{x,\varrho}|^p \, dy \le 2^p \int_{\Omega(x,\varrho)} |u - \xi|^p \, dy$$

$$+ 2^p |u_{x,\varrho} - \xi|^p \le 2^{p+1} \int_{\Omega(x,\varrho)} |u - \xi|^p \, dy \,.$$

[5] Needless-to-say, it will be sufficient that these inclusions be valid for ϱ small enough.

Taking $\xi = u_{\Omega\cap I_{\beta\varrho}}$ and recalling (2.17), we at once get

$$\int_{\Omega(x,\varrho)} |u - u_{x,\varrho}|^p \, dy \le 2^{p+1} \int_{\Omega\cap I(x,\beta\varrho)} |u - u_{\Omega\cap I(x,\beta\varrho)}|^p \, dy \, .$$

It follows that the quantity (2.16) can be estimated by the analogous quantity, obtained by substituting to the cubes the neighborhoods I_ϱ. Interchanging the roles of these two families, we get the opposite estimate, and in conclusion the isomorphism of the relative spaces.

The same conclusion holds of course for Morrey's spaces. In particular it is possible to work with cubes with the edges in given directions, for instance parallel to the axes, as we shall do systematically in what follows. □

Remark 2.2 The quantity $[u]_{p,\lambda}$ is a seminorm in $\mathcal{L}^{p,\lambda}$, equivalent to

$$\sup_{\substack{x_0\in\Omega \\ \varrho>0}} \varrho^{-\lambda} \inf_{\xi\in\mathbf{R}^N} \int_{\Omega(x_0,\varrho)} |u - \xi|^p \, dx \, . \tag{2.18}$$

In fact, it follows from the above that

$$\int_{\Omega(x_0,\varrho)} |u - u_{x_0,\varrho}|^p \, dx \le 2^{p+1} \int_{\Omega(x_0,\varrho)} |u - \xi|^p \, dx \tag{2.19}$$

for every $\xi \in \mathbf{R}^N$. □

It follows easily that

$$|||u|||_{p,\lambda} = \|u\|_p + [u]_{p,\lambda}$$

is a norm, with which $\mathcal{L}^{p,\lambda}(\Omega, \mathbf{R}^N)$ is a Banach space. Moreover, the immersion (2.15) remains valid for the spaces $\mathcal{L}^{p,\lambda}$.

Taking $\xi = 0$ in (2.19), we conclude that the immersion of $L^{p,\lambda}$ in $\mathcal{L}^{p,\lambda}$ is continuous. Actually, if $0 \le \lambda < n$ the two spaces are equivalent, provided $\partial\Omega$ is regular enough.

Definition 2.5 *We say that Ω has no external cusps if there exists a constant $A > 0$ such that for every $x_0 \in \bar{\Omega}$ and for every ϱ, $0 < \varrho \le \operatorname{diam}\Omega$, we have*

$$|\Omega(x_0, \varrho)| \ge A|Q_\varrho| \, .$$

It is clear that an open set with Lipschitz-continuous boundary has neither external nor internal cusps. If Ω has no external cusps, the norm

$$\left\{ \sup_{\substack{x_0\in\Omega \\ \varrho>0}} |\Omega(x_0,\varrho)|^{-\frac{\lambda}{n}} \int_{\Omega(x_0,\varrho)} |u|^p \, dx \right\}^{\frac{1}{p}}$$

is equivalent to (2.14). It is also possible to replace $\varrho^{-\lambda}$ with $|\Omega(x_0,\varrho)|^{-\frac{\lambda}{n}}$ in (2.16) and (2.18), obtaining equivalent seminorms.

Lemma 2.2 *Assume that Ω has no external cusps, and let u be a function in $\mathcal{L}^{p,\lambda}(\Omega)$, and let $\tau = \frac{\lambda-n}{np}$. For every $x_0 \in \bar{\Omega}$ and for every ϱ, R, with $0 < \varrho < R < \operatorname{diam}\Omega$, we have*

$$|u_{x_0,R} - u_{x_0,\varrho}| \le c[u]_{p,\lambda} |\Omega(x_0,\varrho)|^{\tau} \quad \textit{if } \tau < 0\,, \tag{2.20}$$

$$|u_{x_0,R} - u_{x_0,\varrho}| \le c[u]_{p,\lambda} |\Omega(x_0,R)|^{\tau} \quad \textit{if } \tau > 0\,. \tag{2.21}$$

Proof. Let $\varrho \le r < s \le R$. We have[6]

$$|u_s - u_r| \le \fint_{\Omega_r} |u - u_s| dx \le \left(\fint_{\Omega_r} |u - u_s|^p \, dx \right)^{\frac{1}{p}}$$

and hence

$$|u_s - u_r| \le |\Omega_s|^{\frac{\lambda}{pn}} |\Omega_r|^{-\frac{1}{p}} [u]_{p,\lambda}\,.$$

Since Ω has no external cusps, there holds

$$A\left(\frac{s}{r}\right)^n |\Omega_r| \le |\Omega_s| \le A^{-1}\left(\frac{s}{r}\right)^n |\Omega_r| \tag{2.22}$$

and therefore

$$|u_s - u_r| \le c[u]_{p,\lambda} \left(\frac{s}{r}\right)^{\frac{\lambda}{p}} |\Omega_r|^{\tau}\,. \tag{2.23}$$

Choose now $r_i = R2^{-i}$ and $\Omega_i = \Omega_{r_i}$. Writing (2.23) with $r = r_i$ and $s = r_{i-1}$ and summing over i one obtains

$$|u_R - u_{r_k}| \le c[u]_{p,\lambda} \sum_{i=1}^{k} |\Omega_i|^{\tau}\,.$$

From (2.22) it follows that

$$A^{-1}2^{-ni}|\Omega_0| \ge |\Omega_i| \ge A2^{n(k-i)}|\Omega_k|\,,$$

---

[6] We have omitted the non-essential indication of the point x_0.

and hence, if $\tau < 0$,

$$|u_R - u_{r_k}| \le c[u]_{p,\lambda}|\Omega_k|^\tau\,,$$

whereas if $\tau > 0$,

$$|u_R - u_{r_k}| \le c[u]_{p,\lambda}|\Omega_R|^\tau\,.$$

Finally, choosing k in such a way that $r_k \le \varrho < r_{k-1}$, we obtain from (2.23) the estimate

$$|u_\varrho - u_{r_k}| \le c[u]_{p,\lambda}|\Omega_k|^\tau\,,$$

if $\tau < 0$, and

$$|u_\varrho - u_{r_k}| \le c[u]_{p,\lambda}|\Omega_\varrho|^\tau\,,$$

if $\tau > 0$. Comparing with the preceding estimates, and taking into account (2.22), we get the required result. □

It is now a simple matter to prove:

Proposition 2.3 *If Ω is a bounded open set without external cusps, and if $0 \le \lambda < n$, $L^{p,\lambda}(\Omega)$ is isomorphic to $\mathcal{L}^{p,\lambda}(\Omega)$.*

Proof. It will be sufficient to prove that $\|u\|_{p,\lambda} \le c|||u|||_{p,\lambda}$.

For $\varrho < \mathrm{diam}(\Omega)$ we have

$$\left(|\Omega_\varrho|^{-\frac{\lambda}{n}}\int_{\Omega_\varrho}|u|^p\,dx\right)^{\frac{1}{p}} \le \left(|\Omega_\varrho|^{-\frac{\lambda}{n}}\int_{\Omega_\varrho}|u-u_\varrho|^p\,dx\right)^{\frac{1}{p}} + |\Omega_\varrho|^{-\tau}|u_\varrho|\,.$$

Since $\tau < 0$, taking $R = \mathrm{diam}(\Omega)$ we get from the preceding lemma:

$$|\Omega_\varrho|^{-\tau}|u_\varrho| \le |\Omega|^{-\tau}|u_R| + c[u]_{p,\lambda} \le c(\|u\|_p + [u]_{p,\lambda})$$

and the proposition follows immediately. □

The isomorphism between $L^{p,\lambda}$ and $\mathcal{L}^{p,\lambda}$ does not hold if $\lambda \ge n$. This is evident when $\lambda > n$, since in this case the only function in $L^{p,\lambda}$ is the null function, whereas $\mathcal{L}^{p,\lambda}$ contains all the Hölder-continuous functions in $\bar{\Omega}$, with exponent $\alpha = \frac{\lambda-n}{p}$.

For such functions we have in fact:

$$|u(x) - u_\varrho| \le [u]_{0,\alpha}(2\varrho)^\alpha$$

for every $x \in \Omega_\varrho$, and therefore

$$[u]_{p,\lambda} \le c[u]_{0,\alpha}\,.$$

Regarding the case $\lambda = n$, we remark that the function $-\log x$ belongs to $\mathcal{L}^{1,1}((0,1))$ but not to $L^{1,1} = L^\infty((0,1))$.

The interest of Campanato's spaces lies mainly in the following result.

Theorem 2.9 *Let Ω be a bounded open set without internal cusps, and let $n < \lambda \leq n+p$. The space $\mathcal{L}^{p,\lambda}(\Omega)$ is isomorphic to $C^{0,\alpha}(\bar{\Omega})$, with $\alpha = \frac{\lambda - n}{p}$.*

Proof. We have already seen that $C^{0,\alpha} \subset \mathcal{L}^{p,\lambda}$. To show the opposite inclusion, let us start from inequality (2.21), from which it follows that the limit

$$\lim_{R\to 0} u_{x,R} =: v(x)$$

exists uniformly in $x \in \Omega$. The function $x \to u_{x,R}$ being continuous, even $v(x)$ will be continuous, and the same is true for $u(x)$, which by the LEBESGUE theorem coincides almost everywhere with v.

Passing to the limit in (2.21) for $\varrho \to 0$, and writing $2R$ instead of R, we get

$$|u_{x,2R} - u(x)| \leq c[u]_{p,\lambda} |\Omega(x, 2R)|^\tau . \tag{2.24}$$

Let now x, $y \in \Omega$, and let $R = |x-y|$. We have

$$|u(x) - u(y)| \leq |u(x) - u_{x,2R}| + |u_{x,2R} - u_{y,2R}| + |u_{y,2R} - u(y)| .$$

On the other hand

$$|u_{x,2R} - u_{y,2R}| \leq |u(z) - u_{x,2R}| + |u(z) - u_{y,2R}|$$

and integrating on $z \in \Omega(x, 2R) \cap \Omega(y, 2R) \supset \Omega(x, R) \cup \Omega(y, R)$:

$$|u_{x,2R} - u_{y,2R}| \leq |\Omega(x,R)|^{-1} \int_{\Omega(x,2R)} |u(z) - u_{x,2R}| dz$$
$$+ |\Omega(y,R)|^{-1} \int_{\Omega(y,2R)} |u(z) - u_{y,2R}| dz .$$

A straightforward application of Hölder's inequality, taking into account (2.22), gives

$$|u(x) - u(y)| \leq c[u]_{p,\lambda} |x-y|^\alpha .$$

Finally, setting $2R = \operatorname{diam} \Omega$ in (2.24), we find

$$|u(x)| \leq |u_{x,2R}| + |u(x) - u_{x,2R}| \leq c|||u|||_{p,\lambda}$$

and therefore in conclusion

$$\|u\|_{C^{0,\alpha}} \le c|||u|||_{p,\lambda},$$

which proves the result. □

We conclude this section with a proof of the isomorphism of Campanato (and therefore Morrey) spaces with respect to diffeomorphisms of the open set Ω.

We recall that if $g : A \to B$ is a homeomorphism between two open sets A and B, and if $\mathcal{M}(A)$ and $\mathcal{M}(B)$ are the spaces of measurable functions respectively in A and B, the induced map g_* of $\mathcal{M}(B)$ in $\mathcal{M}(A)$ is defined by $g_* u = u \circ g$.

Proposition 2.4 *Let Ω and Λ be two bounded open sets in $\mathbf{R}^n$, both without exterior cusps, and let g be the restriction to Λ of a bilipschitzian*[7] *homeomorphism of $\Lambda_1 \supset \bar{\Lambda}$ onto $\Omega_1 \supset \bar{\Omega}$. Then, g_* is an isomorphism between $\mathcal{L}^{p,\lambda}(\Omega)$ and $\mathcal{L}^{p,\lambda}(\Lambda)$.*

Proof. If $u : \Omega \to \mathbf{R}$, we have $U =: u \circ g : \Lambda \to \mathbf{R}$. We must prove that $u \in \mathcal{L}^{p,\lambda}(\Omega)$ if and only if $U \in \mathcal{L}^{p,\lambda}(\Lambda)$.

Since g is Lipschitz-continuous together with its inverse, there will exist a constant $L > 0$ such that for every x, $x_0 \in \Lambda_1$

$$L^{-1}|g(x) - g(x_0)| \le |x - x_0| \le L|g(x) - g(x_0)|.$$

Setting $y_0 = g(x_0)$, and taking

$$R < R_0 =: \min\{\operatorname{dist}(\Omega, \partial\Omega_1), \operatorname{dist}(\Lambda, \partial\Lambda_1)\},$$

we have

$$g(\Lambda(x_0, R)) \subset \Omega(y_0, LR);\ g^{-1}(\Omega(y_0, R)) \subset \Lambda(x_0, LR).$$

If $R < R_0$ we have therefore for every $\xi \in \mathbf{R}^N$:

$$\int_{\Omega(y_0,R)} |u - \xi|^p\, dy = \int_{g^{-1}(\Omega(y_0,R))} |U - \xi|^p |J| dx \le c \int_{\Lambda(x_0,LR)} |U - \xi|^p\, dx$$

from which the inequality

$$|||u|||_{p,\lambda} \le c|||U|||_{p,\lambda}$$

follows at once. Interchanging g and g^{-1} we get the opposite inequality, and the theorem is proved. □

[7] That is Lipschitz-continuous together with its inverse.

2.4 The Lemmas of John and Nirenberg

We gather in this section two results concerning estimates for the measures of level sets of a function u, subject to suitable conditions. The first of these is evidently connected with Campanato's space $\mathcal{L}^{1,n}$, and will be useful in the proof of the isomorphism of the spaces $\mathcal{L}^{p,n}$.

The proof makes use of the following theorem, that is also interesting in itself, and will be useful in other occasions.

Theorem 2.10 (CALDERON–ZYGMUND [1]) *Let Q_0 be a cube in $\mathbf{R}^n$, and let $g(x)$ be a positive function in $L^1(Q_0)$. Let L be a real number such that*

$$\fint_{Q_0} g\,dx =: \frac{1}{|Q_0|}\int_{Q_0} g\,dx \leq L\,.$$

There exists a sequence (possibly a finite number) of pairwise disjoint cubes $Q_i \subset Q_0$, with faces parallel to those of Q_0, such that

(i) $L < \fint_{Q_i} g\,dx \leq 2^n L$.

(ii) $g \leq L$ *a.e. in* $Q_0 - \cup Q_i$.

Proof. We call final a cube Q for which $\fint_Q g\,dx > L$. Let us cut Q_0 (which by assumption is not final) into 2^n equal cubes, each with side one half of that of Q_0. If any of these cubes is not final, we cut it again in 2^n equal cubes, and we continue as before. Let Q_i, $i = 1, 2, \dots$ be the family of final cubes.

We have obviously $\fint_{Q_i} g\,dx > L$. On the other hand Q_i must arise by division from a cube W (of double side) which is not final. We have thus $\fint_W g\,dx \leq L$, and hence $\fint_{Q_i} g\,dx \leq 2^n \fint_W g\,dx \leq 2^n L$. This proves (i).

In order to prove (ii), we remark that every point $x \in Q_0 - \cup Q_i$ is the intersection of a decreasing sequence of cubes W_s, none of which is final. It follows that for every s there holds

$$\fint_{W_s} g\,dx \leq L$$

so that, passing to the limit for $s \to \infty$, we get $g(x) \leq L$ for almost every $x \in Q_0 - \cup Q_i$. □

Let now $u : Q_0 \to \mathbf{R}^N$ be a summable function in Q_0. Denoting by Q a generic cube, with sides parallel to those of Q_0, we set

$$u_* =: [u]_{*,Q_0} = \sup_Q \fint_Q |u - u_Q|dx\,. \tag{2.25}$$

Finally, we denote by $BMO(Q_0) = BMO(Q_0, \mathbf{R}^N)$[8] the space of functions $u \in L^1(Q_0, \mathbf{R}^N)$ for which the quantity u_* is finite.

Proposition 2.5 *The space BMO is isomorphic to* $\mathcal{L}^{1,n}$.

Proof. We have obviously $[u]_* \leq [u]_{1,n}$. On the other hand, if Z is an arbitrary cube of $\mathbf{R}^n$, with center in Q_0, there exists a cube $Q \subset Q_0$, with sides parallel to those of Q_0, such that $P =: Z \cap Q_0 \subset Q$ and $|Q| \leq c|P|$, with c depending only on n. It follows that

$$\fint_P |u - u_P| dx \leq 2 \fint_P |u - u_Q| dx \leq 2 \frac{|Q|}{|P|} \fint_Q |u - u_Q| dx \leq c \fint_Q |u - u_Q| dx$$

and therefore $[u]_{1,n} \leq c[u]_*$. □

For $v \in BMO(Q)$ and $\sigma > 0$, we set

$$\Upsilon_{\sigma,Q}(v) =: \{x \in Q : |v(x) - v_Q| > \sigma\}.$$

Theorem 2.11 (John–Nirenberg I [1]) *There exists two positive constants A and α such that for every $u \in BMO(Q_0, \mathbf{R}^N)$ and $\sigma > 0$ we have*

$$|\Upsilon_{\sigma,Q_0}(u)| \leq A \exp\left(\frac{-\alpha\sigma}{[u]_*}\right) |Q_0| \qquad (2.26)$$

Proof. Writing $u/[u]_*$ instead of u, and $\sigma/[u]_*$ instead of σ, we can assume $[u]_* = 1$. We can also assume that $u_{Q_0} = 0$.

For any cube Q, and for $\sigma > 0$, we set

$$\varphi(\sigma) = \sup\left\{\frac{|\Upsilon_{\sigma,Q}(u)|}{|Q|},\ [u]_{*,Q} = 1,\ u_Q = 0\right\}.$$

The function φ does not depend on the cube Q, since both the ratio $|\Upsilon_\sigma|/|Q|$, and the quantity $[u]_{*,Q}$ are invariant under omotheties. Moreover

$$\varphi(\sigma) = \sup_{[u]_{*,Q} \leq 1} \frac{|\Upsilon_{\sigma,Q}(u)|}{|Q|}.$$

If $[v]_{*,Q} \leq 1$, we have

$$|\Upsilon_{\sigma,Q}(v)| \leq \frac{1}{\sigma} \int_Q |v - v_Q| dx \leq \frac{|Q|}{\sigma}$$

---

[8] From the initials of *Bounded Mean Oscillation.* Sometimes the space BMO is called $\mathcal{E}^0$.

and therefore $\varphi(\sigma) \leq \frac{1}{\sigma}$. We shall prove that for $s > 1$ and $\sigma > 2^n s$ we have

$$\varphi(\sigma) \leq \frac{1}{s}\varphi(\sigma - 2^n s).$$

For that, let w be a function in $BMO(Q_0)$, with $[w]_{*,Q_0} = 1$ and zero mean value, and let us apply the Calderon–Zygmund theorem to the function $|w|$. We have

$$⨍_{Q_0} |w|dx \leq [w]_* = 1 < s.$$

and hence there exists cubes $Q_k \subset Q_0$ such that

$$s \leq ⨍_{Q_k} |w|dx < 2^n s$$

$$|w| \leq s \text{ in } Q_0 - \cup Q_k .$$

In particular, from the first inequality it follows that

$$\sum_k |Q_k| \leq \frac{1}{s}\int_{Q_0} |w|dx .$$

Denoting by w_k the average of w in Q_k, we have $|w_k| < 2^n s$, and therefore if $x \in Q_k$ and $|w(x)| > \sigma$ we deduce $|w(x) - w_k| \geq |w(x)| - |w_k| > \sigma - 2^n s$. Since $|w| \leq s < \sigma$ in $Q_0 - \cup Q_k$, we get

$$|\Upsilon_{\sigma,Q_0}| \leq \sum_k |\Upsilon_{\sigma-2^n s,Q_k}| .$$

On the other hand $[w]_{*,Q_k} \leq [w]_{*,Q_0} = 1$, and hence

$$|\Upsilon_{\sigma-2^n s,Q_k}| \leq \varphi(\sigma - 2^n s)|Q_k| .$$

It follows that

$$|\Upsilon_{\sigma,Q_0}(w)| \leq \varphi(\sigma-2^n s)\sum_k |Q_k| \leq \frac{1}{s}\varphi(\sigma-2^n s)\int_{Q_0} |w|dx \leq \frac{|Q_0|}{s}\varphi(\sigma-2^n s)$$

and therefore

$$\varphi(\sigma) \leq \frac{1}{s}\varphi(\sigma - 2^n s) \tag{2.27}$$

for every $s > 1$ and $\sigma > 2^n s$.

Choose now $s = e$ and set $\alpha = 2^{-n}e^{-1}$. If $0 < \sigma < 2^n e$ we have

$$|\Upsilon_\sigma| \le |Q_0| \le e^{1-\alpha\sigma}|Q_0|$$

that is (2.26) with $A = e$.

Assume now that $2^n ek \le \sigma < 2^n e(k+1)$, $k \ge 1$. From (2.27) we get by induction

$$\varphi(\sigma) \le e^{-k}\varphi(\sigma - k2^n e).$$

On the other hand $0 < \sigma - 2^n ek < 2^n e$, and therefore

$$\varphi(\sigma - 2^n ek) \le Ae^{-\alpha(\sigma - 2^n ek)} = Ae^{k-\alpha\sigma}$$

which in combination with the preceding estimate again gives (2.26). □

From the above theorem, we have the following:

Corollary 2.2 *If $u \in BMO(Q_0, \mathbf{R}^N)$, then $u \in L^p(Q_0, \mathbf{R}^N)$ for every $p \ge 1$, and for every cube Q parallel to Q_0 it holds that*

$$\fint_Q |u - u_Q|^p\, dx \le c[u]_*^p.$$

Proof. We have actually

$$\begin{aligned}\int_Q |u - u_Q|^p\, dx &= p\int_0^\infty \sigma^{p-1}|\Upsilon_{\sigma,Q}|d\sigma \\ &\le pA\int_0^\infty \sigma^{p-1}\exp\left(-\frac{\alpha\sigma}{[u]_*}\right)|Q|d\sigma \\ &= pA\left(\frac{[u]_*}{\alpha}\right)^p |Q|\int_0^\infty t^{p-1}e^{-t}\, dt \le c|Q|[u]_*^p. \end{aligned}$$ □

In particular, if $u \in BMO(Q_0, \mathbf{R}^N)$, then $u \in \mathcal{L}^{p,n}(Q_0, \mathbf{R}^N)$, and

$$[u]_{p,n} \le c[u]_*.$$

On the other hand, we have trivially $[u]_* \le [u]_{1,n} \le [u]_{p,n}$, and hence BMO is isomorphic to $\mathcal{L}^{p,n}$ for every $p \ge 1$. As a consequence we have the following:

Corollary 2.3 *The spaces $\mathcal{L}^{p,n}(Q_0, \mathbf{R}^N)$ are all isomorphic among themselves.*

The same is true for $\mathcal{L}^{p,n}(\Omega)$, as long as the boundary of Ω is regular enough (for instance, Lipschitz-continuous).

The next lemma is also due to John and Nirenberg, and has been widely used in the theory of elliptic equations.

Lemma 2.3 (JOHN–NIRENBERG II) [1] *Let $u : Q_0 \to \mathbf{R}^N$ be a summable function in Q_0, and assume that there exist two constants $K > 0$ and $p > 1$ such that for every partition of Q_0 in countably many cubes Q_j, pairwise without common interior points, we have*

$$\sum_{j=1}^{\infty} |Q_j| \left(\fint_{Q_j} |u - u_{Q_j}| dx \right)^p \leq K^p . \tag{2.28}$$

In this case, denoting by $[u]_p$ the smallest constant for which (2.28) holds, we have

$$|\Upsilon_{\sigma,Q_0}| \leq A \left(\frac{[u]_p}{\sigma} \right)^p \tag{2.29}$$

with a constant A depending only on n and p.

Proof. For $1 < p < +\infty$, let q be the conjugate exponent of p: $q^{-1} + p^{-1} = 1$, and

$$\lambda_k = \sum_{j=0}^{k} q^{-j} = \frac{1 - q^{-k-1}}{1 - q^{-1}} = p(1 - q^{-k-1}) .$$

We have $\lambda_k = \lambda_{k-1} + q^{-k}$, and hence $q^k \lambda_k = q^k \lambda_{k-1} + 1$. Define moreover

$$\tau_k = \frac{1}{2^{n+k(n+1)} q^k \lambda_k} .$$

We remark that we can always assume $u_{Q_0} = 0$, and we begin by proving that if v has zero average in some cube Q, and if for some integer k we have

$$\fint_Q |v| dx \leq \sigma \tau_k , \tag{2.30}$$

then

$$|\Upsilon_{\sigma,Q}(v)| \leq A_k \left(\frac{\lambda_k [v]_p}{\sigma} \right)^{\lambda_k} \left(\frac{1}{[v]_p} \int_Q |v| dx \right)^{q^{-k}} \tag{2.31}$$

with $A_0 = 1$ and

$$A_k = \prod_{j=1}^{k} (q^j 2^{n+j(n+1)})^{q^{-j}} .$$

We set for simplicity $V_\sigma = \Upsilon_{\sigma,Q}(v) = \{x \in Q : |v(x)| > \sigma\}$. Since

$$|V_\sigma| \leq \frac{1}{\sigma}\int_Q |v|dx$$

(2.31) holds for $k = 0$. We assume now that (2.30) implies (2.31) for $k-1$, and we prove that the same is true for k.

Assume then that (2.30) is satisfied. By Theorem 2.10, we can conclude that there exists a sequence of subcubes Q_m of Q, pairwise without common internal points, and such that

$$\sigma\tau_k \leq \fint_{Q_m} |v|dx < 2^n\sigma\tau_k\,, \tag{2.32}$$

$$|v(x)| \leq \sigma\tau_k < \sigma \text{ in } Q_0 - \cup_m Q_m\,. \tag{2.33}$$

Define $w_m = v - v_{Q_m}$ in Q_m. The function w_m has obviously zero average on Q_m. Moreover, by the definition of $[v]_p$, we have

$$\sum_{m=1}^{\infty}[w_m]_p^p \leq [v]_p^p\,. \tag{2.34}$$

Now

$$1 - 2^n\tau_k > 1 - \frac{1}{q^k\lambda_k} = \frac{\lambda_{k-1}}{\lambda_k} \tag{2.35}$$

and hence

$$2^{n+1}\tau_k = \frac{2^{-n-(k-1)(n+1)}}{q^{k-1}\lambda_{k-1}}\frac{\lambda_{k-1}}{q\lambda_k} < \tau_{k-1}(1-2^n\tau_k)\,.$$

Consequently, taking (2.32) into account:

$$\fint_{Q_m}|w_m|dx \leq 2\fint_{Q_m}|v|dx \leq 2^{n+1}\sigma\tau_k \leq \sigma(1-2^n\tau_k)\tau_{k-1}\,,$$

so that we can write (2.31) for w_m, with $k-1$ instead of k and $\sigma(1-2^n\tau_k)$ instead of σ. Setting

$$W_{\sigma,k,m} = \{x \in Q_m : |w_m(x)| > \sigma(1-2^n\tau_k)\}$$

it follows from (2.35) that

$$|W_{\sigma,k,m}| \leq A_{k-1}\left(\frac{[w_m]_p\lambda_k}{\sigma}\right)^{\lambda_{k-1}}\left(\frac{1}{[w_m]_p}\int_{Q_m}|w_m|dx\right)^{q^{1-k}}.$$

On the other hand, from (2.32) it follows that $|v_{Q_m}| \leq 2^n \tau_k \sigma$, and therefore, if $x \in Q_m$ and $|v(x)| > \sigma$, we get $|w_m(x)| \geq |v(x)| - |v_{Q_m}| > \sigma(1 - 2^n \tau_k)$. It follows that $V_\sigma \cap Q_m \subset W_{\sigma,k,m}$ and hence, taking (2.33) into account, we obtain

$$|V_\sigma| \leq \sum_m |W_{\sigma,k,m}| \leq A_{k-1} \lambda_k^{\lambda_{k-1}} \sigma^{-\lambda_{k-1}} \sum_m [w_m]_p^{\lambda_k - 2} a_m^{q^{1-k}}, \tag{2.36}$$

where

$$a_m = \int_{Q_m} |w_m| dx .$$

In order to evaluate the last sum, we use Hölder inequality. We have

$$\sum_m [w_m]_p^{\lambda_k - 2} a_m^{q^{1-k}} \leq \left(\sum_m [w_m]_p^p \right)^{1 - q^{1-k}} \left(\sum_m a_m \right)^{q^{1-k}}$$

and moreover

$$\sum_m a_m = \sum_m |Q_m|^{\frac{1-p}{p}} a_m |Q_m|^{\frac{p-1}{p}} \leq \left(\sum_m |Q_m|^{1-p} a_m^p \right)^{\frac{1}{p}} \left(\sum_m |Q_m| \right)^{\frac{1}{q}}$$
$$\leq \left(\sum_m [w_m]_p^p \right)^{\frac{1}{p}} \left(\sum_m |Q_m| \right)^{\frac{1}{q}} .$$

The last term can be estimated by means of (2.32). We have

$$\sum_m |Q_m| \leq \frac{1}{\sigma \tau_k} \sum_m \int_{Q_m} |v| dx \leq \frac{1}{\sigma \tau_k} \int_Q |v| dx .$$

Introducing all these inequalities into (2.36), and taking into account (2.34), we get

$$|V_\sigma| \leq A_{k-1} \lambda_k^{\lambda_{k-1}} \tau_k^{-q^{-k}} \left(\frac{[v]_p}{\sigma} \right)^{\lambda_k} \left(\frac{1}{[v]_p} \int_Q |v| dx \right)^{q^{-k}} .$$

To obtain the desired inequality (2.31) we need only adjust the constant A_k. We have $\tau_k^{-1} = q^k \lambda_k 2^{n+k(n+1)}$, and therefore:

$$\lambda_k^{\lambda_{k-1}} \tau_k^{-q^{-k}} \leq \lambda_k^{\lambda_k} [q^k 2^{n+k(n+1)}]^{q^{-k}}$$

from which we get at once (2.31) for k.

Coming back to the function u, and observing that the constants A_k and λ_k are bounded, we can conclude that if

$$\fint_{Q_0} |u|dx \le \sigma\tau_k \, ,$$

then

$$|\Upsilon_\sigma| \le c\left(\frac{[u]_p}{\sigma}\right)^{\lambda_k}\left(\frac{1}{[u]_p}\int_{Q_0}|u|dx\right)^{q^{-k}} .$$

The last relation can also be written in the form:

$$|\Upsilon_\sigma| \le c\left(\frac{[u]_p}{\sigma}\right)^{p}\left(\frac{\sigma^{\frac{p}{q}}|Q_0|}{[u]_p^{\frac{p}{q}}}\fint_{Q_0}|u|dx\right)^{q^{-k}} . \tag{2.37}$$

Let now $\fint_{Q_0}|u|dx < 2^{-n}\sigma$, and let k be such that

$$\sigma\tau_{k+1} < \fint_{Q_0}|u|dx \le \sigma\tau_k \, .$$

In this case (2.37) holds, and estimating σ by $\tau_{k+1}^{-1}\fint_{Q_0}|u|dx$, we get

$$|\Upsilon_\sigma| \le c\left(\frac{[u]_p}{\sigma}\right)^{p}\tau_{k+1}^{-pq^{-k-1}}\left(\frac{|Q_0|^{\frac{1}{p}}}{[u]_p}\fint_{Q_0}|u|dx\right)^{pq^{-k}} .$$

The last factor is less than unity, and $\tau_{k+1}^{-pq^{-k-1}}$ is less than a fixed constant, so that in conclusion we get (2.29), provided $\sigma > 2^n\fint_{Q_0}|u|dx$.

In the opposite case, we have

$$\sigma \le 2^n\fint_{Q_0}|u|dx \le 2^n[u]_p|Q_0|^{-\frac{1}{p}}$$

and therefore

$$|\Upsilon_\sigma| \le |Q_0| \le \left(\frac{2^n[u]_p}{\sigma}\right)^p$$

so that (2.29) holds in this case too. □

2.5 Interpolation

We have already remarked that if u is a function in $L^p(\Omega)$, and if for $\sigma > 0$ we set

$$U_\sigma = \{x \in \Omega : |u(x)| > \sigma\}$$

it holds that

$$\lambda(u,\sigma) =: |U_\sigma| \le \left(\frac{\|u\|_p}{\sigma}\right)^p .$$

Of course the opposite is not true; the function $u(x) = |x|^{-n/p}$ satisfies $|U_\sigma| \le c\sigma^{-p}$ in the unit ball B, but does not belong to $L^p(B)$.

Definition 2.6 *We say that a measurable function $u : \Omega \to \mathbf{R}^N$ belongs to the space $L^p(\Omega)$-weak (that we shall denote by $L^p_w(\Omega)$), if there exists a constant K such that for every $\sigma > 0$:*

$$|U_\sigma| \le \left(\frac{K}{\sigma}\right)^p . \tag{2.38}$$

It follows from (2.4) that $L^p \subset L^p_w$. Moreover, if Ω has finite measure, we have $L^p_w(\Omega) \subset L^s(\Omega)$ for every $s < p$. Actually, we have from (2.5):

$$\begin{aligned}\int_\Omega |u|^s\,dx &= s\int_0^\infty \sigma^{s-1}|U_\sigma|d\sigma \le |\Omega| + s\int_1^\infty \sigma^{s-1}|U_\sigma|d\sigma \\ &\le |\Omega| + sK^p \int_1^\infty \sigma^{s-1-p}\,d\sigma \\ &= |\Omega| + \frac{sK^p}{p-s} .\end{aligned}$$

The weak L^p spaces can be characterized in terms of integrals over arbitrary sets.

Definition 2.7 *For $0 \le \vartheta \le 1$ let us denote by $M^\vartheta(\Omega)$ the space of measurable functions $u : \Omega \to \mathbf{R}^N$ such that*

$$\|u\|_\vartheta =: \sup |E|^{-\vartheta}\int_E |u|dx < +\infty ,$$

the supremum being taken with respect to all the measurable sets $E \subset \Omega$, with $|E| > 0$.

Sometimes the spaces M^ϑ are called Lorentz spaces. It is easily seen that $\|u\|_\vartheta$ is a norm, and that M^ϑ is complete. We have $M^0 = L^1$ and $M^1 = L^\infty$; moreover $M^\vartheta \subset L^{1,n\vartheta}$, since in the latter we integrate only over cubes and their intersections with Ω. We have:

Proposition 2.6 *Let $1 < p < +\infty$, and let $\vartheta = \frac{1}{q} = 1 - \frac{1}{p}$. Then, $L^p_w = M^\vartheta$.*

Proof. For $u \in L^p_w$ we have

$$\int_E |u|dx = \int_0^\infty |U_\sigma \cap E| d\sigma \le |E|^\vartheta + \int_{|E|^{\vartheta-1}}^\infty |U_\sigma| d\sigma$$

$$\le |E|^\vartheta + K^p \int_{|E|^{\vartheta-1}}^\infty \sigma^{-p}\, d\sigma \le c|E|^\vartheta .$$

Conversely, if $u \in M^\vartheta$ we have

$$\sigma|U_\sigma| \le \int_{U_\sigma} |u|dx \le \|u\|_\vartheta |U_\sigma|^\vartheta$$

and hence

$$|U_\sigma| \le \left(\frac{\|u\|_\vartheta}{\sigma}\right)^p . \qquad \square$$

We introduce now some notation. In order to avoid non-essential complications, we shall suppose that Ω has finite measure, so that $L^s(\Omega) \subset L^p(\Omega)$ for $p \le s \le +\infty$. If T is a mapping of $L^p(\Omega, \mathbf{R}^N)$ into $L^p(\Lambda, \mathbf{R}^M)$, we shall say that T is *quasilinear* if there exists a constant Q such that for almost every $x \in \Omega$:

$$|T(u+v)(x)| \le Q(|Tu(x)| + |Tv(x)|) .$$

The first result is an interpolation theorem between s and $+\infty$.

Theorem 2.12 (MARCINKIEWICZ I) *Let $1 \le s < +\infty$, and let T be a quasilinear mapping from $L^s(\Omega, \mathbf{R}^N)$ into the space of measurable functions in Λ, with values in $\mathbf{R}^\nu$. Assume that T maps $L^s(\Omega, \mathbf{R}^N)$ into $L^s_w(\Lambda, \mathbf{R}^\nu)$ and $L^\infty(\Omega, \mathbf{R}^N)$ into $L^\infty(\Lambda, \mathbf{R}^\nu)$, with the estimates:*

$$\lambda(Tf, \sigma) \le \left(\frac{A_s\|f\|_s}{\sigma}\right)^s , \tag{2.39}$$

$$\|Tg\|_\infty \le A_\infty \|g\|_\infty \tag{2.40}$$

for every $f \in L^s$, $g \in L^\infty$ and $\sigma > 0$. Then, T maps $L^p(\Omega, \mathbf{R}^N)$ into $L^p(\Lambda, \mathbf{R}^\nu)$ for every $p > s$, and we have

$$\|Tu\|_p \le c A_s^{\frac{s}{p}} A_\infty^{\frac{p-s}{p}} \|u\|_p \tag{2.41}$$

for every $u \in L^p$.

Proof. Let $u \in L^p$ and let $\gamma > 0$. Setting

$$f(x) = \begin{cases} u(x) & \text{if } |u(x)| > \gamma\sigma\,, \\ 0 & \text{if } |u(x)| \leq \gamma\sigma\,, \end{cases}$$

$$g(x) = \begin{cases} 0 & \text{if } |u(x)| > \gamma\sigma\,, \\ u(x) & \text{if } |u(x)| \leq \gamma\sigma\,, \end{cases}$$

we have $f \in L^s$, $g \in L^\infty$ and $u = f + g$, so that $|Tu(x)| \leq Q(|Tf(x)| + |Tg(x)|)$. In order that $|Tu(x)| > \sigma$, at least one of the two quantities $|Tf(x)|$ and $|Tg(x)|$ must be greater than $\sigma/2Q$, and hence

$$\lambda(Tu, \sigma) \leq \lambda\left(Tf, \frac{\sigma}{2Q}\right) + \lambda\left(Tg, \frac{\sigma}{2Q}\right). \tag{2.42}$$

From (2.39) we have

$$\lambda(Tf, \frac{\sigma}{2Q}) \leq \left(\frac{2QA_s\|f\|_s}{\sigma}\right)^s \leq \left(\frac{2QA_s}{\sigma}\right)^s \int_{U_{\gamma\sigma}} |u|^s\, dx\,. \tag{2.43}$$

On the other hand

$$\|Tg\|_\infty \leq A_\infty \|g\|_\infty \leq A_\infty \gamma\sigma$$

so that, by choosing

$$\gamma = \frac{1}{2QA_\infty}$$

we get

$$\lambda\left(Tg, \frac{\sigma}{2Q}\right) = 0\,. \tag{2.44}$$

Setting $a_s = (2QA_s)^s$, it follows from (2.42) and (2.43)

$$\begin{aligned} &\int_\Lambda |Tu|^p\, dx \\ &\quad \leq p\int_0^\infty \sigma^{p-1}\lambda\left(Tf, \frac{\sigma}{2Q}\right) d\sigma \leq pa_s \int_0^\infty \sigma^{p-s-1}\, d\sigma \int_{U_{\gamma\sigma}} |u|^s\, dx \\ &\quad = pa_s \int_0^\infty \sigma^{p-s-1}\left(s\int_{\gamma\sigma}^\infty \tau^{s-1}\lambda(u,\tau)d\tau + (\gamma\sigma)^s\lambda(u,\gamma\sigma)\right) d\sigma \end{aligned}$$

$$= pa_s\left(s\int_0^\infty \tau^{s-1}\lambda(u,\tau)d\tau\int_0^{\frac{\tau}{\gamma}}\sigma^{p-s-1}\,d\sigma+\gamma^s\int_0^\infty\sigma^{p-1}\lambda(u,\gamma\sigma)d\sigma\right)$$

$$= a_s\gamma^{s-p}\frac{p}{p-s}\int_\Omega |u|^p\,dx\,,$$

from which we get the desired inequality (2.41), taking into account the definitions of a_s and of γ. □

We consider now the interpolation between two finite exponents s and r.

Theorem 2.13 (MARCINKIEWICZ II) *Let $1 \le s < r < +\infty$, and let T be a quasilinear mapping of $L^s(\Omega, \mathbf{R}^N)$ into the space of measurable functions in Λ, with values in $\mathbf{R}^\nu$. Assume that T maps $L^s(\Omega,\mathbf{R}^N)$ into $L^s_w(\Lambda,\mathbf{R}^\nu)$ and $L^r(\Omega,\mathbf{R}^N)$ into $L^r_w(\Lambda,\mathbf{R}^\nu)$, with the estimates:*

$$\lambda(Tf,\sigma)\le\left(\frac{A_s\|f\|_s}{\sigma}\right)^s, \tag{2.45}$$

$$\lambda(Tg,\sigma)\le\left(\frac{A_r\|g\|_r}{\sigma}\right)^r \tag{2.46}$$

for every $f\in L^s$, $g\in L^r$ and $\sigma>0$.

Then, for every p between s and r, T maps $L^p(\Omega,\mathbf{R}^N)$ into $L^p(\Lambda,\mathbf{R}^N)$, and we have

$$\|Tu\|_p\le cA_s^\epsilon A_r^{1-\epsilon}\|u\|_p \tag{2.47}$$

for every $u\in L^p$, where $\epsilon\in(0,1)$ is such that

$$\frac{1}{p}=\frac{\epsilon}{s}+\frac{1-\epsilon}{r}.$$

Proof. Proceeding as above we get (2.42) and (2.43), whereas instead of (2.44) we get

$$\lambda\left(Tg,\frac{\sigma}{2Q}\right)\le\left(\frac{2QA_r\|g\|_r}{\sigma}\right)^r=\left(\frac{2QA_r}{\sigma}\right)^r\int_{\Omega-U_{\gamma\sigma}}|u|^r\,dx\,.$$

It follows that

$$\int_\Lambda |Tu|^p\,dx\le p(2QA_s)^s\int_0^\infty\sigma^{p-s-1}\,d\sigma\int_{U_{\gamma\sigma}}|u|^s\,dx$$

$$+\,p(2QA_r)^r\int_0^\infty\sigma^{p-r-1}\,d\sigma\int_{\Omega-U_{\gamma\sigma}}|u|^r\,dx\,.$$

The first integral can be estimated as above by

$$(2QA_s)^s\gamma^{s-p}\frac{p}{p-s}\int_\Omega |u|^p\,dx\,.$$

For the second, we have

$$\int_0^\infty \sigma^{p-r-1}\int_{\Omega-U_{\gamma\sigma}}|u|^r\,dx\,d\sigma$$

$$=\int_0^\infty \sigma^{p-r-1}\left(r\int_0^{\gamma\sigma}\tau^{r-1}\lambda(u,\tau)d\tau-(\gamma\sigma)^r\lambda(u,\gamma\sigma)\right)d\sigma$$

$$=r\int_0^\infty \tau^{r-1}\lambda(u,\tau)d\tau\int_{\frac{\tau}{\gamma}}^\infty \sigma^{p-r-1}\,d\sigma-\gamma^r\int_0^\infty \sigma^{p-1}\lambda(u,\gamma\sigma)d\sigma$$

$$=\frac{p}{r-p}\gamma^{r-p}\int_0^\infty \sigma^{p-1}\lambda(u,\sigma)d\sigma=\frac{1}{r-p}\gamma^{r-p}\int_\Omega |u|^p\,dx\,.$$

We have therefore,

$$\int_\Lambda |Tu|^p\,dx\le\left(\frac{1}{p-s}\gamma^{s-p}(2QA_s)^s+\frac{1}{r-p}\gamma^{r-p}(2QA_r)^r\right)\int_\Omega |u|^p\,dx\,,$$

and the conclusion follows if we choose

$$\gamma=\frac{1}{2Q}A_s^{\frac{s}{r-s}}A_r^{-\frac{r}{r-s}}\,. \qquad \square$$

We can now prove a last interpolation theorem, that we shall use later.

Theorem 2.14 (STAMPACCHIA [3]) *Let Q_0 be a cube in $\mathbf{R}^n$, and let T be a linear mapping of $L^s(Q_0,\mathbf{R}^N)$ into $L^s(Q_0,\mathbf{R}^\nu)$, mapping $L^\infty(Q_0,\mathbf{R}^N)$ into $BMO=\mathcal{L}^{s,n}(Q_0,\mathbf{R}^\nu)$, with the estimates:*

$$\|Tf\|_s\le A_s\|f\|_s \tag{2.48}$$

$$[Tg]_*\le A_\infty\|g\|_\infty \tag{2.49}$$

for every $f\in L^s$ and $g\in L^\infty$.

Then, T maps $L^p(Q_0,\mathbf{R}^N)$ into $L^p(Q_0,\mathbf{R}^\nu)$ for every $p>s$, and we have

$$\|Tu-(Tu)_{Q_0}\|_p\le A_p\|u\|_p\,, \tag{2.50}$$

where

$$A_p=cA_s^{\frac{s}{p}}A_\infty^{\frac{p-s}{p}}\,. \tag{2.51}$$

Proof. (CAMPANATO [5]) Let $\{Q_k\}$ be a partition of Q_0 in cubes pairwise without common internal points. Let $\mathcal{T}$ be the mapping that to every $u \in L^s$ associates the function $\mathcal{T}u$ that in each cube Q_k takes the constant value:

$$⨍_{Q_k} |Tu - (Tu)_k| dx \,.$$

The mapping $\mathcal{T}$ is quasilinear (with $Q = 1$) and, taking into account the definition of BMO, it maps $L^\infty(Q_0, \mathbf{R}^N)$ into $L^\infty(Q_0, \mathbf{R}^\nu)$. Moreover

$$\begin{aligned}\int_{Q_0} |\mathcal{T}u|^s \, dx &= \sum_k |Q_k| \left(⨍_{Q_k} |Tu - (Tu)_k| \, dx \right)^s \\ &\le 2^s \sum_k |Q_k| ⨍_{Q_k} |Tu|^s \, dx \\ &= 2^s \int_{Q_0} |Tu|^s \, dx \le (2A_s)^s \int_{Q_0} |u|^s \, dx\end{aligned}$$

so that $\mathcal{T}$ maps L^s into L^s, with the appropriate estimate.

By Marcinkiewicz theorem I, $\mathcal{T}$ maps L^q into L^q for every $q > s$, with the estimate

$$\|\mathcal{T}u\|_q \le c A_s^{\frac{s}{q}} A_\infty^{\frac{q-s}{q}} \|u\|_q =: A_q \|u\|_q \,.$$

From the definition of $\mathcal{T}$ we get

$$\sum_k |Q_k| \left(⨍_{Q_k} |Tu - (Tu)_k| dx \right)^q \le A_q^q \|u\|_q^q$$

for every partition Q_k. Applying Lemma 2.3, we can conclude that for every $\sigma > 0$,

$$|\{x \in Q_0 : |Tu(x) - (Tu)_{Q_0}| > \sigma\}| \le c \left(\frac{A_q \|u\|_q}{\sigma} \right)^q . \tag{2.52}$$

Consequently, the mapping $\Theta : u \to Tu - (Tu)_{Q_0}$ maps L^s into L^s, with

$$\|\Theta u\|_s \le 2A_s \|u\|_s$$

and L^q into L^q_w, with the estimate (2.52). By the second theorem of Marcinkiewicz we conclude that Θ maps L^p into L^p, $s < p < q$, and

$$\|\Theta u\|_p \le A_p \|u\|_p \,.$$

A simple computation shows that the constant A_p is given by (2.51).

Finally, if we want to estimate the norm $\|Tu\|_p$, it will be sufficient to estimate the quantity $|(Tu)_{Q_0}|$. We have

$$|(Tu)_{Q_0}| \leq \fint_{Q_0} |Tu| dx \leq \left(\fint_{Q_0} |Tu|^s \, dx\right)^{\frac{1}{s}}$$

$$\leq K_s \left(\fint_{Q_0} |u|^s \, dx\right)^{\frac{1}{s}} \leq K_s \left(\fint_{Q_0} |u|^p \, dx\right)^{\frac{1}{p}}$$

and hence

$$\|Tu\|_p \leq \|\Theta u\|_p + |Q_0|^{\frac{1}{p}} |(Tu)_{Q_0}| \leq (cK_s^{\frac{s}{p}} K_\infty^{1-\frac{s}{p}} + K_s)\|u\|_p . \qquad \square$$

Remark 2.3 The above theorem remains valid in open sets Ω more general than the cubes, and in particular if Ω is the image of a cube Q_0 under a map g, restriction to Q_0 of a bilipschitzian homeomorphism of an open set $C \supset Q_0$ in $\Lambda \supset \Omega$.[9]

In fact, let T be a linear continuous mapping from $L^s(\Omega, \mathbf{R}^N)$ into $L^s(\Omega, \mathbf{R}^\nu)$ as well as from $L^\infty(\Omega)$ into $BMO(\Omega)$, and define

$$T_* = g_* T g_*^{-1} .$$

The mapping T_* is linear, maps $L^s(Q_0)$ into $L^s(Q_0)$ and by Proposition 2.4, $L^\infty(Q_0)$ into $BMO(Q_0)$, with the appropriate estimates. From the preceding theorem, T_* maps $L^p(Q_0)$ into $L^p(Q_0)$ for every $p \geq s$. By consequence, $T = g_*^{-1} T_* g_*$ maps $L^p(\Omega, \mathbf{R}^N)$ into $L^p(\Omega, \mathbf{R}^\nu)$, with the estimate

$$\|Tu\|_p \leq c\|u\|_p .$$

$\square$

2.6 The Hausdorff Measure

The Lebesgue measure is essentially n-dimensional, and cannot distinguish between different sets of zero measure, nor determine the dimension of such sets. For that purpose, several measures have been introduced; the one we shall describe has proved rather useful in the theory of partial differential equations.

---

[9]This happens for instance if Ω is a ball.

Definition 2.8 *Let E be a subset of $\mathbf{R}^n$, and let $k \geq 0$ and $\delta > 0$ be two real numbers. Having set*

$$H_\delta^k(E) = \omega_k 2^{-k} \inf \left\{ \sum_{j=1}^{\infty} (\operatorname{diam} S_j)^k;\ E \subseteq \bigcup_{j=1}^{\infty} S_j;\ \operatorname{diam} S_j < \delta \right\},$$

where $\omega_k = \Gamma(\frac{1}{2})^k / \Gamma(\frac{k}{2}+1)$, we define the k-dimensional Hausdorff measure *of E:*

$$H^k(E) = \lim_{\delta \to 0} H_\delta^k(E) = \sup_{\delta > 0} H_\delta^k(E). \tag{2.53}$$

The constant ω_k in the above definition[10] is chosen in such a way that when E is a regular k-dimensional surface, its Hausdorff measure $H^k(E)$ coincides with the k-dimensional measure elementarily defined. In particular, if E is a measurable set in $\mathbf{R}^n$, we have $H^n(E) = |E|$.

It follows immediately from the definition that $H^0(E)$ counts the number of points of E, and that for $k > n$ we have $H^k(\mathbf{R}^n) = 0$, and hence $H^k(E) = 0$ for every $E \subset \mathbf{R}^n$. Moreover, if $H^k(E) > 0$ for some k, then for every $h < k$ we have $H^h(E) = +\infty$; whereas from $H^k(E) < +\infty$ it follows that $H^r(E) = 0$ for every $r > k$. Consequently, there exists a unique real number d with the property that $H^s(E) = 0$ for every $s > d$, and $H^r(E) = +\infty$ for every $s < d$. This number d is called the Hausdorff *dimension* of the set E, and is denoted by $\dim_H(E)$, or simply by $\dim(E)$.

The next lemma will be quite useful later.

Lemma 2.4 *Let $\mathcal{G}$ be an arbitrary family of cubes, such that*[11] $M =: \sup_{Q \in \mathcal{G}} r(Q) < +\infty$. *There exists a countable (or finite) subfamily $\Gamma =: \{\Pi_i\}$ of pairwise disjoint cubes, such that*

$$\bigcup_{i=1}^{\infty} \hat{\Pi}_i \supset \bigcup \mathcal{G}$$

where $\hat{\Pi}$ indicates the cube concentric with Π and with quintuple side.

Proof. For every integer h, we set

$$\mathcal{G}_h = \{P \in \mathcal{G} : 2^{-h} M < r(P) \leq 2^{1-h} M\}.$$

Let $\tilde{\mathcal{G}}_1$ be a maximal subfamily of pairwise disjoint cubes of $\mathcal{G}_1$, that is such that

[10] Note that when k is integer, ω_k is the measure of the unit ball of dimension k.

[11] If P is a cube, we indicate by $r(P)$ its radius, that is half its side.

(i) for every $P, Q \in \tilde{\mathcal{G}}_1$, $P \cap Q = \emptyset$.
(ii) for every $Q \in \mathcal{G}_1$ there exists $P \in \tilde{\mathcal{G}}_1$ with $P \cap Q \neq \emptyset$.

The family $\tilde{\mathcal{G}}_1$ is at most countable, since, being $r(P) > M/2$ for every $P \in \tilde{\mathcal{G}}_1$, at most a finite number of cubes of $\tilde{\mathcal{G}}_1$ can intersect any compact set.

Assume now that $\tilde{\mathcal{G}}_1, \tilde{\mathcal{G}}_2, \ldots, \tilde{\mathcal{G}}_{h-1}$ have been defined, and among the cubes of $\mathcal{G}_h$ which do not intersect any of the cubes of $\Gamma_{h-1} =: \cup_{i=1}^{h-1} \tilde{\mathcal{G}}_i$ let us choose a maximal family $\tilde{\mathcal{G}}_h$ of pairwise disjoint cubes. We have

(i) for every $P, Q \in \Gamma_h$, $P \cap Q = \emptyset$.
(ii) for every $Q \in \mathcal{G}_h$ there exists $P \in \Gamma_h$, with $P \cap Q \neq \emptyset$.

As above, $\tilde{\mathcal{G}}_h$ is at most countable.

The family $\Gamma = \cup_{i=1}^{\infty} \tilde{\mathcal{G}}_i$ is what we are looking for. In fact it is at most countable, being the union of countably many families of countable or finite sets. Moreover, if Q is a cube of $\mathcal{G}$, whence of $\mathcal{G}_h$ for some h, and if $Q \notin \Gamma$, there will exist a cube $P \in \Gamma_h$ with $P \cap Q \neq \emptyset$. Since $2r(P) \geq r(Q)$, we have therefore $Q \subset \hat{P}$, and the lemma is proved. □

Remark 2.4 If we write $1 + \epsilon$ instead of 2 in the definition of $\mathcal{G}_h$, we see immediately that the lemma continues to hold with cubes $\hat{\Pi}$ with $r(\hat{\Pi}) = (3 + 2\epsilon) r(\Pi)$ instead of $5r(\Pi)$. □

A first consequence of the above lemma is given in the following proposition:

Proposition 2.7 *Let A be an open set in $\mathbf{R}^n$, and let μ be a positive Radon measure in A, with $\mu(A) < +\infty$. For $0 < \alpha < n$ let*

$$E^{\alpha} = \left\{ x \in A : \limsup_{\varrho \to 0^+} \varrho^{-\alpha} \mu(Q(x, \varrho)) > 0 \right\}. \tag{2.54}$$

Then

$$\dim(E^{\alpha}) \leq \alpha \,. \tag{2.55}$$

Proof. Setting

$$E_s = \left\{ x \in A : \limsup_{\varrho \to 0^+} \varrho^{-\alpha} \mu(Q(x, \varrho)) > \frac{1}{s} \right\},$$

it will be sufficient to prove that $H^{\alpha+\epsilon}(E_s) = 0$ for every integer s and every $\epsilon > 0$.

For every $\delta > 0$ and every $x \in E_s$ there exists a cube $Q(x, \varrho) \subset A$, with $\varrho < \delta$, and such that

$$\mu(Q(x, \varrho)) > \frac{\varrho^\alpha}{2s}.$$

By the preceding lemma, it is possible to find a sequence (possibly a finite number) of pairwise disjoint cubes $\Pi_i = Q(x_i, \varrho_i)$ such that

$$\bigcup_{i=1}^{\infty} \hat{\Pi}_i \supset E_s.$$

We have therefore[12]

$$H_\delta^{\alpha+\epsilon}(E_s) \leq c \sum_{i=1}^{\infty} \varrho_i^{\alpha+\epsilon} \leq c\delta^\epsilon \sum_{i=1}^{\infty} \mu(\Pi_i) \leq c\delta^\epsilon \mu(E_s^\delta \cap A) \leq c\delta^\epsilon \mu(A),$$

and the conclusion follows immediately letting δ go to zero. □

Remark 2.5 In particular, for $H^{\alpha+\epsilon}$ almost every $x \in A$ we have

$$\lim_{\varrho \to 0^+} \varrho^{-\alpha} \mu(Q(x, \varrho)) = 0. \tag{2.56}$$

Moreover, if

$$\lim_{\delta \to 0^+} \mu(E_s^\delta \cap A) = 0,$$

we can conclude that (2.56) holds for H^α almost every $x \in A$. □

Remark 2.6 The assumption $\mu(A) < +\infty$ can be replaced by $\mu(A \cap Q_r) < +\infty$ for every $r > 0$. Actually, repeating the argument for $E_s \cap Q_r$, we obtain $H^{\alpha+\epsilon}(E_s \cap Q_r) = 0$ for every $\epsilon > 0$, and for every integer s and r, and hence $H^{\alpha+\epsilon}(E^\alpha) = 0$. □

2.7 Notes and Comments

The spaces of Morrey and Campanato have been introduced by the latter in [1] and [2], and have proved an important tool in the proof of the Hölder continuity of the solutions of partial differential equations.

These spaces can be generalized in several directions. In the first place, one can introduce a metric δ different from the Euclidean distance, and

[12] If $A \subset \mathbf{R}^n$, we indicate by A^δ the envelope of radius δ of A:

$$A^\delta = \{x \in \mathbf{R}^n : \text{dist}(x, A) < \delta\}.$$

topologically (but not metrically) equivalent to the latter. As an example, we mention the parabolic metric; denoting by (x,t), $x \in \mathbf{R}^n$, $t \in \mathbf{R}$ a point of $\mathbf{R}^{n+1}$, we set

$$\delta((x,t),(y,s)) = |x-y| + \sqrt{|t-s|}$$

or equivalently

$$\delta((x,t),(y,s)) = \max\left\{ \max_{1\le i\le n} |x_i - y_i|, \sqrt{|t-s|}\right\}.$$

The balls $Q(0,R)$ in the last metric are parallelepipeds in $\mathbf{R}^{n+1}$, having as base a cube of $\mathbf{R}^n$ of side $2R$ and $2\sqrt{R}$ as height.

More generally, given a metric δ, one can define the spaces $\mathcal{L}^{p,\lambda}(\Omega,\delta)$ by means of the same (2.16) defining the spaces $\mathcal{L}^{p,\lambda}$, the only difference being that Ω_ϱ denotes now the intersection of Ω with the ball of radius ϱ in the metric δ.

These spaces have been studied by DA PRATO [1] and BAROZZI [1], and have proved useful in the study of the regularity of the solutions of parabolic (CAMPANATO [3], [4]) and quasi-elliptic equations (GIUSTI [1]).

A second generalization of Campanato spaces is given by the spaces $\mathcal{L}^{p,\lambda}_k(\Omega)$. Denoting by $\mathcal{P}_k$ the class of polynomials of degree $\le k$, we set

$$[u]^p_{k,p,\lambda} = \sup_{\substack{x_0\in\Omega\\ \varrho>0}} \varrho^{-\lambda} \inf_{P\in\mathcal{P}_k} \int_{\Omega_\varrho} |u-P|^p\, dx$$

and we define $\mathcal{L}^{p,\lambda}_k(\Omega)$ as the space of functions in $L^p(\Omega)$ such that $[u]_{k,p,\lambda} < +\infty$. $\mathcal{L}^{p,\lambda}_k(\Omega)$ is a Banach space with the norm

$$\|u\|_{k,p,\lambda} = \|u\|_p + [u]_{k,p,\lambda}\,.$$

By means of methods similar to those of section 3, one can prove (CAMPANATO [1]) that

(i) if $\lambda < n + kp$, $\mathcal{L}^{p,\lambda}_k$ is isomorphic to $\mathcal{L}^{p,\lambda}_{k-1}$;

(ii) if $n + kp < \lambda \le n + (k+1)p$, $\mathcal{L}^{p,\lambda}_k$ is isomorphic to $C^{k,\alpha}$, with $\alpha = \frac{\lambda - n}{p} - k$;

(iii) if $\lambda > n + (k+1)p$, $\mathcal{L}^{p,\lambda}_k = \mathcal{P}_k$.

We remark that, in spite of the similarity in their definitions, the spaces $L^{p,\lambda}$ and M^ϑ are essentially different. For instance, whereas M^ϑ coincides with L^p_w ($p = (1-\vartheta)^{-1}$), and hence the functions of M^ϑ are summable with any exponent less than p, a function of $L^{1,\lambda}$ need not belong to any

L^p with $p > 1$, even if λ is very close to n. An example can be found in PICCININI [1].[13]

We could also define the spaces $M^{p,\vartheta}$, of functions in L^p such that

$$\sup |E|^{-\vartheta} \int_E |u|^p \, dx < +\infty \, .$$

A simple application of Hölder's inequality shows that $M^{p,\vartheta} \subset M^\sigma$, with $\sigma = 1 - \frac{1-\vartheta}{p}$. Actually these two spaces are isomorphic, since we have

$$u \in M^{p,\vartheta} \Leftrightarrow |u|^p \in M^\vartheta \Leftrightarrow |u|^p \in L_w^{\frac{1}{1-\vartheta}}$$
$$\Leftrightarrow u \in L_w^{\frac{p}{1-\vartheta}} \Leftrightarrow u \in M^\sigma \, .$$

The lemmas of John and Nirenberg have a "strong" (or rather "weak") version, which consists in eliminating the mean value both in the definition (2.25) and in the assumption (2.28). In the first case we get nothing interesting, since

$$\sup \fint_E |u| dx < +\infty$$

if and only if u is bounded.

The second case corresponds to a theorem by RIESZ [1]: if for every partition of Q_0 in subcubes Q_j it holds that

$$\sum_{j=1}^{\infty} |Q_j| \left(\fint_{Q_j} |u| dx \right)^p \leq K^p$$

then $u \in L^p(Q_0)$, and *vice versa*.

[13]For $\lambda = n - 1$, an example is given by the function $f(x) = x_1^{-1}(\log |x_1|)^{-2}$ in the cube $Q_{1/2}$ of $\mathbf{R}^n$.

Chapter 3

Sobolev Spaces

3.1 Partitions of Unity

Lemma 3.1 *Let Ω be an open set of $\mathbf{R}^n$, and let $\Phi = \{\Omega_j\}$, $j = 1, 2, \ldots$ be a countable covering of Ω with bounded open sets $\Omega_j \subset\subset \Omega$. There exists an open set $\tilde{\Omega}_1$, $\tilde{\Omega}_1 \subset\subset \Omega_1$, such that the family $\{\tilde{\Omega}_1, \Omega_2, \Omega_3, \ldots\}$ is again a covering of Ω.*

Proof. For any open set A we define

$$A_t = \{x \in A : \operatorname{dist}(x, \partial A) > t\}. \tag{3.1}$$

If A is bounded, we have $A_t \subset\subset A$ for every $t > 0$.

Consider now the family

$$\Phi_t = \{\Omega_{1t}, \Omega_2, \Omega_3, \ldots\}.$$

We shall prove that there exists a $t > 0$ such that Φ_t is a covering of Ω. Otherwise, for every integer k there would exist a point x_k belonging to Ω but not to the union of the open sets of $\Phi_{1/k}$. All the points x_k lie in Ω_1, and since this set is relatively compact, a subsequence x_{k_i} will converge to a point $x_0 \in \partial\Omega_1$.

On the other hand Φ is a covering of Ω, and hence x_0 must belong to some Ω_j, with $j \neq 1$. This is impossible, since none of the points x_{k_i} lies in Ω_j. □

An open covering Φ of Ω is *pointwise finite* if every $x \in \Omega$ belongs to a finite number of members of Φ; it is *locally finite* if every compact set $K \subset \Omega$ meets at most a finite number of open sets in Φ.

Proposition 3.1 *Let $\Phi = \{\Omega_j\}$ be a pointwise finite covering of Ω, with $\Omega_j \subset\subset \Omega$. For every j there exists an open set $\tilde{\Omega}_j \subset\subset \Omega_j$ such that*

$$\Omega = \bigcup_{j=1}^{\infty} \tilde{\Omega}_j .$$

Proof. Using repeatedly the preceding lemma we obtain a sequence of coverings

$$\begin{aligned} \Phi^1 &= \{\tilde{\Omega}_1, \Omega_2, \Omega_3, \ldots\}, \\ \Phi^2 &= \{\tilde{\Omega}_1, \tilde{\Omega}_2, \Omega_3, \ldots\}, \\ &\vdots \\ \Phi^k &= \{\tilde{\Omega}_1, \ldots, \tilde{\Omega}_k, \Omega_{k+1}, \ldots\}. \end{aligned}$$

The family $\tilde{\Phi} = \{\tilde{\Omega}_j\}$ is the required covering. For otherwise there would exist a point $x_0 \in \Omega$ which does not belong to any of the sets $\tilde{\Omega}_j$. Since Φ is pointwise finite, there exists an integer N such that $x_0 \notin \Omega_j$ for $j > N$. But in this case x_0 could not belong to any open set of Φ^N, and this could not be a covering of Ω. □

We remark that if A and B are two open sets, and $A \subset\subset B$, there exists a function $\gamma \in C_0^\infty(B)$, with $0 \le \gamma \le 1$ and $\gamma \equiv 1$ in A. Actually, let 2δ be the distance between A and ∂B, and let $\epsilon < \delta$. The function

$$\gamma = \varphi_\epsilon * \chi ,$$

where χ is the characteristic function of $A^\delta = \{x \in \mathbf{R}^n : \operatorname{dist}(x, A) < \delta\}$, has the required properties.

Theorem 3.1 *Let $\Phi = \{\Omega_j\}$ be a locally finite covering of Ω, with $\Omega_j \subset\subset \Omega$. For each j there exists a function $\{\alpha_j\}$ such that*

(i) $\alpha_j \in C_0^\infty(\Omega_j)$,
(ii) $0 \le \alpha_j \le 1$,
(iii) $\sum_{j=1}^{\infty} \alpha_j = 1$ *in* Ω.

The family $\{\alpha_j\}$ is called a *partition of unity* relative to the covering Φ.

Proof. Let $\{\tilde{\Omega}_j\}$ be the covering given by the preceding proposition, and for every j let γ_j be a function in $C_0^\infty(\Omega_j)$ with $\gamma_j \equiv 1$ in $\tilde{\Omega}_j$. Since Φ is locally finite, for every compact set $K \subset \Omega$ only a finite number of γ_j are not identically zero in K. There follows that the series

$$\sum_{j=1}^{\infty} \gamma_j$$

converges uniformly with all its derivatives in every compact set $K \subset \Omega$, so that its sum γ is infinitely differentiable in Ω.

On the other hand, the family $\tilde{\Phi}$ is itself a covering of Ω, and hence $\gamma \geq 1$ in Ω. The functions

$$\alpha_i(x) = \frac{\gamma_i(x)}{\gamma(x)}$$

give the required partition of unity. □

Remark 3.1 If we accept the possibility that some of the functions α_i are identically zero, the conclusion of the theorem remains valid also for a generic covering. In fact on one side from every covering of Ω it is possible to extract a countable subcovering; and on the other, given a countable covering $\Phi = \{\Omega_i\}$, it is always possible to find a locally finite covering $\tilde{\Phi} = \{\tilde{\Omega}_i\}$, with $\tilde{\Omega}_i \subset \Omega_i$.

In order to prove the last statement, let $\emptyset = K_0 \subset K_1 \subset K_2 \subset \cdots \subset \Omega$ be a sequence of compact sets with the property that every compact set $K \subset \Omega$ is contained in some K_m, and for every h let n_h be the integer such that

$$K_h \subset \Omega_1 \cup \Omega_2 \cup \cdots \cup \Omega_{n_h},$$

$$K_h \not\subset \Omega_1 \cup \Omega_2 \cup \cdots \cup \Omega_{n_h - 1}.$$

For $n_h < j \leq n_{h+1}$ ($n_0 = 0$) define

$$\tilde{\Omega}_j = \Omega_j - K_h .$$

The family $\tilde{\Phi} = \{\tilde{\Omega}_j\}$ is what we are looking for, since we have

$$\bigcup_{j=1}^{m} \tilde{\Omega}_j = \bigcup_{j=1}^{m} \Omega_j . \tag{3.2}$$

The above relation is trivial for $m \leq n_1$. Assume now it holds for $m \leq n_h$, and let $n_h < m \leq n_{h+1}$. We have then

$$\begin{aligned}\bigcup_{j=1}^{m} \tilde{\Omega}_j &= \bigcup_{j=1}^{n_h} \Omega_j \cup \bigcup_{j=n_h+1}^{m} (\Omega_j - K_h) \\ &= \bigcup_{j=1}^{n_h} \Omega_j \cup \left(\bigcup_{j=n_h+1}^{m} \Omega_j - K_h \right) \\ &= \bigcup_{j=1}^{m} \Omega_j\end{aligned}$$

since $K_h \subset \bigcup_{j=1}^{n_h} \Omega_j$. □

In what follows, we shall use also partitions of unity relative to coverings $\{\Omega_j\}$ such that

$$\Omega \subset\subset \bigcup_{j=1}^{\infty} \Omega_j .$$

This case cannot be treated as the preceding one, though the conclusion is the same.

Theorem 3.2 *Let $\Phi = \{\Omega_j\}$ be a covering of Ω, with $\Omega \subset\subset \cup\Phi$. There exists a partition of unity relative to Φ.*

Proof. Since in particular $\bar{\Omega} \subset \cup\Phi$, from Φ we can extract a finite covering $\{\Omega_1, \Omega_2, \ldots, \Omega_N\}$ of $\bar{\Omega}$. Reasoning as above, we can show that there exists a covering $\{\tilde{\Omega}_1, \tilde{\Omega}_2, \ldots, \tilde{\Omega}_N\}$ of $\bar{\Omega}$, with $\tilde{\Omega}_j \subset\subset \Omega_j$. Let γ_i be a function in $C_0^\infty(\Omega_i)$, with $\gamma_i \equiv 1$ in $\tilde{\Omega}_i$, and define

$$\begin{aligned}\alpha_1 &= \gamma_1 , \\ \alpha_2 &= \gamma_2(1 - \gamma_1) , \\ &\vdots \\ \alpha_N &= \gamma_N(1 - \gamma_1)(1 - \gamma_2) \ldots (1 - \gamma_N) .\end{aligned}$$

By induction,

$$1 - \sum_{i=1}^{k} \alpha_i = (1 - \gamma_1)(1 - \gamma_2) \ldots (1 - \gamma_k)$$

for every $k = 1, 2, \ldots, N$.

For $x \in \tilde{\Omega}_j$ we have $\gamma_j = 1$, and therefore $1 - \sum_{i=1}^{N} \alpha_i = 0$. Since the open sets $\tilde{\Omega}_i$ are a covering of Ω, we conclude that

$$\sum_{i=1}^{N} \alpha_i(x) = 1$$

for every $x \in \Omega$. □

3.2 Weak Derivatives

Let us begin by introducing some notation. A *multi-index* α is a n-vector $(\alpha_1, \alpha_2, \ldots, \alpha_n)$ whose components are non-negative integers. If α and β are two multi-indices, we say that $\alpha \le \beta$ if $\alpha_i \le \beta_i$ for every $i = 1, 2, \ldots, n$. If at least one of these inequalities is strict, we shall say that $\alpha < \beta$. Moreover, we define

$$|\alpha| = \alpha_1 + \alpha_2 + \cdots + \alpha_n\,,$$

$$\alpha! = \alpha_1!\alpha_2!\ldots\alpha_n!\,,$$

$$\alpha + \beta = (\alpha_1 + \beta_1, \alpha_2 + \beta_2, \ldots, \alpha_n + \beta_n)$$

and if $\alpha \le \beta$:

$$\binom{\beta}{\alpha} = \binom{\beta_1}{\alpha_1}\binom{\beta_2}{\alpha_2}\cdots\binom{\beta_n}{\alpha_n} = \frac{\beta!}{\alpha!(\beta-\alpha)!}\,.$$

For $x \in \mathbf{R}^n$ we set

$$x^\alpha = x_1^{\alpha_1} x_2^{\alpha_2} \ldots x_n^{\alpha_n}$$

and if $f(x)$ is an infinitely differentiable function:

$$D^\alpha f(x) = \frac{\partial^{|\alpha|} f(x)}{\partial x_1^{\alpha_1} \partial x_2^{\alpha_2} \ldots \partial x_n^{\alpha_n}}\,.$$

With these definitions, the formulae involving partial derivatives of functions of n variables become very compact; for instance, Taylor's formula can be written in the form

$$f(x) = \sum_{|\alpha| \le k} \frac{D^\alpha f(x_0)}{\alpha!}(x - x_0)^\alpha + R_k(x; x_0)\,,$$

where R_k is the rest of order k, whereas the formula for the derivative of a product becomes

$$D^\alpha(fg) = \sum_{\gamma \leq \alpha} \binom{\alpha}{\gamma} D^\gamma f D^{\alpha-\gamma} g .$$

We introduce now some extensions of the classical concept of derivative.

Definition 3.1 *Let $u \in L^p_{\text{loc}}(\Omega)$, $p \geq 1$, and let α be a multi-index. We say that u has a* weak derivative *(or a* derivative in the sense of distributions*) of order α in $L^p_{\text{loc}}(\Omega)$ if there exists a function $v_\alpha \in L^p_{\text{loc}}(\Omega)$ such that for every function $\varphi \in C_0^\infty(\Omega)$*

$$\int_\Omega u D^\alpha \varphi \, dx = (-1)^{|\alpha|} \int_\Omega v_\alpha \varphi \, dx . \tag{3.3}$$

The function v_α is usually denoted by the standard symbol $D^\alpha u$. This notation is appropriate, since we have the following:

Proposition 3.2 *Weak derivatives are unique.*

Proof. If v_α and w_α are both weak derivatives of order α of the same function u, we have for every $\varphi \in C_0^\infty(\Omega)$:

$$\int_\Omega (v_\alpha - w_\alpha)\varphi \, dx = 0 .$$

Let now $A \subset\subset \Omega$. Since $C_0^\infty(A)$ is dense in $L^1(A)$, it is possible to find a sequence $\varphi_j \in C_0^\infty(A)$, with $|\varphi_j| \leq 2$, converging almost everywhere to the function $H(v_\alpha - w_\alpha)$, where H is the Haeviside function:

$$H(t) = \begin{cases} 1 & \text{if } t > 0 , \\ 0 & \text{if } t = 0 , \\ -1 & \text{if } t < 0 . \end{cases}$$

Writing φ_j instead of φ in the preceding equation, and passing to the limit as $j \to \infty$, we obtain:

$$\int_A |v_\alpha - w_\alpha| dx = 0$$

and therefore, since A is arbitrary, $v_\alpha = w_\alpha$ in Ω. □

From the preceding result it follows at once that for $u \in C^{|\alpha|}(\Omega)$ the weak derivative $D^\alpha u$ coincides with the standard one. However, one has to be careful in general, since a function (as for instance $H(t)$) can be differentiable almost everywhere, and yet may not possess a weak derivative.[1]

A second way to define derivatives of functions in L^p_{loc} is the following.

Definition 3.2 *Let $u \in L^p_{\text{loc}}(\Omega)$, and let α be a multi-index. We say that u has* strong derivative *of order α in $L^p_{\text{loc}}(\Omega)$ if there exists a sequence of functions $u_j \in C^{|\alpha|}(\Omega)$ such that for every open set $A \subset\subset \Omega$:*

(i) *$u_j \to u$ in $L^p(A)$,*
(ii) *$D^\alpha u_j$ is a Cauchy sequence in $L^p(A)$.*

The functions $D^\alpha u_j$ will converge in $L^p_{\text{loc}}(\Omega)$ to a function v_α, which we call the *strong derivative* of u of order α.

If $\varphi \in C_0^\infty(\Omega)$ we have

$$\begin{aligned}(-1)^{|\alpha|} \int_\Omega v_\alpha \varphi \, dx &= (-1)^{|\alpha|} \lim_{j\to\infty} \int_\Omega \varphi D^\alpha u_j \, dx \\ &= \lim_{j\to\infty} \int_\Omega u_j D^\alpha \varphi \, dx = \int_\Omega u D^\alpha \varphi \, dx\end{aligned}$$

so that the strong derivative v_α of u coincides with its weak derivative. As a consequence, strong derivatives are unique.

Proposition 3.3 *A function u has strong derivative of order α in $L^p_{\text{loc}}(\Omega)$ if and only if for every open set $A \subset\subset \Omega$ there exists a sequence u_j satisfying* (i) *and* (ii).

Proof. The necessity of the condition is trivial. In order to prove that it is sufficient, consider a sequence $\{\Omega_k\}$ of open sets such that $\Omega_k \subset\subset \Omega_{k+1} \subset\subset \Omega$ and

$$\bigcup_{j=1}^\infty \Omega_j = \Omega .$$

For every Ω_k there exists a sequence $\{u_j^{(k)}\}_j$ such that

(i) $u_j^{(k)} \to u$ in $L^p(\Omega_k)$,
(ii) $D^\alpha u_j^{(k)} \to v_\alpha^{(k)}$ in $L^p(\Omega_k)$.

---

[1] More precisely, the derivative of $H(t)$ in the sense of distributions is the Dirac measure δ, and hence it does not coincide with its pointwise derivative, which is zero. In other words, the fact that a function f is almost everywhere differentiable, with derivative f' in L^p_{loc}, does not imply that f' be the weak derivative of f.

It is clear that $v_\alpha^{(k)} = D^\alpha u$ in Ω_k. Moreover, from the uniqueness of strong derivatives there follows that $v_\alpha^{(k)} = v_\alpha^{(k+h)}$ in Ω_k. If for $x \in \Omega_k$ we define $v_\alpha(x) = v_\alpha^{(k)}(x)$, the function v_α belongs to $L^p_{\text{loc}}(\Omega)$, and is the weak derivative of u: $v_\alpha = D^\alpha u$.

On the other hand, for every integer k it is possible to find an index j_k such that

$$\|u_{j_k}^{(k)} - u\|_{p,\Omega_k} + \|D^\alpha u_{j_k}^{(k)} - v_\alpha\|_{p,\Omega_k} < \frac{1}{k}.$$

Let now $\gamma_k \in C_0^\infty(\Omega_{k+1})$, with $\gamma_k = 1$ on Ω_k. The sequence $w_k = \gamma_k u_{j_{k+1}}^{(k+1)}$ satisfies (i) and (ii) of Definition 3.2, and hence u has strong derivative $D^\alpha u$ in L^p_{loc}. □

We have already remarked that strong derivatives are weak derivatives. The converse is also true.

Theorem 3.3 *Weak derivatives are strong derivatives.*

Proof. By the preceding proposition, it will be sufficient to show that for every $A \subset\subset \Omega$ there exists a sequence u_k satisfying conditions (i) and (ii). Let $2d = \min\{1, \text{dist}(A, \partial\Omega)\}$, and let $\chi(x)$ be the characteristic function of the set $A^d = \{x \in \Omega : \text{dist}(x, A) < d\}$. For $\epsilon < d$ we set

$$w_\epsilon = (\chi u) * \varphi_\epsilon .$$

Let us show first that if $D^\alpha u$ is the weak derivative of order α of u, then for every $x \in A$:

$$D^\alpha w_\epsilon = D^\alpha u_\epsilon = (D^\alpha u)_\epsilon .$$

Let $x \in A$ and $|z| \le \epsilon < d$. We have $x - z \in A^d$ and therefore $\chi(x - z) = 1$, whence

$$w_\epsilon(x) = \int_{|z|<\epsilon} \varphi_\epsilon(z)\chi(x-z)u(x-z)dz = u_\epsilon(x) .$$

For such points x we have then

$$D^\alpha w_\epsilon(x) = D^\alpha u_\epsilon(x) = \int D_x^\alpha \varphi_\epsilon(x-y)u(y)dy$$
$$= (-1)^{|\alpha|} \int D_y^\alpha \varphi_\epsilon(x-y)u(y)dy .$$

By the definition of weak derivative, and recalling that for every x the function $y \to \varphi_\epsilon(x-y)$ belongs to $C_0^\infty(\Omega)$, the last integral is equal to

$$\int \varphi_\epsilon(x-y) D^\alpha u(y) dy = (D^\alpha u)_\epsilon(x) \,.$$

The proof of the theorem is now simple. Since u and $D^\alpha u$ both belong to $L^p(A)$, it follows from Theorem 2.8 that $u_\epsilon \to u$ and $D^\alpha u_\epsilon \to D^\alpha u$ in $L^p(A)$, and therefore $D^\alpha u$ is the strong derivative of u. □

3.3 The Sobolev Spaces $W^{k,p}$

Definition 3.3 *We shall denote by $W^{k,p}(\Omega)$ the space of functions having weak derivatives up to the order k in $L^p(\Omega)$.*

$W^{k,p}(\Omega)$ is a Banach space with the norm

$$\|u\|_{k,p} = \left\{ \sum_{|\alpha| \le k} \int_\Omega |D^\alpha u|^p \, dx \right\}^{\frac{1}{p}} . \tag{3.4}$$

With $C^{k,p}(\Omega)$ we indicate the space of the functions $u \in C^k(\Omega)$ whose norm (3.4) is finite: $C^{k,p} = C^k \cap W^{k,p}$. We have the following

Theorem 3.4 (Meyers and Serrin [1]) *For every $u \in W^{k,p}(\Omega)$ there exists a sequence of functions $u_i \in C^{k,p}(\Omega)$ converging to u in the norm* (3.4).

Proof. It is sufficient to show that for every $u \in W^{k,p}(\Omega)$ and for every $\tau > 0$ there exists a function $w \in C^{k,p}(\Omega)$ with $\|u - w\|_{k,p} < \tau$.

We remark in the first place that this is true if u has compact support in Ω; for in this case the functions $u_\epsilon = u * \varphi_\epsilon$ belong to $C_0^\infty(\Omega)$ (hence to $C^{k,p}(\Omega)$) for every ϵ small enough, and converge to u in the norm (3.4). Consider now a general function $u \in W^{k,p}(\Omega)$, and set

$$\Omega_{(i)} = \left\{ x \in \Omega : \operatorname{dist}(x, \partial\Omega) > \frac{1}{i} \right\} ;$$

$$\Omega_{(0)} = \emptyset ;$$

$$A_i = \Omega_{(i+1)} - \Omega_{(i-1)} , \quad i \ge 3 ;$$

$$A_2 = \Omega_{(3)} .$$

The open sets A_i, $i \geq 2$, form a locally finite covering of Ω. By Theorem 3.1 there exists a partition of unity $\{\alpha_i\}$ relative to that covering. It is easily seen that for every i we have

$$\alpha_i u \in W^{k,p}(\Omega)\,; \quad \operatorname{supp}(\alpha_i u) \subset A_i \subset\subset \Omega$$

so that, according to the above, for every integer $i \geq 2$ it will be possible to find an $\epsilon_i > 0$ such that:

$$\operatorname{supp} \varphi_{\epsilon_i} * (\alpha_i u) \subset \Omega_{(i+2)} - \Omega_{(i-2)}\,, \tag{3.5}$$

$$\|\varphi_{\epsilon_i} * (\alpha_i u) - \alpha_i u\|_{k,p} < \tau 2^{-i}\,. \tag{3.6}$$

We define now

$$w = \sum_{i=2}^{\infty} \varphi_{\epsilon_i} * (\alpha_i u)\,. \tag{3.7}$$

By the definition of the sets Ω_i, in every compact set $K \subset \Omega$ at most a finite number of terms of the series will be different from zero; and hence the function w is infinitely differentiable in Ω. Moreover, we have:

$$\|w - u\|_{k,p} = \left\| \sum_{i=2}^{\infty} \{\varphi_{\epsilon_i} * (\alpha_i u) - \alpha_i u\} \right\|_{k,p}$$

$$\leq \sum_{i=2}^{\infty} \|\varphi_{\epsilon_i} * (\alpha_i u) - \alpha_i u\|_{k,p} < \tau$$

and the theorem is proved. □

Remark 3.2 We remark that the function w constructed above is of class $C^\infty(\Omega)$. Moreover w and u have the same trace on $\partial\Omega$ (see later, Secs. 3.7 and 3.8). □

The preceding theorem can be rephrased by saying that $W^{k,p}(\Omega)$ is the closure of $C^{k,p}(\Omega)$ in the norm of $W^{k,p}$, or else that for any open set Ω, $C^{k,p}(\Omega)$ is dense in $W^{k,p}(\Omega)$. The matter is different if one wants to approximate functions in $W^{k,p}(\Omega)$ with functions of class $C^k(\bar{\Omega})$ (that is continuous together with their derivatives up to the boundary of Ω). Generally speaking, this is not possible unless the boundary of Ω is regular enough (see later, Theorem 3.6).

A consequence of Theorem 3.4 is the following.

Proposition 3.4 *Let $f(t)$ be a function of class $C^1(\mathbf{R})$ with bounded derivative, and let $u \in W^{1,p}(\Omega)$. Then, $f \circ u$ belongs to $W^{1,p}(\Omega)$, and $D(f \circ u) = f' \circ u\, Du$.*

Proof. The theorem is trivial for functions u of class $C^{1,p}(\Omega)$. Let now u_k be a sequence in $C^{1,p}$, converging to u in the norm of $W^{1,p}(\Omega)$. Passing possibly to a subsequence, we may assume that $u_k \to u$ almost everywhere. We have

$$|f(u_k) - f(u)| \leq L|u_k - u|$$

and therefore $f(u_k) \to f(u)$ in L^p. On the other hand

$$\begin{aligned} Df(u_k) - f'(u)Du &= f'(u_k)Du_k - f'(u)Du \\ &= f'(u_k)[Du_k - Du] + [f'(u_k) - f'(u)]Du\,. \end{aligned}$$

The first term on the right-hand side can be estimated by $L|Du_k - Du|$, and therefore it tends to zero in L^p. The second term tends to zero almost everywhere (by the continuity of f') and is bounded by $2L|Du|$; by Lebesgue's theorem of dominated convergence, it tends to zero in L^p. □

The above proposition remains valid when the function $f(t)$ is only Lipschitz-continuous (see later, Sec. 3.9). However, for our purposes we shall need only the particular case $f(t) = |t|$, which often is taken for granted.

Proposition 3.5 *If $u \in W^{1,p}(\Omega)$, then $|u| \in W^{1,p}(\Omega)$, and*

$$D|u| = H(u)Du$$

where $H(t)$ is the Heaviside function:

$$H(t) = \begin{cases} 1 & \text{if } t > 0\,, \\ 0 & \text{if } t = 0\,, \\ -1 & \text{if } t < 0\,. \end{cases}$$

Proof. The function $\eta_\epsilon = \sqrt{\epsilon^2 + (u+\epsilon)^2}$ tends to $|u|$ in L^p as $\epsilon \to 0$; moreover by the preceding proposition

$$D\eta_\epsilon = \frac{u+\epsilon}{\eta_\epsilon} Du\,.$$

Let now $\epsilon \to 0$. For almost every x we have

$$\lim_{\epsilon \to 0^+} \frac{u+\epsilon}{\eta_\epsilon} = \sigma(u)$$

and

$$\lim_{\epsilon \to 0^-} \frac{u+\epsilon}{\eta_\epsilon} = \vartheta(u)$$

where

$$\sigma(t)=\begin{cases}1 & \text{if } t>0\,,\\ \dfrac{1}{\sqrt{2}} & \text{if } t=0\,,\\ -1 & \text{if } t<0\,,\end{cases} \quad \text{and} \quad \vartheta(t)=\begin{cases}1 & \text{if } t>0\,,\\ -\dfrac{1}{\sqrt{2}} & \text{se } t=0\,,\\ -1 & \text{if } t<0\,.\end{cases}$$

Both the functions $\sigma(u)Du$ and $\vartheta(u)Du$ are strong derivatives of $|u|$; since strong derivatives are unique, they must coincide almost everywhere, so that in particular $Du(x)=0$ for almost every $x\in E=\{x\in\Omega : u(x)=0\}$. But then we also have $D|u|=H(u)Du$. □

If we define

$$u^+(x)=\frac{|u|+u}{2} \text{ and } u^-=\frac{|u|-u}{2}\,,$$

we have

$$Du^+=\begin{cases}Du & \text{if } u>0\,,\\ 0 & \text{if } u\le 0\,,\end{cases} \tag{3.8}$$

$$Du^-=\begin{cases}0 & \text{if } u\ge 0\,,\\ -Du & \text{if } u<0\,.\end{cases} \tag{3.9}$$

Theorem 3.5 *Let Ω and Λ be two open sets in $\mathbf{R}^n$ and let $g:\bar\Lambda\to\bar\Omega$ be a diffeomorphism. Then, the induced mapping g_*, defined by $g_*u =: u\circ g$, is an isomorphism between $W^{1,p}(\Omega)$ and $W^{1,p}(\Lambda)$.*

Proof. Let $U(x)=g_*u(x)=u(g(x))$, and let us assume first that $u\in C^{1,p}(\Omega)$. We have $U\in C^{1,p}(\Lambda)$, and[2]

$$DU=\left(\frac{\partial g}{\partial x}\right)^t Du\circ g$$

and hence, setting

$$M=\left\|\left(\frac{\partial g}{\partial x}\right)^t\right\| =: \sup_{x\in\Lambda}\sup_{|\xi|=1}\left|\left(\frac{\partial g}{\partial x}\right)^t\xi\right|$$

we get

$$|DU|^p\le M^p|Du\circ g|^p\,.$$

[2]We denote by A^t the transposed matrix of A.

But then

$$\int_\Lambda |DU|^p\,dx \le M^p \int_\Lambda |Du\circ g|^p\,dx$$

$$\le M^p N \int_\Lambda |Du\circ g|^p \left|\det \frac{\partial g}{\partial x}\right| dx = M^p N \int_\Omega |Du|^p\,dy\,,$$

where

$$N^{-1} =: \inf_\Lambda \left|\det \frac{\partial g}{\partial x}\right| > 0\,.$$

Interchanging the role of u and U, we conclude that

$$c_1 \int_\Omega |Du|^p\,dy \le \int_\Lambda |DU|^p\,dx \le c_2 \int_\Omega |Du|^p\,dy\,, \tag{3.10}$$

where c_1 and c_2 are positive constants independent of u.

A similar argument holds for the integral of $|u|^p$, and leads to the estimates:

$$c_1 \int_\Omega |u|^p\,dy \le \int_\Lambda |U|^p\,dx \le c_2 \int_\Omega |u|^p\,dy\,. \tag{3.11}$$

From the estimates (3.10) and (3.11) we conclude that g_* can be extended to a linear mapping from $W^{1,p}(\Omega)$ to $W^{1,p}(\Lambda)$, continuous with its inverse, and the theorem is proved. □

We remark that the conclusion of the theorem holds whenever g is a bi-Lipschitz mapping (namely, Lipschitz-continuous with its inverse) from $\bar\Lambda$ onto $\bar\Omega$, provided $|\det \frac{\partial g}{\partial x}| \ge \alpha > 0$, as we shall always assume. It is not difficult to prove that if $g \in C^k(\bar\Lambda)$, g_* is an isomorphism between $W^{k,p}(\Omega)$ and $W^{k,p}(\Lambda)$.

Proposition 3.6 *Let $u \in L^p(\Omega)$ and let $\{\Omega_j\}_{1\le j\le N}$ be a finite covering of $\bar\Omega$. Assume that for every j the function u belongs to $W^{k,p}(\Omega\cap\Omega_j)$. Then, $u \in W^{k,p}(\Omega)$.*

Proof. For $|\beta| \le k$ and $j = 1, 2, \ldots, N$, let v_j^β be the weak derivative of order β of u in $\Omega_j \cap \Omega$:

$$\int v_j^\beta \varphi\,dx = (-1)^{|\beta|} \int u D^\beta \varphi\,dx \quad \forall\,\varphi \in C_0^\infty(\Omega_j \cap \Omega)\,.$$

If the support of φ is contained in $\Omega_i \cap \Omega_j$, we have

$$\int v_i^\beta \varphi\,dx = (-1)^{|\beta|} \int u D^\beta \varphi\,dx = \int v_j^\beta \varphi\,dx$$

and hence

$$v_j^\beta = v_i^\beta \quad \text{in} \quad \Omega_i \cap \Omega_j \cap \Omega\,.$$

We have, therefore, a function $v^\beta \in L^p(\Omega)$ such that

$$v^\beta(x) = v_i^\beta(x) \quad \forall\, x \in \Omega_i \cap \Omega\,.$$

Let now $\varphi \in C_0^\infty(\Omega)$, and let $\{\alpha_i\}$ be a partition of unity relative to the covering Ω_i. We have

$$\begin{aligned}\int u D^\beta \varphi\, dx &= \sum_{i=1}^N \int u D^\beta(\alpha_i \varphi) dx \\ &= (-1)^{|\beta|} \int \sum_{i=1}^N v^\beta \alpha_i \varphi\, dx = (-1)^{|\beta|} \int v^\beta \varphi\, dx\end{aligned}$$

and hence $v^\beta = D^\beta u$. □

We can now discuss the problem of approximating functions u in $W^{k,p}(\Omega)$ with functions regular in $\bar{\Omega}$. For simplicity, we shall only treat the case $k = 1$.

Setting

$$B^+ = \{x \in \mathbf{R}^n : |x| < 1, x_n > 0\} = B \cap \mathbf{R}^n_+\,,$$

let u be a function in $W^{1,p}(B^+)$, whose support does not intersect the round part $\partial^+ B = \partial B \cap \mathbf{R}^n_+$ of the boundary of B^+.

For $s > 0$ let us define

$$\tau_s u(x) = u(x_1, x_2, \ldots, x_n + s)\,.$$

Moreover, let u_ϵ be a regularization of u (extended as 0 outside B^+). For ϵ small enough, the functions $\tau_{2\epsilon} u_\epsilon(x)$ belong to $C^\infty(\overline{B^+})$. We have

$$\int_{B^+} |\tau_{2\epsilon} u_\epsilon - \tau_{2\epsilon} u|^p\, dx = \int_{\mathbf{R}^n_{2\epsilon}} |u_\epsilon - u|^p\, dx\,,$$

where we have set

$$\mathbf{R}^n_{2\epsilon} = \{x \in \mathbf{R}^n : x_n > 2\epsilon\}\,.$$

Moreover, since $D(\tau_s u) = \tau_s Du$, we have

$$\int_{B^+} |D(\tau_{2\epsilon}(u_\epsilon - u))|^p\, dx = \int_{\mathbf{R}^n_{2\epsilon}} |D(u_\epsilon - u)|^p\, dx$$

and hence

$$\lim_{\epsilon \to 0^+} \|\tau_{2\epsilon}(u_\epsilon - u)\|_{1,p} = 0 .$$

On the other hand from Theorems 2.1 (Lusin) and 2.3 (the absolute continuity of the integral) it follows that

$$\lim_{\epsilon \to 0^+} \|\tau_{2\epsilon} u - u\|_{1,p} = 0$$

and therefore

$$\lim_{\epsilon \to 0^+} \|\tau_{2\epsilon} u_\epsilon - u\|_{1,p} = 0$$

so that the function u is the limit of functions of class $C^\infty(\overline{B^+})$.

More generally, we have

Theorem 3.6 *Let Ω be a bounded open set with boundary of class C^1. The space $C^1(\bar{\Omega})$ is dense in $W^{1,p}(\Omega)$.*

Proof. For every point $x_0 \in \partial\Omega$ there exists a neighborhood A and a diffeomorphism $g : B \to A$ mapping B^+ in $A \cap \Omega$ and the flat part P of ∂B^+ $(P = \partial B^+ \cap \{x : x_n = 0\})$ on $A \cap \partial\Omega$. A finite number of such sets $A_1, A_2, \dots, A_N$ cover $\partial\Omega$; adding possibly an additional open set $A_0 \subset\subset \Omega$ we get a finite covering of $\bar{\Omega}$. Let $\{\alpha_i\}$ be a partition of unity relative to that covering, and let g_i be the diffeomorphism relative to the set A_i.

For $u \in W^{1,p}(\Omega)$, the functions

$$U_i = (g_i)_*(\alpha_i u) = (\alpha_i u) \circ g_i$$

belong to $W^{1,p}(B^+)$, and have support non-intersecting $\partial^+ B$. As we have seen, for every $\vartheta > 0$ we can find a function $Z_i \in C^\infty(\overline{B^+})$, whose support does not intersect $\partial^+ B$, and such that

$$\|Z_i - U_i\|_{1,p} < \vartheta .$$

Setting

$$z_i = (g_i^{-1})_* Z_i = Z_i \circ g_i^{-1}$$

we have $z_i \in C^1(\overline{\Omega \cap A_i})$ and

$$\|z_i - \alpha_i u\|_{1,p} < c\vartheta .$$

Let now z_0 be a function in $C_0^1(A_0)$, with $\|z_0 - \alpha_0 u\|_{1,p} < \vartheta$, and let

$$z = \sum_{i=0}^{N} z_i .$$

We have $z \in C^1(\bar{\Omega})$, and

$$\|z - u\|_{1,p} \le \sum_{i=0}^{N} \|z_i - \alpha_i u\|_{1,p} < (1 + Nc)\vartheta .$$

The theorem is thus proved. If instead we only want to approximate u with functions in Lip(Ω), it will be sufficient to assume that $\partial\Omega$ is Lipschitz-continuous. □

3.4 Imbedding Theorems

Lemma 3.2 *Let $f_1, f_2, \ldots, f_N$ be functions in $L^N(\Omega)$, $\Omega \subset \mathbf{R}^n$. Then,*

$$\int_\Omega \prod_{i=1}^{N} |f_i| dx \le \prod_{i=1}^{N} \left(\int_\Omega |f_i|^N \, dx \right)^{\frac{1}{N}} . \tag{3.12}$$

Proof. For $N = 2$, (3.12) reduces to a Schwartz inequality. We shall assume that it holds for N, and we shall prove it for $N + 1$.

By Hölder's inequality (2.7) we have

$$\int_\Omega \prod_{i=1}^{N+1} |f_i| dx \le \left(\int_\Omega |f_{N+1}|^{N+1} \, dx \right)^{\frac{1}{N+1}} \left(\int_\Omega \prod_{i=1}^{N} |f_i|^{\frac{N+1}{N}} \, dx \right)^{\frac{N}{N+1}} ,$$

and the conclusion follows at once applying (3.12) to the functions $g_i = |f_i|^{\frac{N+1}{N}}$. □

Lemma 3.3 (Gagliardo [2]) *Let $f_1, f_2, \ldots, f_n$ be non-negative functions in $\mathbf{R}^n$, and assume that for every i the function f_i does not depend on the variable x_i. Then,*

$$\int \prod_{i=1}^{n} f_i \, dx \le \prod_{i=1}^{n} \left(\int f_i^{n-1} \, d\hat{x}_i \right)^{\frac{1}{n-1}} , \tag{3.13}$$

where

$$d\hat{x}_i = dx_1 \ldots dx_{i-1} \, dx_{i+1} \ldots dx_n .$$

Proof. For $n = 2$, (3.13) is trivial. Let us assume it holds for $n \geq 2$ and let us prove it for $n + 1$.

We have from Hölder's inequality:

$$\int \prod_{i=1}^{n+1} f_i \, dx \, dx_{n+1} = \int f_{n+1} \, dx \int \prod_{i=1}^{n} f_i \, dx_{n+1}$$

$$\leq \left(\int f_{n+1}^n \, dx \right)^{\frac{1}{n}} \left\{ \int \left(\int \prod_{i=1}^{n} f_i \, dx_{n+1} \right)^{\frac{n}{n-1}} dx \right\}^{\frac{n-1}{n}},$$

where $dx = dx_1 \, dx_2 \ldots dx_n$.

From the preceding lemma we get

$$\int \prod_{i=1}^{n+1} f_i \, dx \, dx_{n+1} \leq \left(\int f_{n+1}^n \, dx \right)^{\frac{1}{n}} \left\{ \int \prod_{i=1}^{n} \left(\int f_i^n \, dx_{n+1} \right)^{\frac{1}{n-1}} dx \right\}^{\frac{n-1}{n}}.$$

Setting now

$$g_i = \left(\int f_i^n \, dx_{n+1} \right)^{\frac{1}{n-1}}$$

and applying (3.13) to the functions g_i, we get at once

$$\int \prod_{i=1}^{n} g_i \, dx \leq \left(\prod_{i=1}^{n} \int g_i^{n-1} \, d\hat{x}_i \right)^{\frac{1}{n-1}}$$

and hence the conclusion. □

We can now prove a first imbedding result.

Theorem 3.7 (Sobolev I) *Let $u \in C_0^\infty(\mathbf{R}^n)$ and for $p < n$ let $p^* = \frac{np}{n-p}$. Then,*

$$\|u\|_{p^*} \leq c(n,p) \|Du\|_p \,, \tag{3.14}$$

where $c(n,p)$ is a constant depending only on n and p.

Proof. We shall consider first the case $p = 1$. We have

$$u(x) = \int_{-\infty}^{x_i} D_i u(x_1, \ldots, x_{i-1}, t, x_{i+1}, \ldots, x_n) dt$$

and hence

$$|u(x)| \leq \int_{-\infty}^{+\infty} |D_i u(x)| dx_i \,,$$

so that

$$|u(x)|^{\frac{n}{n-1}} \leq \prod_{i=1}^{n} \left\{ \int_{-\infty}^{+\infty} |D_i u(x)| dx_i \right\}^{\frac{1}{n-1}} =: \prod_{i=1}^{n} f_i \,.$$

We can apply the preceding lemma to the functions f_i, getting easily

$$\int |u|^{\frac{n}{n-1}} \, dx \leq \prod_{i=1}^{n} \left\{ \int |D_i u| dx \right\}^{\frac{1}{n-1}} \leq \left\{ \int |Du| dx \right\}^{\frac{n}{n-1}}$$

and therefore (3.14) for $p = 1$.

In the general case we apply (3.14) to the function $v = |u|^{r+1}$, $r > 0$. Remarking that $|Dv| = (r+1)|u|^r |Du|$, we get

$$\begin{aligned} \left\{ \int |u|^{\frac{(r+1)n}{n-1}} \, dx \right\}^{\frac{n-1}{n}} &\leq (r+1) \int |u|^r |Du| dx \\ &\leq (r+1) \left\{ \int |Du|^p \, dx \right\}^{\frac{1}{p}} \left\{ \int |u|^{\frac{rp}{p-1}} \, dx \right\}^{\frac{p-1}{p}} . \end{aligned}$$

If we choose now $r = \frac{n(p-1)}{n-p}$, we get $\frac{(r+1)n}{n-1} = \frac{rp}{p-1} = \frac{np}{n-p} = p^*$, and therefore (3.14) with $c(n,p) = \frac{p(n-1)}{n-p}$. □

Theorem 3.8 (SOBOLEV II) *Let $u \in C_0^\infty(\mathbf{R}^n)$, and let $p > n$. Setting $\alpha = 1 - \frac{n}{p}$ we have:*

$$[u]_\alpha \leq c(n,p) \|Du\|_p \,, \tag{3.15}$$

$$\sup |u| \leq c(n,p) (\operatorname{diam}(\operatorname{supp} u))^\alpha \|Du\|_p \,, \tag{3.16}$$

where $c(n,p)$ is a constant depending only on n and p.

Proof. Let x and y be two points in $\mathbf{R}^n$, and let $\delta = |x - y|$. Let

$$S = B(x,\delta) \cap B(y,\delta)$$

and let $z \in S$. We have

$$|u(x) - u(y)| \leq |u(x) - u(z)| + |u(y) - u(z)|$$

and integrating over S:

$$|S| |u(x) - u(y)| \leq \int_S |u(x) - u(z)| dz + \int_S |u(y) - u(z)| dz \,.$$

The two integrals can be estimated in the same way. Let us consider for instance the first of them. Since

$$u(z) - u(x) = \int_0^1 \frac{d}{dt} u(x + t(z - x))dt = \int_0^1 \langle z - x, Du(x + t(z - x))\rangle dt\,,$$

we have for every $z \in S$:

$$|u(x) - u(z)| \leq \delta \int_0^1 |Du(x + t(z - x))|dt$$

and integrating:

$$\int_S |u(x) - u(z)|dz \leq \int_{B(x,\delta)} |u(x) - u(z)|dz$$

$$\leq \delta \int_0^1 dt \int_{B(x,\delta)} |Du(x + t(z - x))|dz\,.$$

After a change of variables $w = x + t(z - x)$ we get

$$\int_S |u(x) - u(z)|dz \leq \int_0^1 t^{-n}\, dt \int_{B(x,t\delta)} |Du(w)|\, dw\,.$$

Using Hölder's inequality the right-hand side can be estimated by

$$\omega_n^{1-\frac{1}{p}} \delta^{n+1-\frac{n}{p}} \int_0^1 t^{-\frac{n}{p}}\, dt \left(\int_{B(x,t\delta)} |Du|^p\, dw\right)^{\frac{1}{p}}$$

and therefore in conclusion by

$$\frac{\omega_n^{1-\frac{1}{p}} \delta^{n+1-\frac{n}{p}}}{1 - \frac{n}{p}} \|Du\|_p$$

where we have indicated with ω_n the measure of the unit ball in $\mathbf{R}^n$. The same quantity gives a bound for the second integral. On the other hand

$$|S| = c(n)\delta^n$$

and hence, remembering that $\delta = |x - y|$:

$$|u(x) - u(y)| \leq c(n,p)\|Du\|_p |x - y|^\alpha$$

from which (3.15) follows.

If we choose $y \in \operatorname{supp}(u)$ such that $u(y) = 0$, the preceding inequality leads immediately to (3.16). □

The theorems of Sobolev extend immediately to the closure of $C_0^\infty(\Omega)$.

Definition 3.4 *By $W_0^{k,p}(\Omega)$ we indicate the closure of $C_0^\infty(\Omega)$ in the norm of $W^{k,p}(\Omega)$. In other words, a function u belongs to $W_0^{k,p}(\Omega)$ if and only if there exists a sequence of functions $u_k \in C_0^\infty(\Omega)$ such that*

$$\lim_{k\to\infty} \|u_k - u\|_{k,p} = 0 .$$

It is evident that $W_0^{k,p}(\Omega)$ is a Banach space, with the norm (3.4).

Theorem 3.9 *Let Ω be an open set in $\mathbf{R}^n$, and let $u \in W_0^{1,p}(\Omega)$. Then:*

(i) *If $p < n$, u belongs to L^{p^*}, $p^* = \frac{np}{n-p}$, and*

$$\|u\|_{p^*} \leq c\|Du\|_p . \tag{3.17}$$

(ii) *If $p > n$, u belongs to $C^{0,\alpha}(\bar{\Omega})$, $\alpha = 1 - \frac{n}{p}$, and*

$$[u]_\alpha \leq c\|Du\|_p . \tag{3.18}$$

Moreover, if Ω is bounded, we have

$$\sup|u| \leq c(\mathrm{diam}(\Omega))^\alpha \|Du\|_p . \tag{3.19}$$

Proof. Let us prove for instance (i). Let u_j be a sequence in $C_0^\infty(\Omega)$, converging to u in $W_0^{1,p}(\Omega)$. In particular, $\{Du_j\}$ will be a Cauchy sequence in $L^p(\Omega)$, and since

$$\|u_j - u_k\|_{p^*} \leq c\|Du_j - Du_k\|_p$$

u_j will be a Cauchy sequence in $L^{p^*}(\Omega)$. Since $u_j \to u$ in $L^p(\Omega)$, we will have also $u_j \to u$ in $L^{p^*}(\Omega)$. Writing (3.14) for u_j and passing to the limit for $j \to \infty$ we get the conclusion.

The proof of (ii) is similar, and we leave it to the reader. □

A consequence of the above theorem is that, differently from what happens for the spaces L^p, $C_0^\infty(\Omega)$ is not dense in $W^{1,p}(\Omega)$. For instance, the characteristic function of Ω, that obviously belongs to $W^{1,p}(\Omega)$, cannot be approximated in $W^{1,p}$ with functions with compact support, since it does not satisfy inequality (3.17).

In some sense, the functions in $W_0^{1,p}(\Omega)$ "take the value zero" at the boundary of Ω. As a consequence, two functions u and v of $W^{1,p}(\Omega)$ "have the same boundary value" if their difference belongs to $W_0^{1,p}(\Omega)$.[3]

[3]These questions will be treated in greater detail later.

A corollary of the preceding theorem, that can be also proved directly, is the following:

Corollary 3.1 *Let Ω be an open set in $\mathbf{R}^n$ with finite measure, and let $u \in W_0^{1,p}(\Omega)$. Then,*

$$\|u\|_p \leq c(n,p)|\Omega|^{\frac{1}{n}}\|Du\|_p .$$

Proof. We distinguish the two cases $1 \leq p < \frac{n}{n-1}$ and $p \geq \frac{n}{n-1}$. In the first case we have

$$\|u\|_p \leq |\Omega|^{\frac{1}{p}-\frac{1}{p^*}}\|u\|_{p^*} \leq c|\Omega|^{\frac{1}{n}}\|Du\|_p .$$

If instead $p \geq \frac{n}{n-1}$, setting $p_* = \frac{np}{n+p}$, it results $(p_*)^* = p$, and again:

$$\|u\|_p \leq c\|Du\|_{p_*} \leq c|\Omega|^{\frac{1}{n}}\|Du\|_p . \qquad \square$$

In particular, for $u \in W_0^{1,p}(\Omega)$ the norm (3.4) of $W^{1,p}(\Omega)$ is equivalent to

$$|u|_{1,p} = \left\{\int_\Omega |Du|^p\, dx\right\}^{\frac{1}{p}} .$$

More generally, if $|\Omega| < +\infty$, the quantity

$$|u|_{k,p} = \left\{\sum_{|\alpha|=k}\int_\Omega |D^\alpha u|^p\, dx\right\}^{\frac{1}{p}}$$

is a norm in $W_0^{1,p}(\Omega)$, equivalent to the standard norm $\|u\|_{k,p}$.

We shall consider now the space $W^{1,p}(\Omega)$. In general, without suitable assumptions on the boundary of Ω, the imbedding $W^{1,p}(\Omega) \hookrightarrow L^{p^*}(\Omega)$ does not hold. For instance, taking $n = 2$ and

$$\Omega = \left\{x \in \mathbf{R}^2 : 0 < x < 1,\ |y| < \exp\left(-\frac{1}{x^2}\right)\right\},$$

the function $f(x,y) = x^3\exp(\frac{1}{x^2})$ belongs to $W^{1,1}(\Omega)$, but does not belong to any $L^p(\Omega)$, with $p > 1$.

The imbedding $W^{1,p}(\Omega) \hookrightarrow L^{p^*}(\Omega)$ holds if $\partial\Omega$ is Lipschitz-continuous. In order to prove this statement, we shall begin by establishing some results which are of interest by themselves.

Lemma 3.4 *Let $v(x)$ be a function in $W^{1,p}(B^+)$, and let*

$$V(x) = \sigma(v)(x) = \begin{cases} v(x) & \text{if } x \in B^+ ,\\ v(x_1, x_2, \ldots, x_{n-1}, -x_n) & \text{if } x \in B^- = B - \overline{B^+} . \end{cases}$$

Then, the function $V(x)$ belongs to $W^{1,p}(B)$.

Proof. By Theorem 3.6 and the remark immediately following it, there will exist a sequence $\{v_k\}$ of Lipschitz-continuous functions in $\overline{B^+}$, converging to v in $W^{1,p}(B^+)$. The corresponding functions $V_k = \sigma(v_k)$ are Lipschitz-continuous in B, converge to V in $L^p(B)$, and $\{DV_k\}$ is a Cauchy sequence in $L^p(B)$. Denoting by g the limit of DV_k, we have

$$\int V D\varphi\, dx = \lim_{k\to\infty} \int V_k D\varphi\, dx = \lim_{k\to\infty} -\int \varphi DV_k\, dx = -\int g\varphi\, dx$$

for every $\varphi \in C_0^\infty(B)$, and hence $V \in W^{1,p}(B)$ and $DV = g$. □

It is clear that $DV = Dv$ in B^+, whereas in B^- we have

$$D_i V = \sigma(D_i v) \quad \text{if } i < n\,, \quad D_n V = -\sigma(D_n v)$$

and hence

$$\|\sigma(v)\|_{W^{1,p}(B)} = 2^{\frac{1}{p}} \|v\|_{W^{1,p}(B^+)}\,.$$

Theorem 3.10 *Let Ω be a bounded open set in $\mathbf{R}^n$ with Lipschitz-continuous boundary. There exists an open set $\Sigma \supset\supset \Omega$ and for every $u \in W^{1,p}(\Omega)$ a function $U \in W_0^{1,p}(\Sigma)$ such that $U(x) \equiv u(x)$ in Ω and*

$$\|U\|_{1,p,\Sigma} \le c\|u\|_{1,p,\Omega}$$

with the constant c depending only on p, n and Ω.

Proof. As in the proof of Theorem 3.6, there exists a finite open covering $A_0, A_1, \ldots, A_N$, with $A_0 \subset\subset \Omega$ and $A_i \cap \partial\Omega \neq \emptyset$ for $1 \le i \le N$, and for every $i \ge 1$ a homeomorphism $g_i : B \to A_i$, Lipschitz-continuous with its inverse, mapping B^+ in $A_i \cap \Omega$.

Let $\{\alpha_i\}$ be a partition of unity relative to that covering. If $u \in W^{1,p}(\Omega)$, the functions $v_i = (g_i)_*(\alpha_i u)$ belong to $W^{1,p}(B^+)$, and are zero in a neighborhood of $\partial^+ B =: \partial B \cap \mathbf{R}^n_+$. By the preceding lemma, the functions $V_i(x) = \sigma(v_i)(x)$ belong to $W_0^{1,p}(B)$, and

$$\|V_i\|_{1,p,B} = 2^{\frac{1}{p}} \|v_i\|_{1,p,B^+} \le c\|\alpha_i u\|_{1,p,\Omega}\,.$$

Setting

$$U_i = (g_i^{-1})_* V_i = V_i \circ g_i^{-1}\,,$$

we have $U_i = \alpha_i u$ in Ω and

$$\|U_i\|_{1,p,A_i} \le c\|\alpha_i u\|_{1,p,\Omega}\,.$$

The function

$$U = \alpha_0 u + \sum_{i=1}^{N} U_i$$

is the required one. In the first place, U belongs to $W_0^{1,p}(\Sigma)$, with $\Sigma = \cup_{i=0}^{N} A_i$; moreover for $x \in \Omega$ we have

$$U(x) = \sum_{i=0}^{N} \alpha_i(x) u(x) = u(x).$$

Finally,

$$\|U\|_{1,p,\Sigma} \leq c \sum_{i=0}^{N} \|\alpha_i u\|_{1,p,\Omega} \leq cM(1+N)\|u\|_{1,p,\Omega},$$

where

$$M = \max_{0 \leq i \leq N} (\|\alpha_i\|_\infty + \|D\alpha_i\|_\infty). \qquad \square$$

The immersion theorem is now an immediate corollary.

Theorem 3.11 *Let Ω be a bounded open set in $\mathbf{R}^n$ with Lipschitz-continuous boundary, and let $u \in W^{1,p}(\Omega)$. Then:*

(i) *If $p < n$, u belongs to L^{p^*}, $p^* = \frac{np}{n-p}$, and we have*

$$\|u\|_{p^*,\Omega} \leq c\|u\|_{1,p,\Omega}. \tag{3.20}$$

(ii) *If $p > n$, u belongs to $C^{0,\alpha}(\bar{\Omega})$, $\alpha = 1 - \frac{n}{p}$, and*

$$\|u\|_\alpha \leq c\|u\|_{1,p,\Omega}. \tag{3.21}$$

Proof. Let U be the function given by the preceding theorem. Remarking that

$$\|u\|_{p^*,\Omega} = \|U\|_{p^*,\Omega} \leq \|U\|_{p^*,\Sigma},$$

$$\|u\|_{\alpha,\Omega} = \|U\|_{\alpha,\Omega} \leq \|U\|_{\alpha,\Sigma},$$

the conclusion follows immediately from Theorem 3.9 applied to the function U. $\square$

Let now $u \in W^{k,p}(\Omega)$. Applying the preceding theorem to the derivatives of order $k-1$, then to those of order $k-2$ and so on, we get the following result:

Theorem 3.12 *Let Ω be a bounded open set in $\mathbf{R}^n$, with Lipschitz-continuous boundary $\partial\Omega$, and let $u \in W^{k,p}(\Omega)$. Then:*

(i) *If $n > kp$, $u \in L^{\frac{np}{n-kp}}(\Omega)$, and we have*

$$\|u\|_{\frac{np}{n-kp}} \leq c\|u\|_{k,p} . \tag{3.22}$$

(ii) *If $n < kp$ and $k - \frac{n}{p}$ is not an integer, then, denoting by h its integer part and by $\alpha = k - h - \frac{n}{p}$ its fractionary part, we have $u \in C^{h,\alpha}(\bar{\Omega})$, with*

$$\|u\|_{C^{h,\alpha}} \leq c\|u\|_{k,p} . \tag{3.23}$$

(iii) *Finally, if $n < kp$, and $k - \frac{n}{p}$ is an integer, then $u \in C^{k-\frac{n}{p}-1,\alpha}(\bar{\Omega})$ for every $\alpha < 1$, and (3.23) holds with c depending on α.*

3.5 Compactness

In the preceding section we have proved, among other things, that the immersion $W^{1,p} \hookrightarrow L^{p^*}$ is continuous. If Ω is bounded and $q < p^*$, we have $L^{p^*} \hookrightarrow L^q$, and hence the immersion $W^{1,p} \hookrightarrow L^q$ is also continuous. We will show now that the last immersion is compact.

Definition 3.5 *Let X and Y be two complete metric spaces, and let T be a map from X into Y. We say that T is* compact *if:*

(i) *T is continuous,*
(ii) *T maps bounded sets in X into relatively compact sets in Y.*

We note that if X and Y are two Banach spaces and if T is a linear map, condition (ii) implies (i).

Lemma 3.5 *Let Ω be a bounded open set in $\mathbf{R}^n$, and let Z be a bounded subset of $L^q(\Omega)$, such that the mollified functions u_ϵ tend to u in $L^q(\Omega)$, uniformly for $u \in Z$. Then, Z is relatively compact in $L^q(\Omega)$.*

Proof. It will be sufficient to prove that Z is totally bounded, that is that for every $\delta > 0$ there exists a δ-net, i.e. a finite covering of Z made of sets with diameter less than δ.

Let $\delta > 0$. By assumption, there exists $\epsilon_0 > 0$ such that for every $u \in Z$ we have

$$\|u_{\epsilon_0} - u\|_q < \frac{\delta}{4} .$$

Setting

$$Z_0 = \{u_{\epsilon_0} : u \in Z\},$$

it will be sufficient to prove the existence of a $\frac{\delta}{2}$-net for Z_0, or else that Z_0 is relatively compact in $L^q(\Omega)$. This goal will be achieved if we prove that Z_0 is relatively compact in the topology of uniform convergence, since the last topology is stronger than that of L^q.

By Ascoli–Arzelà's theorem, it will be sufficient to prove that Z_0 is bounded in $C^1(\bar{\Omega})$. We have:

$$u_{\epsilon_0}(x) = \int_\Omega \varphi_{\epsilon_0}(x-y)u(y)dy\,,$$

$$Du_{\epsilon_0}(x) = \int_\Omega D\varphi_{\epsilon_0}(x-y)u(y)dy$$

and therefore

$$|u_{\epsilon_0}| \leq M|\Omega|^{1-\frac{1}{q}}\|u\|_q\,,$$

$$|Du_{\epsilon_0}| \leq N|\Omega|^{1-\frac{1}{q}}\|u\|_q\,,$$

where $M = \sup|\varphi_{\epsilon_0}|$ and $N = \sup|D\varphi_{\epsilon_0}|$.

Since Z is bounded in L^q, Z_0 is then bounded in $C^1(\bar{\Omega})$, and the lemma is proved. □

We can now prove the following:

Theorem 3.13 (Rellich) *Let Ω be a bounded open set in $\mathbf{R}^n$, with Lipschitz-continuous boundary $\partial\Omega$, and let $1 \leq p < n$ and $1 \leq q < p^* = \frac{np}{n-p}$. The immersion*

$$W^{1,p}(\Omega) \hookrightarrow L^q(\Omega)$$

is compact.

Proof. Let Z be a bounded subset of $W^{1,p}(\Omega)$. It is clear that Z is bounded in $L^q(\Omega)$; we have actually

$$\|u\|_q \leq |\Omega|^{\frac{1}{q}-\frac{1}{p^*}}\|u\|_{p^*} \leq c\|u\|_{1,p}\,.$$

In order to apply the preceding lemma, it will suffice to verify that $u_\epsilon \to u$ in L^q, uniformly for $u \in Z$. We shall begin with the case $q = 1$. We have

$$\int_\Omega |u_\epsilon - u|dx = \int_\Omega \left| \int \varphi_\epsilon(x-y)[u(y)-u(x)]dy \right| dx$$
$$\leq \int \varphi(w)dw \int_\Omega |u(x-\epsilon w) - u(x)|dx \,.$$

Let now $\tau > 0$ and let $\Sigma \subset\subset \Omega$ be an open set such that $|\Omega - \Sigma| < \tau^{\frac{p^*}{p^*-1}}$. We have then

$$\int_{\Omega-\Sigma} |u(x)|dx \leq |\Omega - \Sigma|^{1-\frac{1}{p^*}} \|u\|_{p^*} < \tau c \|u\|_{1,p}$$

and similarly

$$\int_{\Omega-\Sigma} |u(x-\epsilon w)|dx < \tau c \|u\|_{1,p} \,,$$

so that

$$\int_\Omega |u(x-\epsilon w) - u(x)|dx < 2\tau c\|u\|_{1,p} + \int_\Sigma |u(x-\epsilon w) - u(x)|dx \,.$$

If $\epsilon < \text{dist}(\Sigma, \partial\Omega)$, we have $x - \epsilon w \in \Omega$ for every $w \in B$; whence:

$$|u(x-\epsilon w) - u(x)| = \left| \int_0^1 \langle Du(x - t\epsilon w), \epsilon w \rangle dt \right|$$
$$\leq \epsilon \int_0^1 |Du(x - t\epsilon w)|dt$$

for almost every $x \in \Sigma$. We have therefore

$$\int_\Sigma |u(x-\epsilon w) - u(x)|dx \leq \epsilon \int_0^1 dt \int_\Sigma |Du(x - t\epsilon w)|dx \leq \epsilon \int_\Omega |Du|dx$$

and in conclusion:

$$\int_\Omega |u_\epsilon - u|dx < (2\tau c + \epsilon|\Omega|^{1-\frac{1}{p}})\|u\|_{1,p} \,. \tag{3.24}$$

From the above inequality it follows at once that $u_\epsilon \to u$, uniformly for $u \in Z$, and hence the conclusion of the theorem for $q = 1$.

In the general case $q < p^*$ we note that

$$\int_\Omega |u_\epsilon - u|^q \, dx = \int_\Omega |u_\epsilon - u|^{\frac{p^*(q-1)}{p^*-1}} |u_\epsilon - u|^{\frac{p^*-q}{p^*-1}} \, dx$$

and by the Hölder inequality:

$$\int_\Omega |u_\epsilon - u|^q \, dx \le \left\{ \int_\Omega |u_\epsilon - u|^{p^*} \, dx \right\}^{\frac{q-1}{p^*-1}} \left\{ \int_\Omega |u_\epsilon - u| dx \right\}^{\frac{p^*-q}{p^*-1}} .$$

From the inequality $\|u_\epsilon\|_{p^*} \le \|u\|_{p^*}$ and from Theorem 3.11 it follows that

$$\|u_\epsilon - u\|_{p^*} \le c\|u\|_{1,p}$$

and hence, by (3.24):

$$\|u_\epsilon - u\|_q \le c\|u\|_{1,p}(\tau + \epsilon)^{\frac{p^*-q}{p^*-1}}$$

so that we can apply again the preceding lemma. □

It is obvious that if $p \ge n$, the immersion $W^{1,p}(\Omega) \hookrightarrow L^q(\Omega)$ is compact for every q. More generally, if $kp < n$, $q < \frac{np}{n-kp}$, and Ω has boundary of class C^1, the immersion of $W^{k,p}(\Omega)$ in $L^q(\Omega)$ is compact, since by Theorem 3.12 the first derivatives belong to L^r, with $r = \frac{np}{n-kp+p}$, and $q < r^* = \frac{np}{n-kp}$.

Remark 3.3 It follows from Rellich's theorem that if u_k converges weakly to u in $W^{1,p}$ (or in other words if $u_k \rightharpoonup u$ and $Du_k \rightharpoonup Du$ in L^p), then $u_k \to u$ strongly in L^q, for every $q < p^*$. In fact, the sequence u_k is bounded in $W^{1,p}$, and hence relatively compact in L^q, so that from any of its subsequences it is possible to extract a subsequence convergent in L^q to a function v. On the other hand $u_k \rightharpoonup u$ in L^r, and hence $v = u$, so that the whole sequence u_k converges to u strongly in L^q. □

3.6 Inequalities

Theorem 3.14 (POINCARÉ's Inequality) *Let $\Omega \subset \mathbf{R}^n$ be a bounded connected open set, with Lipschitz-continuous boundary $\partial\Omega$. There exists a constant $c(n, p, \Omega)$ such that for every $u \in W^{1,p}(\Omega)$*

$$\int_\Omega |u - u_\Omega|^p \, dx \le c \int_\Omega |Du|^p \, dx \, , \tag{3.25}$$

where

$$u_\Omega =: \fint_\Omega u \, dx =: \frac{1}{|\Omega|} \int_\Omega u \, dx \tag{3.26}$$

is the average of u in Ω.

Proof. Since (3.25) does not change if we add a constant to u, we can assume that $u_\Omega = 0$.

If the theorem were false, it would be possible to find a sequence of functions $u_k \in W^{1,p}(\Omega)$, with $u_{k\Omega} = 0$, and such that

$$\int_\Omega |u_k|^p \, dx = 1 \,, \tag{3.27}$$

$$\int_\Omega |Du_k|^p \, dx \leq \frac{1}{k} \,. \tag{3.28}$$

By Rellich's theorem a subsequence will converge to a function $u \in L^p(\Omega)$, with $\|u\|_p = 1$ by (3.27). On the other hand the sequence $\|Du_k\|_p$ tends to zero, and therefore $u \in W^{1,p}(\Omega)$ and $Du = 0$. Since Ω is connected, u will be constant in Ω,[4] and having zero average (remember that all the functions u_k have zero average), it will be identically zero. This contradicts the fact that $\|u\|_p = 1$. □

A joint application of the preceding inequality and of Theorem 3.11 gives immediately the following

Theorem 3.15 (Sobolev–Poincaré's *inequality*) *With the assumptions of the preceding theorem, if $p < n$, we have*

$$\|u - u_\Omega\|_{p^*} \leq c(n, p, \Omega)\|Du\|_p \,. \tag{3.29}$$

A proof similar to that of Theorem 3.14 gives an inequality of the type (3.17) for functions in $W^{1,p}(\Omega)$ (not necessarily zero on $\partial\Omega$), provided they are zero on a set of positive measure.

Theorem 3.16 *Let Ω be a bounded connected open set in $\mathbf{R}^n$, with Lipschitz-continuous boundary. For every $u \in W^{1,p}(\Omega)$, $p < n$, taking the value zero in a set A of positive measure, we have*

$$\|u\|_{p^*,\Omega} \leq c \left(\frac{|\Omega|}{|A|}\right)^{\frac{1}{p^*}} \|Du\|_{p,\Omega} \,, \tag{3.30}$$

where c is twice the constant in Sobolev–Poincaré's inequality (3.29).

[4]The proof of that assertion can be made by remarking that all the mollified functions u_ϵ have zero gradient in Ω_ϵ, and therefore are constant in Ω_ϵ. The same is true for their limit u, and since $\epsilon > 0$ is arbitrary, u is constant in Ω. Recalling the proof of Proposition 3.5, we can conclude that if $u \in W^{1,p}$ and u is constant in K, then $Du = 0$ in K; and conversely if $Du = 0$ in K, then u is constant in every connected component of K.

Proof. We have

$$|u_\Omega||A|^{\frac{1}{p^*}} \leq \|u - u_\Omega\|_{p^*},$$

and hence

$$\|u\|_{p^*} \leq \|u - u_\Omega\|_{p^*} + |u_\Omega||\Omega|^{\frac{1}{p^*}} \leq 2\left(\frac{|\Omega|}{|A|}\right)^{\frac{1}{p^*}} \|u - u_\Omega\|_{p^*}.$$

The conclusion follows immediately from (3.29). □

It is possible to prove that (3.30) holds whenever u is zero on a set of positive *p-capacity*. We shall not enter in these problems, that are outside the scope of this work; instead we shall discuss the dependence on Ω of the constants entering in the various inequalities, when Ω is a cube of $\mathbf{R}^n$. For functions in $W_0^{1,p}$, we get immediately from the Sobolev inequality (3.17)

$$\|u\|_q \leq |Q_R|^{\frac{1}{q} - \frac{1}{p^*}} \|u\|_{p^*} \leq c R^{\frac{n}{q} - \frac{n}{p^*}} \|Du\|_p \tag{3.31}$$

for every $q < p^*$, and in particular

$$\int_{Q_R} |u|^p \, dx \leq c(n,p) R^p \int_{Q_R} |Du|^p \, dx; \;\; u \in W_0^{1,p}(Q_R). \tag{3.32}$$

In general, we have the following:

Theorem 3.17 *For every function $u \in W^{1,p}(Q_R)$ it holds that*

$$\|u - u_R\|_{p^*, Q_R} \leq c(n,p) \|Du\|_{p, Q_R} \tag{3.33}$$

and hence

$$\int_{Q_R} |u - u_R|^p \, dx \leq c(n,p) R^p \int_{Q_R} |Du|^p \, dx, \tag{3.34}$$

where $u_R = u_{Q_R}$.

Moreover, if the function u is zero on a set $A \subset Q_R$ of positive measure, we have

$$\|u\|_{p^*, Q_R} \leq c(n,p) \left(\frac{|Q_R|}{|A|}\right)^{\frac{1}{p^*}} \|Du\|_{p, Q_R}, \tag{3.35}$$

$$\int_{Q_R} |u|^p \, dx \leq c(n,p) \left(\frac{|Q_R|}{|A|}\right)^{\frac{1}{p^*}} R^p \int_{Q_R} |Du|^p \, dx. \tag{3.36}$$

Proof. If $u \in W^{1,p}(Q_R)$, the function $w(x) = u(Rx)$ belongs to $W^{1,p}(Q)$. Writing inequality (3.29) for w, we have

$$\|w - w_1\|_{p^*, Q} \leq c(n,p,Q) \|Dw\|_{p, Q}.$$

On the other hand it is easily seen that $w_1 = u_R$ and $Dw(x) = RDu(Rx)$, so that we have (3.33) with $c(n,p) = c(n,p,Q)$. Inequality (3.34) can be proved likewise. In a similar way, using Theorem 3.16, one proves (3.35) and (3.36). □

Remark 3.4 It is evident that we can replace the left-hand side of (3.33) and (3.34) respectively with the quantities:

$$\inf_{\xi} \|u - \xi\|_{p^*, Q_R}$$

and

$$\inf_{\xi} \int_{Q_R} |u - \xi|^p \, dx \, .$$

The resulting inequalities are equivalent to (3.33) and (3.34), in virtue of (2.19).

Moreover the average on Q_R can be replaced with that on any cube $Q_{\alpha R}$, with $\alpha < 1$, since

$$\begin{aligned}\int_{Q_R} |u - u_{\alpha R}|^s \, dx &\le c \left\{ \int_{Q_R} |u - u_R|^s \, dx + |Q_R| |u_R - u_{\alpha R}|^s \right\} \\ &\le c \left\{ \int_{Q_R} |u - u_R|^s \, dx + \alpha^{-n} \int_{Q_{\alpha R}} |u - u_R|^s \, dx \right\} \\ &\le c(1 + \alpha^{-n}) \int_{Q_R} |u - u_R|^s \, dx \, . \end{aligned} \tag{3.37}$$

Finally, we remark that the preceding theorem remains valid if we replace the cubes Q_R with balls of radius R, or more generally with any family of sets deriving by homothety from one of them. Of course, the constant will depend on the family in question, but not on R. □

Poincaré's inequality has as a consequence an interesting relation in the spaces $\mathcal{L}^{p,\lambda}$.

Proposition 3.7 *Let Ω be a bounded open set with Lipschitz-continuous boundary, and let u be a function in $W^{1,p}(\Omega, \mathbf{R}^N)$ with $Du \in L^{p,\lambda}$. Then, $u \in \mathcal{L}^{p,\lambda+p}(\Omega, \mathbf{R}^N)$.*

Proof. It will suffice to prove that, at least for R small enough, we have the estimate

$$\int_{\Omega_R} |u - u_R|^p \, dx \le cR^p \int_{\Omega_{\vartheta R}} |Du|^p \, dx \tag{3.38}$$

for some $\vartheta \ge 1$.

That inequality holds with $\vartheta = 1$ if $Q_R \subset \Omega$ (see (3.34)); it remains only to examine the case when Q_R meets $\partial\Omega$.

Let us assume first that Ω is the half-space $\mathbf{R}^n_+$. In this case, setting as above

$$U(x) = \sigma(u)(x) = \begin{cases} u(x) & \text{if } x_n \geq 0\,, \\ u(x_1, \ldots, x_{n-1}, -x_n) & \text{if } x_n < 0\,, \end{cases}$$

we have for every cube $Q_R = Q(x_0, R)$ with center in the upper half-space:

$$\begin{aligned} \int_{\Omega_R} |u - U_R|^p \, dx &\leq \int_{Q_R} |U - U_R|^p \, dx \\ &\leq cR^p \int_{Q_R} |DU|^p \, dx \leq 2cR^p \int_{\Omega_R} |Du|^p \, dx \end{aligned}$$

and (3.38) is proved in this case.

In general, for every point x_0 of $\partial\Omega$ there exists a neighborhood W of x_0 and a bi-Lipschitz map g from the unit ball B onto W, mapping the half-ball B^+ onto $W \cap \Omega$. A finite number $W_1, \ldots, W_N$ of such neighborhoods will cover $\partial\Omega$, and there will exist a number $R_1 > 0$ such that for $R \leq R_1$ every cube Q_R meeting $\partial\Omega$ is contained in one of these. Let L be the greatest Lipschitz constant of the functions g relative to these neighborhoods and of their inverse, and let $R_0 = L^{-2}R_1$. If $R \leq R_0$ and $Q_R \subset W_k$, setting $U = u \circ g_k$, we have

$$\int_{\Omega_R} |u - \xi|^p \, dy \leq c \int_{Q_{LR} \cap \mathbf{R}^n_+} |U - \xi|^p \, dx$$

and therefore, with a suitable choice of ξ, we have

$$\int_{\Omega_R} |u - \xi|^p \, dy \leq cR^p \int_{Q_{LR} \cap \mathbf{R}^n_+} |DU|^p \, dx \leq cR^p \int_{\Omega_{L^2 R}} |Du|^p \, dy$$

from which the conclusion follows easily. □

In particular, if $\lambda > n - p$, the function u is Hölder-continuous, a result known as the DIRICHLET *growth theorem* (see MORREY [3], Theorem 3.5.2).

Finally, always in the spirit of Poincaré's inequality, we can prove the following theorem concerning functions $u \in W^{2,p}$.

Theorem 3.18 *Let Ω be a bounded open set with Lipschitz-continuous boundary. For every $\epsilon > 0$ there exists a constant $c(\epsilon)$ such that for every $u \in W^{2,p}(\Omega)$*

$$\int_\Omega |Du|^p\,dx \le \epsilon \int_\Omega |D^2u|^p\,dx + c(\epsilon)\int_\Omega |u|^p\,dx\,. \tag{3.39}$$

Proof. If for some ϵ_0 such a constant would not exist, we could find a sequence $u_k \in W^{2,p}$ such that

$$\int_\Omega |Du_k|^p\,dx > \epsilon_0 \int_\Omega |D^2u_k|^p\,dx + k\int_\Omega |u_k|^p\,dx\,.$$

Writing $u_k\|Du_k\|_p^{-1}$ instead of u_k, we can suppose that the left-hand side of the preceding relation be equal to unity. The sequence u_k is bounded in $W^{2,p}$, and hence, passing possibly to a subsequence, we can assume that $u_k \to u$ in $W^{1,p}$. We have

$$\int_\Omega |Du|^p\,dx = 1\,.$$

On the other hand, $k\int_\Omega |u_k|^p\,dx < 1$, and hence $u_k \to 0$ in L^p. We have therefore $u = 0$, contradicting the preceding relation. □

3.7 Traces

We have already remarked that in some sense it is possible to speak of the boundary values of a function $u \in W^{1,p}(\Omega)$, or at least to say when two such functions have the same boundary values. We could therefore define the *boundary value* of a function $u \in W^{1,p}(\Omega)$ in an abstract way, as the equivalence class of all the functions v such that $u - v \in W_0^{1,p}(\Omega)$. To this formal definition we prefer the following discussion that, though incomplete, has nevertheless the merit of introducing the *trace* of a function $u \in W^{1,p}(\Omega)$ as a function φ defined on the boundary $\partial\Omega$.

We begin by discussing the case of a cylinder $C_{R,T} = D_R \times (0,T)$, where D_R is a ball of radius R in $\mathbf{R}^{n-1}$. A generic point of $C_{R,T}$ will be denoted by $x = (\bar{x}, t)$, with $\bar{x} \in D_R$ and $t \in (0,T)$. If $u(x)$ is a function of class C^1 in $C_{R,T}$ and if $0 < s < t < T$, we have

$$u(\bar{x}, t) - u(\bar{x}, s) = \int_s^t D_n u(\bar{x}, \tau)d\tau$$

and hence

$$\int_{D_R} |u(\bar{x},t) - u(\bar{x},s)|^p \, d\bar{x} = \int_{D_R} \left| \int_s^t D_n u(\bar{x},\tau) d\tau \right|^p d\bar{x}$$
$$\leq (t-s)^{p-1} \int_{C_{R,t}} |Du|^p \, dx \, .$$

By approximation, the preceding relation holds for every function $u \in W^{1,p}(C_{R,T})$, and for almost every s and t. From it we deduce that when $h \to 0$ the function $u(\bar{x},h)$ tends in the strong topology of $L^p(D_R)$ to a function $\varphi(\bar{x})$. Such a function is called the *trace* of u on D_R. Letting s go to zero in the preceding relation, we get

$$\int_{D_R} |u(\bar{x},t) - \varphi(\bar{x})|^p \, d\bar{x} \leq t^{p-1} \int_{C_{R,t}} |Du|^p \, dx \qquad (3.40)$$

for almost every $t \in (0,T)$.

A consequence of the above inequality is the following:

Proposition 3.8 *Let $p > 1$ and let u_k be a sequence converging weakly in $W^{1,p}(C_{R,T})$ to a function u. Then, the traces φ_k of the functions u_k converge in $L^p(D_R)$ to the trace φ of u.*

Proof. For $0 < t < T$ let $u^t(\bar{x}) = u(\bar{x},t)$. We have

$$\|\varphi_k - \varphi\|_p \leq \|\varphi_k - u_k^t\|_p + \|u_k^t - u^t\|_p + \|u^t - \varphi\|_p$$

and hence by (3.40):

$$\|\varphi_k - \varphi\|_p^p \leq c(p)(\|\varphi_k - u_k^t\|_p^p + \|u_k^t - u^t\|_p^p + \|u^t - \varphi\|_p^p)$$
$$\leq c(p)\{t^{p-1}(\|Du_k\|_{p,C_{R,t}}^p + \|Du\|_{p,C_{R,t}}^p) + \|u_k^t - u^t\|_p^p\} \, .$$

Integrating with respect to t between 0 and $\vartheta < T$ we get

$$\vartheta\|\varphi_k - \varphi\|_p^p \leq c(p)[\vartheta^p(\|Du_k\|_{p,C_{R,T}}^p + \|Du\|_{p,C_{R,T}}^p) + \|u_k - u\|_{p,C_{R,T}}^p] \, .$$

If we let k go to infinity, and we take into account the strong convergence of u_k in $L^p(C_{R,T})$ and the equiboundedness of the L^p norms of Du_k, we get

$$\limsup_{k\to\infty} \|\varphi_k - \varphi\|_p^p \leq c(p)\vartheta^{p-1}$$

and the conclusion follows from the arbitrariness of ϑ. □

We can now characterize the space $W_0^{1,p}$ by means of the traces.

Theorem 3.19 *Let $u \in W^{1,p}(C_{R,T})$, and assume that* $\text{supp}(u) \cap \partial C_{R,T} = \text{supp}(u) \cap D_R$. *We have $u \in W_0^{1,p}(C_{R,T})$ if and only if the trace of u on D_R is zero.*

Proof. A function $u \in W_0^{1,p}(C_{R,T})$ can be approximated by functions $u_k \in C_0^1$, that have obviously zero trace on D_R. By the preceding proposition, also u has zero trace.

Assume conversely that the trace of u is zero, and for $0 < \tau < T$ let $\eta(t) = \eta_\tau(t)$ be a function of class $C^1((0,T))$ such that $0 \leq \eta \leq 1$, $\eta(t) = 0$ for $t \leq \frac{\tau}{2}$, $\eta(t) = 1$ for $t \geq \tau$ and $|\eta'(t)| \leq \frac{4}{\tau}$. We have from (3.40):

$$\|u - \eta u\|^p_{p,C_{R,\tau}} \leq \int_{C_{R,\tau}} |u|^p \, dx \leq c\tau^p \int_{C_{R,\tau}} |Du|^p \, dx$$

and moreover

$$\|D(u - \eta u)\|^p_{p,C_{R,\tau}} \leq c \left(\int_{C_{R,\tau}} |Du|^p \, dx + \frac{1}{\tau^p} \int_{C_{R,\tau}} |u|^p \, dx \right)$$

$$\leq c \int_{C_{R,\tau}} |Du|^p \, dx \, .$$

It follows that for $\tau \to 0$ we have $\eta_\tau u \to u$ in $W^{1,p}$. On the other hand ηu has support contained in $C_{R,T}$, and hence it can be approximated by functions in $C_0^\infty(C_{R,T})$, so that u belongs to $W_0^{1,p}(C_{R,T})$. □

Let us consider now a generic bounded open set Ω, with boundary of class C^1. Arguing as in Theorem 3.6 we find a finite covering $\{A_i\}$, a partition of unity $\{\alpha_i\}$ relative to that covering, and for every i a diffeomorphism g_i of the unit ball B onto A_i, mapping B^+ onto $A_i \cap \Omega$. The functions

$$U_i = (g_i)_*(\alpha_i u) = (\alpha_i u) \circ g_i$$

belong to $W^{1,p}(B^+)$, and are zero in the curved part of ∂B^+; they can obviously be defined in the whole cylinder $C_{1,1}$ setting them equal to zero outside B^+. Let Φ_i be the trace of U_i; the function

$$\varphi = \sum_{i=1}^{N} \Phi_i \circ g_i^{-1}$$

is the trace of u on $\partial\Omega$. The same construction works when the boundary of Ω is only Lipschitz-continuous.

Both Theorem 3.19 and Proposition 3.8 extend to functions of $W^{1,p}(\Omega)$ without relevant changes. We remark that from Proposition 3.8 it follows that the map $\Gamma : W^{1,p}(\Omega) \to L^p(\partial\Omega)$, mapping every function in $W^{1,p}(\Omega)$ into its trace on $\partial\Omega$, is compact.

The estimate (3.40) becomes in this case[5]

$$\int_{\partial\Omega} |u(x - t\nu(x)) - \varphi(x)|^p \, dH_{n-1}(x) \le ct^{p-1} \int_{\Omega - \Omega_t} |Du|^p \, dx\,, \qquad (3.41)$$

where $\nu(x)$ is the exterior normal to $\partial\Omega$ in x, and we have set as usual

$$\Omega_t = \{x \in \Omega : \operatorname{dist}(x, \partial\Omega) > t\}\,.$$

The trace of a function u will be denoted by the same symbol u. We have the Green's formula:

$$\int_\Omega u D_i \varphi \, dx = -\int_\Omega \varphi D_i u \, dx + \int_{\partial\Omega} \nu_i \varphi u \, dH_{n-1}\,.$$

for every $\varphi \in C^1(\bar{\Omega})$, that can be proved passing to the limit in the same formula for $u \in C^1(\bar{\Omega})$.

Remark 3.5 If the function $u \in W^{1,p}$ is continuous in $\bar{\Omega}$, it is evident that when t tends to 0, $u(x - t\nu(x))$ tends to the value of u on $\partial\Omega$, and hence the trace of u coincides with the restriction of u to $\partial\Omega$.

Actually the linear map T, that to any function $u \in W^{1,p}(\Omega)$ associates its trace on $\partial\Omega$, is completely characterized by the property that it coincides with the restriction to $\partial\Omega$ for functions $u \in C^1(\bar{\Omega})$. To see that, let $u \in W^{1,p}$, and let $u_k \in C^1(\bar{\Omega})$ be a sequence convergent to u in $W^{1,p}$ (see Theorem 3.6). The sequence of traces $\{Tu_k\}$ is then a Cauchy sequence in $L^p(\partial\Omega)$, and therefore it converges to a function $\varphi = Tu$. The limit function does not depend on the sequence u_k; for, if $v_k \to u$ in $W^{1,p}$, we have $u_k - v_k \to 0$, and hence $Tu_k - Tv_k \to 0$ in $L^p(\partial\Omega)$. If we write (3.41) for u_k, and we pass to the limit, we conclude that it holds for φ, so that φ is the trace of u. □

With an argument rather similar to that of Theorem 3.14, we prove the following result:

Theorem 3.20 *Let Ω be a bounded open set, with a boundary $\partial\Omega$ connected and Lipschitz-continuous. There exists a constant $c(p, n, \Omega)$ such that for every $u \in W^{1,p}(\Omega)$*

[5] We denote by H_{n-1} the $(n-1)$-dimensional Hausdorff measure on $\partial\Omega$ (see Sec. 2.6), which coincides with the usual surface measure when $\partial\Omega$ is Lipschitz-continuous.

$$\int_{\partial\Omega} |u(x) - u_{\partial\Omega}|^p \, dH_{n-1} \le c \int_\Omega |Du|^p \, dx \,. \tag{3.42}$$

Moreover, if Ω is the cube Q_R of side $2R$, we have

$$\int_{\partial Q_R} |u(x) - u_{\partial Q_R}|^p \, dH_{n-1} \le c(p,n) R^{p-1} \int_{Q_R} |Du|^p \, dx \,. \tag{3.43}$$

The same inequality, possibly with a different constant c, holds for balls $B_R = \{x \in \mathbf{R}^n : |x| < R\}$.

We conclude this section by recalling that not every function in $L^p(\partial\Omega)$ is the trace of some function in $W^{1,p}(\Omega)$. As a matter of fact, there exists the following characterization of traces of functions in $W^{1,p}$, which we mention for the sake of completeness. Its proof can be found for instance in the book of KUFNER, JOHN and FUČIK [1].

Theorem 3.21 *Let Ω be a bounded open set with Lipschitz-continuous boundary. A necessary and sufficient condition for a function φ in $L^p(\partial\Omega)$ be the trace of a function in $W^{1,p}(\Omega)$ is that*

$$\iint_{\partial\Omega \times \partial\Omega} \frac{|\varphi(x) - \varphi(y)|^p}{|x-y|^{n+p-2}} \, dH_{n-1}(x) dH_{n-1}(y) < +\infty \,. \tag{3.44}$$

3.8 The Values of $W^{1,p}$ Functions

We have often remarked that the functions in $W^{1,p}$ are strictly speaking equivalence classes, and therefore their values are defined up to a set of zero measure. On the other hand, the fact that for such functions it is possible to define a trace, something that cannot be done for instance for functions in L^p, suggests the possibility of defining $W^{1,p}$ functions on some set of zero measure, and even of dimension $n-1$. We can actually do better, as we see in the following theorem.

Theorem 3.22 *Let $\Omega \subset \mathbf{R}^n$ and let $u \in W^{1,p}(\Omega)$. There exists a set $E \subset \Omega$, with $\dim_H(E) \le n-p$, such that for every $x_0 \in \Omega - E$ the limit*

$$\lim_{\varrho \to 0} ⨍_{Q(x_0,\varrho)} u \, dx$$

exists and is finite.

Proof. As usual we set

$$u_{x_0,\varrho} =: ⨍_{Q(x_0,\varrho)} u \, dx =: \frac{1}{|Q_\varrho|} \int_{Q(x_0,\varrho)} u \, dx \,.$$

If Q is the cube with center at the origin and radius 1, we have

$$u_{x_0,\varrho} = \fint_Q u(x_0 + \varrho y)dy\,,$$

so that the function $\varphi(\varrho) = u_{x_0,\varrho}$ is differentiable in $(0, +\infty)$, and

$$\varphi'(\varrho) = \fint_Q y_i D_i u(x_0 + \varrho y)dy\,.$$

Estimating the right-hand side by means of the Hölder inequality, and coming back to the cube of radius ϱ, we get

$$|\varphi'(\varrho)| \le c\left\{\fint_{Q(x_0,\varrho)} |Du|^p\,dx\right\}^{\frac{1}{p}}\,.$$

Let now μ be the measure defined by

$$\mu(B) = \int_B |Du|^p\,dx\,,$$

and consider the corresponding sets E^α defined in (2.54):

$$E^\alpha = \left\{x \in A : \limsup_{\rho\to 0^+} \varrho^{-\alpha}\mu(Q(x,\varrho)) > 0\right\}.$$

If $x_0 \notin E^{n-p+\epsilon}$ we must have

$$\lim_{\varrho\to 0} \varrho^{p-n-\epsilon}\int_{Q(x_0,\varrho)} |Du|^p\,dx = 0\,, \tag{3.45}$$

and therefore

$$|\varphi'(\varrho)| \le c\varrho^{-1+\frac{\epsilon}{p}}\,.$$

From that inequality follows at once that the required limit exists and is finite. Setting then $E = \bigcap_{\epsilon>0} E^{n-p+\epsilon}$, we have $H^{n-p+\epsilon}(E) \le H^{n-p+\epsilon}(E^{n-p+\epsilon}) = 0$, and therefore $\dim_H(E) \le n-p$. □

We note that by Remark 2.5, (3.45) continues to hold for $\epsilon = 0$. The same cannot be asserted for the conclusion of the theorem, since the condition $\epsilon > 0$ is essential for the existence of the limit, as one can see from the function

$$u(x) = \log\log\frac{1}{|x|}$$

in the disc of $\mathbf{R}^2$ of radius $R < 1$.

Since by Lebesgue theorem we have almost everywhere

$$u(x) = \lim_{\varrho \to 0} u_{x,\varrho}\,,$$

the preceding theorem permits to specify the values of functions in $W^{1,p}$. In fact for such functions the limit on the right-hand side exists for every $x \in \Omega$, except at most a set of dimension not larger than $n-p$, and defines a function belonging to the equivalence class of u.

In a similar way we can characterize the traces of functions in $W^{1,p}(\Omega)$. Since our results are local, it will be sufficient to treat the case when Ω is the half-space $\mathbf{R}^n_+$; the general case will follow by flattening locally the boundary.

Let P be the boundary of $\mathbf{R}^n_+$:

$$P = \{x \in \mathbf{R}^n : x_n = 0\}\,.$$

For $x_0 \in P$, we set $Q^+(x_0,\varrho) = Q(x_0,\varrho) \cap \mathbf{R}^n_+$.

Theorem 3.23 *Let $u \in W^{1,p}(\mathbf{R}^n_+)$. There exists a set $F \subset P$, with dimension not larger than $n-p$, such that for every $x_0 \in P - F$ the limit*

$$\lim_{\varrho \to 0} \fint_{Q^+(x_0,\varrho)} u\, dx$$

exists and is finite.

Moreover, for almost every $x_0 \in P$ this limit coincides with the trace $\varphi(x_0)$ of u.

Proof. The first part can be proved as in the preceding theorem, taking $A = \mathbf{R}^n$ and

$$\mu(G) = \int_{G \cap \mathbf{R}^n_+} |Du|^p\, dx\,.$$

In order to show that the limit of the averages coincides with the trace, we remark that, calling $K(y_0,\varrho)$ the $(n-1)$-dimensional cube with center y_0 and side 2ϱ contained in P, we have

$$\begin{aligned} \int_{Q^+(y_0,\varrho)} |u(x) - \varphi(y_0)|^p\, dx &= \int_{K(y_0,\varrho)} dy \int_0^\varrho |u(y,t) - \varphi(y_0)|^p\, dt \\ &\le c \int_{K(y_0,\varrho)} dy \int_0^\varrho |u(y,t) - \varphi(y)|^p\, dt \\ &\quad + c\varrho \int_{K(y_0,\varrho)} |\varphi(y) - \varphi(y_0)|^p\, dy\,. \end{aligned}$$

Using (3.40) we get the estimate

$$\fint_{Q^+(y_0,\varrho)} |u(x) - \varphi(y_0)|^p \, dx$$

$$\leq c\varrho^{p-n} \int_{Q^+(y_0,\varrho)} |Du|^p \, dx + c\varrho^{1-n} \int_{K(y_0,\varrho)} |\varphi(y) - \varphi(y_0)|^p \, dy \, .$$

The first of the two integrals on the right-hand side tends to zero, except possibly for y_0 in a set of zero $(n-p)$-dimensional measure (see the remark at the end of the preceding theorem); whereas the second tends to zero almost everywhere in P by the Lebesgue theorem. This proves the second part of the theorem. □

From the above theorem we obtain two interesting corollaries. The first is what we had stated in Remark 3.2, that is that the functions w and u have the same trace on $\partial\Omega$. Let $R > 0$, and let $k \leq \frac{1}{R} < k+1$. If $x_0 \in \partial\Omega$ and $x \in Q(x_0, R)$, taking into account (3.5), (3.6) and the definition of the functions α_i, we get

$$\left(\int_{Q(x_0,R)} |w - u|^p \, dx \right)^{\frac{1}{p}} \leq \sum_{i=k-2}^{\infty} \|\varphi_{\epsilon_i} * (\alpha_i u) - \alpha_i u\|_p \leq \tau 2^{2-k} \leq c 2^{-\frac{1}{R}}$$

from which we obtain immediately

$$\lim_{\varrho \to 0} \fint_{Q(x_0,\varrho)} (w - u) dx = 0$$

and hence the equality of the traces of u and w.

We have in addition the following:

Corollary 3.2 *Let Ω be an open set with Lipschitz-continuous boundary, and let $u \in W^{2,p}(\Omega)$ have zero trace on $\partial\Omega$. Let $d(x) = \text{dist}(x, \partial\Omega)$, and assume that*

$$|u(x)| \leq \gamma(d(x))$$

in a neighborhood of $\partial\Omega$, where γ is a C^1 function, with $\gamma(0) = 0$. Then, for almost every $x \in \partial\Omega$ we have

$$|Du(x)| \leq c\gamma'(0) \, .$$

Proof. Setting $\Omega_R = \Omega \cap Q_R$, using the Green formula with $\varphi = 1$ and taking into account that $u = 0$ on $\partial\Omega$, we get for R sufficiently small

$$\left|\int_{\Omega_R} Du\,dx\right| = \left|\int_{\Sigma_R} u\nu\,dH_{n-1}\right| \leq \int_{\Sigma_R} \gamma(d)dH_{n-1}\,,$$

with $\Sigma_R = \Omega \cap \partial Q_R$.

On the other hand, if $d < R$ we have $\gamma(d) \leq (\gamma'(0) + \epsilon(R))R$, and therefore

$$\left|\int_{\Omega_R} Du\,dx\right| \leq RH_{n-1}(\Sigma_R)(\gamma'(0) + \epsilon(R))\,.$$

The conclusion follows immediately from the inequality $H_{n-1}(\Sigma_R) \leq cR^{n-1}$, dividing by R^n and letting R tend to zero. □

3.9 Notes and Comments

A general theory of Sobolev spaces can be found in most books devoted to elliptic equations, such as the classical treatises by MORREY [3], LADYŽENSKAYA and URAL'CEVA [1] or GILBARG and TRUDINGER [1]. There are, of course, volumes expressly dedicated to the study of such spaces; among which we mention ADAMS [1] and KUFNER, JOHN and FUČIK [1].

Sobolev spaces were introduced and studied by SOBOLEV [1], and independently by CALKIN [1] and MORREY [1]. Since then, they have been so widely used, that in many cases it is difficult to retrace the exact paternity of a result. For instance, whereas Theorem 3.4 is certainly due to MEYERS and SERRIN [1], Theorem 3.6 is more difficult to attribute with some certainty.

It is possible to define fractional Sobolev spaces, either by interpolation between L^p and $W^{1,p}$ (see LIONS [1] or LIONS and MAGENES [1]), or else by introducing norms similar to (3.44). We shall thus say that $u \in W^{\vartheta,p}(\Omega)$, $0 < \vartheta < 1$, if $u \in L^p$ and

$$\iint_{\Omega\times\Omega} \frac{|u(x) - u(y)|^p}{|x - y|^{n+p\vartheta}}\,dx\,dy < +\infty\,. \tag{3.46}$$

More generally, a function u belongs to $W^{k+\vartheta,p}(\Omega)$ if $u \in W^{k,p}$ and its derivatives of order k belong to $W^{\vartheta,p}(\Omega)$.

For such spaces results similar to those proved for the usual Sobolev spaces hold; in particular the statements of the immersion theorems of

Sobolev (Theorem 3.12) and Rellich (Theorem 3.13) remain valid even if k is not an integer.

In terms of fractional Sobolev spaces, Theorem 3.21 says that a function φ defined on $\partial\Omega$ is the trace of a function in $W^{1,p}(\Omega)$ if and only if it belongs to $W^{1-\frac{1}{p},p}(\partial\Omega)$. Moreover, if $u \in W^{1,p}(\Omega)$ and φ is its trace, we have

$$c_1\|\varphi\|_{W^{1-\frac{1}{p},p}(\partial\Omega)} \leq \inf\{\|u+\eta\|_{W^{1,p}(\Omega)};\ \eta \in W_0^{1,p}(\Omega)\}$$
$$\leq c_2\|\varphi\|_{W^{1-\frac{1}{p},p}(\partial\Omega)}\,.$$

As a consequence of these estimates and of the Sobolev theorem for fractional spaces, we can deduce the continuity of the immersion of $W^{1,p}(\Omega)$ in $L^q(\partial\Omega)$, with $q = \frac{(n-1)p}{n-p}$.

For more general trace theorems (derivatives of arbitrary order on manifolds of codimension greater than one) have proved useful the Besov spaces $B^{\vartheta,p}(\Omega)$, of functions $u \in L^p(\Omega)$ such that

$$\int_\Omega dx \int_{\Omega_x} \frac{|u(x)+u(y)-2u(\frac{x+y}{2})|^p}{|x-y|^{n+p\vartheta}}\,dy < +\infty\,,$$

where $\Omega_x = \{y \in \Omega : \frac{x+y}{2} \in \Omega\}$.

Another definition of fractional derivatives makes use of the Fourier transform. If $u \in L^2(\mathbf{R}^n)$, one can define the function

$$\hat{u}(\xi) =: \frac{1}{(2\pi i)^n}\int u(x)e^{i\langle x,\xi\rangle}\,dx\,.$$

Since

$$\hat{D^\alpha u}(\xi) = i^{|\alpha|}\xi^\alpha\hat{u}(\xi)\,,$$

the function u will belong to $W^{k,2}(\mathbf{R}^n)$ if and only if

$$\int_{\mathbf{R}^n} |\xi|^{2k}|\hat{u}(\xi)|^2\,d\xi < +\infty\,.$$

The form of the last relation permits an immediate extension to the case of fractional exponent, since the order k appears in it only as a parameter, not necessarily integral.

More generally, a function $u \in L^p(\mathbf{R}^n)$ belongs to $L^{s,p}(\mathbf{R}^n)$ if the function $(1+|\xi|^2)^{-s/2}\hat{u}(\xi)$ is the Fourier transform of a function $\tilde{u} \in L^p$. The spaces $L^{s,p}$ do not coincide with the $W^{s,p}$ definite above; we have however

$$L^{s+\epsilon,p}(\mathbf{R}^n) \hookrightarrow W^{s,p}(\mathbf{R}^n) \hookrightarrow L^{s-\epsilon,p}(\mathbf{R}^n)$$

for every $\epsilon > 0$.

More information about fractional Sobolev spaces, as well as about Besov spaces, that are variations of them, can be found in ADAMS [1] and KUFNER, JOHN and FUČIK [1].

The inequalities of Sobolev (3.17) and Sobolev–Poincaré (3.29) are strictly connected with the so-called *isoperimetric inequalities* by means of the *coarea formula* (FLEMING and RISHEL [1]; see also GIUSTI [6]) for functions in $W^{1,1}(\Omega)$:

$$\int_L |Du| dx = \int_{-\infty}^{\infty} H_{n-1}(\partial U_t \cap L) dt \,, \tag{3.47}$$

that holds for any Borel set $L \subset \Omega$, where as usually we have set

$$U_t = \{x \in \Omega : u(x) > t\} \,.$$

The "classical" isoperimetric inequality (DE GIORGI [2]) asserts that the n-ball has the least parameter among bodies of the same measure. It follows that for every set E, and whence in particular for the sets U_t, it holds that

$$|E|^{1-\frac{1}{n}} \leq c(n) H_{n-1}(\partial E)$$

in which the constant $c(n)$ is determined by the fact that we must have the equality when E is an n-dimensional ball.

Let now $u \in C^{0,1}(\Omega)$, with $u \geq 0$.[6] We have

$$|U_t| \leq \left(\frac{\|u\|_\sigma}{t} \right)^\sigma$$

with $\sigma = \frac{n}{n-1}$, and therefore

$$\|u\|_\sigma^\sigma = \sigma \int_0^\infty t^{\frac{1}{n-1}} |U_t| \, dt \leq \sigma \|u\|_\sigma^{\frac{\sigma}{n}} \int_0^\infty |U_t|^{1-\frac{1}{n}} \, dt \,.$$

By applying the isoperimetric inequality above, we get

$$\|u\|_\sigma \leq c \int_0^\infty H_{n-1}(\partial U_t) dt \,.$$

Assume now that u has zero trace. In this case U_t is contained in Ω, and therefore the last quantity is nothing but

[6] This is not a restriction, since we may always divide u into its positive and negative part.

$$\int_\Omega |Du|dx$$

so that the Sobolev inequality is proved when $p = 1$. The general case follows as in Theorem 3.7.

In order to prove the Sobolev–Poincaré inequality we need a slightly more sophisticated isoperimetric inequality (BOMBIERI and GIUSTI [1]): If Ω is a connected set with regular boundary, there exists a constant $c(n, \Omega)$ such that for every $E \subset \Omega$,

$$\min\{|E|, |\Omega - E|\}^{1-\frac{1}{n}} \leq cH_{n-1}(\partial E \cap \Omega).$$

The required inequality follows now choosing a real number λ such that $|A_\lambda| = \frac{1}{2}|\Omega|$, and repeating with small variants the preceding argument for the functions $\max\{u - \lambda, 0\}$ and $\max\{\lambda - u, 0\}$.

When $p = 1$ the above inequalities hold also for functions whose derivatives are measures. For more details see FEDERER [2] or GIUSTI [6].

The coarea formula can be used also to establish a result extending what we have proved in Proposition 3.5.

Proposition 3.9 *Let $A \subset \mathbf{R}$ be a Borel set of zero measure, and let $u \in W^{1,1}(\Omega)$. Then, $Du = 0$ for almost every $x \in u^{-1}(A)$.*

Proof. Since for almost every $t \in \mathbf{R}$ we have

$$\partial U_t = \{x \in \Omega : u(x) = t\},$$

the conclusion follows at once from the coarea formula with $L = u^{-1}(A)$, by remarking that the integral on the right-hand side is made on the set A of zero measure. □

We can prove now the following:

Theorem 3.24 *Let $f(t)$ be a Lipschitz-continuous function in $\mathbf{R}$, and let $u \in W^{1,1}_{\text{loc}}(\Omega)$. Then, the composed function $f \circ u$ belongs to $W^{1,1}_{\text{loc}}(\Omega)$, and $D(f \circ u) = f' \circ u\, Du$.*

Proof. Indicating by e_s the unit vector in the direction of the x_s-axis, for any measurable function v we set

$$\Delta_h v = \Delta_{h,s} v = \frac{v(x + he_s) - v(x)}{h}$$

and we define

$$B_h(u) = \begin{cases} \dfrac{f(u(x)+h\Delta_h u) - f(u(x))}{h\Delta_h u} & \text{if } \Delta_h u \neq 0\,, \\ f'(u(x)) & \text{if } \Delta_h u = 0 \text{ and } u(x) \notin A\,, \\ 0 & \text{if } \Delta_h u = 0 \text{ and } u(x) \in A\,, \end{cases}$$

where A is the set of points where f is not differentiable.

Now let φ be a function with compact support in Ω, and let $|h| < \operatorname{dist}(\operatorname{supp}\varphi, \partial\Omega)$. By the preceding proposition

$$\int_\Omega B_h(x) D_s u\, \varphi\, dx = \int_{\Omega - u^{-1}(A)} B_h(x) D_s u\, \varphi\, dx$$

and therefore

$$\lim_{h\to 0} \int_\Omega B_h(x) D_s u\, \varphi\, dx = \int_\Omega f' \circ u\, D_s u\, \varphi\, dx\,.$$

On the other hand, since f is Lipschitz-continuous in $\mathbf{R}$, we have $|B_h| \leq K$, and since $\Delta_h u \to D_s u$ in $L^1_{\mathrm{loc}}(\Omega)$ (see later, Sec. 8.1), we have also

$$\lim_{h\to 0} \int_\Omega B_h(x) \Delta_h u\, \varphi\, dx = \int_\Omega f' \circ u\, D_s u\, \varphi\, dx\,.$$

The conclusion follows by remarking that

$$B_h(x)\Delta_h u = \frac{f(u(x)+h\Delta_h u) - f(u(x))}{h} = \Delta_h(f\circ u)$$

and that

$$\int \Delta_h(f\circ u)\varphi\, dx = -\int f\circ u\, \Delta_{-h}\varphi\, dx \to -\int f\circ u\, D_s\varphi\, dx\,. \qquad \square$$

The conclusion of the theorem holds even if f is only locally Lipschitz-continuous in $\mathbf{R}$, provided $f\circ u$ and $f'\circ u\, Du$ are locally summable in Ω. For the proof we assume first that f is bounded, and we pass to the limit for $t \to +\infty$ in the relation

$$\int f\circ u_t\, D\varphi\, dx = -\int f' \circ u_t\, Du_t \varphi\, dx$$

in which $u_t = \max\{\min(u,t), -t\}$. When f is not bounded, it will suffice to pass to the limit in the equation

$$\int f_t \circ u\, D\varphi\, dx = -\int f_t' \circ u\, Du\, \varphi\, dx\,.$$

Chapter 4

Convexity and Semicontinuity

4.1 Preliminaries

Once the fundamental results about SOBOLEV spaces have been established, we can pose the problem of lower semicontinuity in the space $W^{1,p}$ for general functionals

$$\mathcal{F}(u) = \mathcal{F}(u, \Omega) =: \int_\Omega F(x, u, Du)\, dx\,. \tag{4.1}$$

Definition 4.1 *Let X be a topological space. A function $\mathcal{F} : X \to \bar{\mathbf{R}}$ is* lower semicontinuous (l.s.c.) *if for every $t \in \mathbf{R}$ the set*

$$\mathcal{F}_t = \{x \in X : \mathcal{F}(x) > t\}$$

is open (or else if the set $\mathcal{G}_t = \{x \in X : \mathcal{F}(x) \leq t\}$ is closed).

In the above definition we have indicated by $\bar{\mathbf{R}}$ the set $\mathbf{R} \cup \{+\infty\}$; in other words the function $\mathcal{F}$ may take the value $+\infty$.[1]

It is easily seen that $\mathcal{F}$ is lower semicontinuous if and only if its *epigraphic*

$$\Sigma(\mathcal{F}) = \{(x, t) \in X \times \mathbf{R} : t \geq \mathcal{F}(x)\}$$

[1] This extension is purely technical, and permits one to avoid non-essential discussions. However, we shall always assume that at least for a point $x \in X$ it holds that $\mathcal{F}(x) < +\infty$.

is closed. Actually, the complement of $\Sigma(\mathcal{F})$ is the set

$$A = \bigcup_{t\in\mathbf{R}} \mathcal{F}_t \times (-\infty, t),$$

which is open if $\mathcal{F}$ is lower semicontinuous. Conversely, if $\Sigma(\mathcal{F})$ is closed, so is $\mathcal{G}_t \times \{t\} = \Sigma(\mathcal{F}) \cap (X \times \{t\})$, and therefore $\mathcal{G}_t$.

For our purposes, a second definition of lower semicontinuity will be more suitable, in terms of convergence of sequences.

Definition 4.2 *Let X be a topological space. We say that the function $\mathcal{F} : X \to \bar{\mathbf{R}}$ is (sequentially)* lower semicontinuous *if for every sequence v_k convergent to some $v \in X$, we have*

$$\mathcal{F}(v) \le \liminf_{k\to\infty} \mathcal{F}(v_k).$$

We note that a lower semicontinuous function $\mathcal{F}$ is sequentially lower semicontinuous. The converse is true if X satisfies the first countability axiom, that is if every point has a countable fundamental system of neighborhoods; in particular if X is metrizable. The proof is left to the reader.

The following result is not difficult to prove.

Lemma 4.1 *If $\mathcal{F}_\alpha$, $\alpha \in I$, is a family of lower semicontinuous functions, then*

$$\mathcal{F}(x) = \sup_{\alpha\in I} \mathcal{F}_\alpha(x)$$

is lower semicontinuous.

Proof. It will be sufficient to remark that

$$\mathcal{F}_t = \{x \in X : \mathcal{F}(x) > t\} = \bigcup_{\alpha\in I} \mathcal{F}_t^\alpha = \bigcup_{\alpha\in I} \{x \in X : \mathcal{F}_\alpha(x) > t\}.$$

In the case of sequential semicontinuity, if $x_k \to x$, we have

$$\begin{aligned} \mathcal{F}(x) &= \sup_\alpha \mathcal{F}_\alpha(x) \le \sup_\alpha \liminf_{k\to\infty} \mathcal{F}_\alpha(x_k) \\ &\le \liminf_{k\to\infty} \sup_\alpha \mathcal{F}_\alpha(x_k) = \liminf_{k\to\infty} \mathcal{F}(x_k). \end{aligned}$$

□

In this book we shall use only sequential lower semicontinuity, that we shall abbreviate LSC, always keeping in mind that the two definitions coincide in metric spaces.

Let now V be a subset of X, in which we want to minimize the functional $\mathcal{F}$. We call *minimizing* a sequence $\{x_k\}$ with values in V, such that

$$\lim_{k\to\infty} \mathcal{F}(x_k) = \inf_V \mathcal{F}\,.$$

The following result generalizes WEIERSTRASS' theorem.

Theorem 4.1 *Let $V \subset X$, and let $\mathcal{F}$ be a LSC function. Assume there exists a minimizing sequence x_k, converging to a point $x_0 \in V$. Then, $\mathcal{F}(x_0)$ is the minimum of $\mathcal{F}$ in V.*

Proof. Indeed, we have

$$\inf_V \mathcal{F} \le \mathcal{F}(x_0) \le \liminf_{k\to\infty} \mathcal{F}(x_k) = \inf_V \mathcal{F}$$

and hence $-\infty < \mathcal{F}(x_0) = \inf_V \mathcal{F}$. □

In the applications to the calculus of variations, neither the space X nor its topology are given *a priori*; they must be chosen according to two contrasting requests: the lower semicontinuity of the functional and the existence of a convergent minimizing sequence. As we have said, these are two concurrent properties. The choice of the topology should be situated at the equilibrium point between these two forces pushing in opposite directions; in particular it will be useful to prove the semicontinuity in the weakest possible topology. Most of this and the next chapter will be concerned with the study of the semicontinuity of functionals in the most general situation.

4.2 Convex Functionals

We shall begin with a semicontinuity theorem in the strong topology of $W^{1,1}(\Omega)$, even if the strong convergence in this space is by far too restrictive to be useful in the applications. However, this result will be a starting point for the subsequent developments.

Definition 4.3 *A* CARATHEODORY function *is a function $F(x,y) : \Omega \times \mathbf{R}^s \to \bar{\mathbf{R}}$ such that*

(i) *$F(\cdot, y)$ is measurable for every $y \in \mathbf{R}^s$,*
(ii) *$F(x, \cdot)$ is continuous for almost every $x \in \Omega$.*

Lemma 4.2 *Let $F(x,y)$ be a* CARATHEODORY *function, and let $y(x)$ be a measurable function. Then, the function $g(x) =: F(x, y(x))$ is measurable in Ω.*

Proof. Assume first that $y(x)$ is a step function, that is

$$y(x) = \sum_{i=1}^{N} \lambda_i \chi_i(x)\,,$$

where λ_i are real numbers, and χ_i are the characteristic functions of pairwise disjoint measurable sets E_i, with $\cup_{i=1}^{N} E_i = \Omega$. For $t \in \mathbf{R}$ we have

$$\{x \in \Omega : g(x) > t\} = \bigcup_{i=1}^{N} \{x \in E_i : F(x, \lambda_i) > t\}\,.$$

By (i) above, all the sets on the right-hand side are measurable, and hence g is measurable.

In the general case, we remark that a measurable function $y(x)$ is the pointwise limit of a sequence y_k of step functions. By (ii) we have

$$F(x, y(x)) = \lim_{k\to\infty} F(x, y_k(x)) \quad \text{a.e. in } \Omega$$

and hence the function $F(x, y(x))$ is measurable, being the pointwise limit of measurable functions. □

We can now prove without difficulty the following:

Theorem 4.2 *Let $F(x, y)$ be a* CARATHEODORY *function, and let $y_k(x)$ be a sequence of functions, strongly convergent in $L^1(\Omega)$ to a function $y(x)$. Then,*

$$\int_\Omega F(x, y(x))\, dx \le \liminf_{k\to\infty} \int_\Omega F(x, y_k(x))\, dx\,.$$

Proof. From y_k we extract a sequence y_k^* such that

$$\lim_{k\to\infty} \int_\Omega F(x, y_k^*(x))\, dx = \liminf_{k\to\infty} \int_\Omega F(x, y_k(x))\, dx\,.$$

Passing possibly to a subsequence, we can assume that $y_k^*(x)$ converges almost everywhere to $y(x)$, in which case $F(x, y_k^*(x))$ will also converge almost everywhere to $F(x, y(x))$. The conclusion then follows from FATOU's lemma. □

If we want to apply the above theorem to the functional (4.1), we must take $y = (u, Du)$, so that we need the strong convergence in $W^{1,1}(\Omega)$; a topology which, as we have remarked, is far too strong. On the other hand, if the functional $\mathcal{F}$ is *convex*, we can prove the following:

Theorem 4.3 *Let $F(x,y)$ be a* CARATHEODORY *function, and assume that for almost every $x \in \Omega$ the function $F(x,\cdot)$ is convex. Then, the functional*

$$\mathcal{F}(y,\Omega) = \int_\Omega F(x,y(x))\,dx$$

is lower semicontinuous in the weak topology of $L^1(\Omega)$.

Proof. Since $\mathcal{F}$ is LSC in the strong topology of L^1, and the latter is a BANACH space, the epigraphic $\Sigma(\mathcal{F})$ of $\mathcal{F}$ is strongly closed. On the other hand, the convexity of F implies that of $\mathcal{F}$, and therefore of $\Sigma(\mathcal{F})$. It follows (see DUNFORD–SCHWARTZ [1] (I), Theorem V. 13) that $\Sigma(\mathcal{F})$ is weakly closed, and hence $\mathcal{F}$ is LSC in the weak topology of $L^1(\Omega)$. □

From this theorem we can already see the role played by the convexity in the passage from strong to weak topology. However, in the case at hand the convexity of the function $F(x,\cdot)$ means that of $F(x,u,z)$ in the couple (u,z), and this assumption is still too strong, to the point that the preceding result is meaningful only when the function F in (4.1) is independent of u.

On the other hand, from the compactness theorems proved in the preceding chapter it follows that the weak convergence of the derivatives implies the strong convergence of the functions. This fact suggests that, whereas the convexity of the function $F(x,u,z)$ with respect to the variable z is somewhat essential for the semicontinuity in the weak topology of $W^{1,p}$, with respect to u is superfluous since the weak convergence of the derivatives implies the strong convergence of the functions. Consequently, we can weaken the hypotheses, and we can assume only the continuity of F with respect to the variable u. This is exactly what we shall do in what follows.

4.3 Semicontinuity

We shall begin with some notation. As is usual, we shall use an arrow $\to$ to denote strong convergence, and a half-arrow $\rightharpoonup$ for the weak convergence (the topology in question, when not explicitly stated, will be clear from the context). If $V(\Omega)$ is a space of functions defined in Ω, we say that $u_k \to u$ (resp. $u_k \rightharpoonup u$) in $V_{\text{loc}}(\Omega)$ if $u_k \to u$ (resp. $u_k \rightharpoonup u$) in $V(\Sigma)$ for every open set $\Sigma \subset\subset \Omega$.

We shall assume that the function $F(x,u,z)$ is defined in $\Omega \times M \times \mathbf{R}^\nu$, where M is a closed set in $\mathbf{R}^N$, possibly coinciding with $\mathbf{R}^N$, and we shall consider the space of functions (u, z), with $z \in L^1(\Omega, \mathbf{R}^\nu)$ and $u \in L^1(\Omega, M)$, that is $u \in L^1(\Omega, \mathbf{R}^N)$ and $u(x) \in M$ for almost all $x \in \Omega$. It is easily seen that $L^1(\Omega, M)$ is closed with respect to strong convergence in $L^1_{\mathrm{loc}}(\Omega)$; if $u_k \in L^1(\Omega, M)$, $u \in L^1(\Omega, \mathbf{R}^N)$, and if $u_k \to u$ in $L^1_{\mathrm{loc}}(\Omega, \mathbf{R}^N)$, then, passing possibly to a subsequence, we can suppose that $u_k \to u$ almost everywhere, so that $u \in L^1(\Omega, M)$.

Our first results concerns the case of a function $F(x, u, z)$ regular enough.

Theorem 4.4 *Let $F(x, u, z)$ be a non-negative function, continuous together with its derivatives with respect to z in $\Omega \times M \times \mathbf{R}^\nu$, and convex in z. Let $u_k, u \in L^1(\Omega, M)$, $z_k, z \in L^1(\Omega, \mathbf{R}^\nu)$, and assume that $u_k \to u$ in $L^1_{\mathrm{loc}}(\Omega)$ and that $z_k \rightharpoonup z$ in $L^1_{\mathrm{loc}}(\Omega)$. Then,*

$$\mathcal{F}(u, z) \leq \liminf_{k \to \infty} \mathcal{F}(u_k, z_k)\,. \tag{4.2}$$

Proof. Assume first that Ω is bounded, and let $D \subset\subset \Omega$. By the theorems of EGOROV and LUSIN, for every $\epsilon > 0$ we can find a compact set $K \subset D$, with $|D - K| < \epsilon$ and such that

(i) $u_k \to u$ uniformly in K,
(ii) u and z are continuous in K,
(iii) $\int_K F(x, u, z)\, dx \geq \int_D F(x, u, z)\, dx - \epsilon$.

Indicating as usual with F_z the vector in $\mathbf{R}^\nu$ whose components are $\frac{\partial F}{\partial z_i}$ and with $\langle\ \rangle$ the scalar product in $\mathbf{R}^\nu$, we have:

$$\begin{aligned}
\int_K F(x, u_k, z_k)\, dx &= \int_K \{F(x, u_k, z) + [F(x, u_k, z_k) - F(x, u_k, z)]\}\, dx \\
&\geq \int_K F(x, u_k, z)\, dx + \int_K \langle F_z(x, u_k, z), z_k - z\rangle\, dx \\
&= \int_K F(x, u_k, z)\, dx + \int_K \langle F_z(x, u, z), z_k - z\rangle\, dx \\
&\quad + \int_K \langle F_z(x, u_k, z) - F_z(x, u, z), z_k - z\rangle\, dx\,.
\end{aligned}$$

Passing to the limit as $k \to \infty$, the first integral on the right-hand side tends to

$$\int_K F(x, u, z)\, dx$$

thanks to the continuity of F and to condition (i). The second integral tends to zero, since $F_z(x, u(x), z(x))$ is continuous, and $z_k \rightharpoonup z$. Finally, the third integral can be estimated by

$$\|z_k - z\|_1 \sup_K |F_z(x, u_k(x), z(x)) - F_z(x, u(x), z(x))|$$

and therefore it also tends to zero by (i) and the continuity of F_z. In conclusion:

$$\liminf_{k\to\infty} \int_K F(x, u_k, z_k)\, dx \geq \int_K F(x, u, z)\, dx \geq \int_D F(x, u, z)\, dx - \epsilon\,.$$

Since $F \geq 0$ and ϵ is arbitrary we have

$$\liminf_{k\to\infty} \int_\Omega F(x, u_k, z_k)\, dx \geq \int_D F(x, u, z)\, dx$$

and taking the supremum over $D \subset \Omega$ we get the conclusion if Ω is bounded.

The general case is obtained remarking that

$$\begin{aligned}\liminf_{k\to\infty} \int_\Omega F(x, u_k, z_k)\, dx &\geq \liminf_{k\to\infty} \int_{\Omega\cap B_r} F(x, u_k, z_k)\, dx \\ &\geq \int_{\Omega\cap B_r} F(x, u, z)\, dx\end{aligned}$$

and passing to the limit as $r \to +\infty$. □

We want now to drop the assumption of continuity of F and F_z. Writing

$$\int_\Omega F(x, u_k, z_k)\, dx = \int_\Omega F(x, u, z_k)\, dx + \int_\Omega [F(x, u_k, z_k) - F(x, u, z_k)]\, dx\,,$$

the first integral on the right is semicontinuous by Theorem 4.3; it will only be necessary to show that the second integral tends to zero, or at least that its liminf is non-negative.

We begin with a lemma, which we shall prove only for the part that will be used later.

Lemma 4.3 *Let Σ be an open set in $\mathbf{R}^n$, with $|\Sigma| < +\infty$, and let z_k be a sequence converging weakly in $L^p(\Sigma)$ $(p \geq 1)$ to a function $z(x)$. For $L > 0$ set*

$$z^L = \begin{cases} z & \textit{if } |z| \leq L\,, \\ 0 & \textit{otherwise}\,. \end{cases}$$

For every integer L there exists a subsequence z_{k_s} and a function v^L such that $z_{k_s}^L \rightharpoonup v^L$ in $L^2(\Sigma)$. Moreover, when L tends to infinity, the sequence v^L tends to z in L^1.

Proof. We shall prove the lemma only for $p > 1$. Since $|z_k^L| \leq L$ and $|\Sigma| < +\infty$, the sequence z_k^L is bounded in $L^2(\Sigma)$, and therefore we can extract from it a subsequence weakly convergent to a function $v^L \in L^2(\Sigma)$ (note that we do not assert that $v^L = z^L$).

Let $\varphi \in L^\infty(\Sigma)$, and let

$$\Sigma_{k,L} = \{x \in \Sigma : |z_k(x)| > L\}.$$

We have

$$\int_\Sigma (z_k - z_k^L)\varphi\, dx = \int_{\Sigma_{k,L}} z_k \varphi\, dx \leq \sup|\varphi| \int_{\Sigma_{k,L}} |z_k|\, dx.$$

From the assumption $p > 1$ we get

$$\int_{\Sigma_{k,L}} |z_k|\, dx \leq |\Sigma_{k,L}|^{1-\frac{1}{p}} \|z_k\|_{p,\Sigma} \tag{4.3}$$

and hence, recalling that

$$|\Sigma_{k,L}| \leq \frac{\|z_k\|_{p,\Sigma}^p}{L^p} \leq cL^{-p},$$

we can conclude that for any fixed $\epsilon > 0$, the integral on the left of (4.3) can be made $< \epsilon$ by taking L large enough, independently of k. Passing to the limit as $k \to \infty$, we get

$$\int_\Sigma (z - v^L)\varphi\, dx \leq \epsilon \sup|\varphi|$$

for $L > L(\epsilon)$. Choosing $\varphi = H(z - v^L)$ (H is the HEAVISIDE function) we find

$$\int_\Sigma |z - v^L|\, dx \leq \epsilon$$

for every $L > L(\epsilon)$, and the lemma is proved for $p > 1$. □

The case $p = 1$ is more complex,[2] since we cannot use the inequality (4.3). In its place, one might use the following lemma, that we state without proof (see DUNFORD and SCHWARTZ [1] (I), Corollary IV.11):

Lemma 4.4 *Let Σ be a bounded open set of $\mathbf{R}^n$, and let $f_k \rightharpoonup f$ in $L^1(\Sigma)$.*

---

[2]We remark, however, that in the applications one has usually $p > 1$.

Then:

(i) *The sequence* $\|f_k\|_1$ *is bounded,*
(ii) *The functions* $A \to \int_A |f_k|\,dx$ *are absolutely continuous, uniformly in* k.

The next step is essentially technical.

Lemma 4.5 *Let* K *be a compact set in* $\mathbf{R}^n$, *and let* $F(x,u,z)$ *be a continuous non-negative function in* $K \times M \times \mathbf{R}^\nu$, *convex in* z *for every* $(x,u) \in K \times M$. *Assume that* $u_h \to u$ *uniformly in* K, *and that* $z_h \rightharpoonup z$ *in* $L^p(K)$, *with* $p \geq 1$. *Then,*

$$\int_K F(x,u,z)\,dx \leq \liminf_{h\to\infty} \int_K F(x,u_h,z_h)\,dx\,.$$

Proof. Passing possibly to a subsequence, we can suppose that the sequence of the integrals on the right-hand side is convergent. Let $R \geq \sup_h \|u_h\|_{\infty,K}$, $M_R = M \cap \overline{B_R}$, and define

$$T = \sup_{K\times M_R} F(x,u,0)\,; \quad \Lambda = \sup_h \|z_h\|_{p,K}\,.$$

Setting $K_{h,L} = \{x \in K : |z_h(x)| > L\}$, we have $|K_{h,L}| \leq (\frac{\Lambda}{L})^p$. From Theorem 4.3 and the preceding lemma we deduce

$$\int_K F(x,u,z)\,dx \leq \liminf_{L\to\infty} \int_K F(x,u,v^L)\,dx$$

and

$$\int_K F(x,u,v^L)\,dx \leq \liminf_{h\to\infty} \int_K F(x,u,z_h^L)\,dx\,.$$

On the other hand, since $F \geq 0$, we have

$$\begin{aligned}\int_K F(x,u,z_h^L)\,dx &= \int_K F(x,u_h,z_h^L)\,dx + \int_K [F(x,u,z_h^L) - F(x,u_h,z_h^L)]\,dx\\ &\leq \int_K F(x,u_h,z_h)\,dx + \int_{K_{h,L}} F(x,u_h,0)\,dx\\ &\quad + \int_K [F(x,u,z_h^L) - F(x,u_h,z_h^L)]\,dx\,.\end{aligned}$$

When $h \to \infty$, the third integral tends to zero, since F is uniformly continuous in the compact set $K \times M_R \times \overline{B_L^\nu}$ and $u_h \to u$ uniformly. The second integral can be estimated by $T(\frac{\Lambda}{L})^p$ independently of h. We have therefore

$$\int_K F(x, u, v^L)\, dx \leq \liminf_{h \to \infty} \int_K F(x, u_h, z_h)\, dx + T\left(\frac{\Lambda}{L}\right)^p$$

and the result follows at once. □

To conclude the proof of the first of our semicontinuity theorems, we need the following lemma that generalizes the theorem of LUSIN.

Lemma 4.6 (SCORZA DRAGONI [1]) *Let Σ be a measurable set, with $|\Sigma| < +\infty$, let $S \subset \mathbf{R}^s$, and let $h(x, y)$ be a function defined in $\Sigma \times S$, measurable in x for every $y \in S$, and uniformly continuous in y for almost all $x \in \Sigma$.*

For every $\delta > 0$ there exists a compact set $K \subset \Sigma$, with $|\Sigma - K| < \delta$, and such that the restriction of $h(x, y)$ to $K \times S$ is continuous.

Proof. For $i \in \mathbf{N}$ we set

$$\omega_i(x) = \sup\left\{ |h(x, y_1) - h(x, y_2)|;\ y_1, y_2 \in S,\ |y_1 - y_2| < \frac{1}{i} \right\}.$$

By assumption, $\omega_i \to 0$ almost everywhere in Σ, and hence by the EGOROV theorem there exists a compact set $K_1 \in \Sigma$, with $|\Sigma - K_1| < \frac{\delta}{2}$, such that $\omega_i \to 0$ uniformly in K_1. In other words, $h(x, \cdot)$, $x \in K_1$ is a family of equicontinuous functions.

Let $\tilde{S} = \{\tilde{y}_1, \tilde{y}_2, \ldots, \tilde{y}_n, \ldots\}$ be a countable dense set in S, and let δ_j be a sequence of positive numbers, with $\sum_{j=1}^{\infty} \delta_j < \frac{\delta}{2}$. By the LUSIN theorem, for every j there exists a compact set $\tilde{K}_j \subset \Sigma$ such that $|\Sigma - \tilde{K}_j| < \delta_j$ and $h(\cdot, \tilde{y}_j)$ is continuous in $\tilde{K}_j$.

Setting $K_2 = \bigcap_j \tilde{K}_j$, all the functions $h(\cdot, \tilde{y}_j)$ are continuous in K_2, and $|\Sigma - K_2| < \frac{\delta}{2}$.

Now let $K = K_1 \cap K_2$, and let (x_n, y_n) be a sequence in $K \times S$, such that $x_n \to x \in K$ and $y_n \to y \in S$. We have

$$|h(x_n, y_n) - h(x, y)| \leq |h(x_n, y_n) - h(x_n, y)| + |h(x_n, y) - h(x, y)|.$$

The first term on the right tends to zero, since the functions $h(x, \cdot)$ are equicontinuous for $x \in K$. As for the second term, we have

$$|h(x_n, y) - h(x, y)| \leq |h(x, \tilde{y}) - h(x, y)| + |h(x_n, \tilde{y}) - h(x, \tilde{y})| \\ + |h(x_n, \tilde{y}) - h(x_n, y)|,$$

where $\tilde{y} \in \tilde{S}$ is chosen in such a way that for every $x \in K$ it holds that

$$|h(x, \tilde{y}) - h(x, y)| < \frac{\epsilon}{2}$$

(once again, this is possible by the equicontinuity of the functions $h(x, \cdot)$).

In this way the first and the third terms are both less than $\frac{\epsilon}{2}$, whereas the second tends to zero when $n \to \infty$. We have therefore

$$\limsup_{n\to\infty} |h(x_n, y_n) - h(x, y)| < \epsilon$$

from which the lemma follows. □

We can now prove the following:

Theorem 4.5 (Semicontinuity) *Let Ω be an open set in $\mathbf{R}^n$, let M be a closed set in $\mathbf{R}^N$, and let $F(x, u, z)$ be a function defined in $\Omega \times M \times \mathbf{R}^\nu$ and such that*

(i) *F is a* CARATHEODORY *function, that is measurable in x for every $(u, z) \in M \times \mathbf{R}^\nu$ and continuous in (u, z) for almost every $x \in \Omega$.*
(ii) *$F(x, u, z)$ is convex in z for almost every $x \in \Omega$ and for every $u \in S$.*
(iii) *$F \geq 0$.*

Let $u_h, u \in L^1(\Omega, M)$, $z_h, z \in L^1(\Omega, \mathbf{R}^\nu)$, and assume that $u_h \to u$ and $z_h \rightharpoonup z$ in $L^1_{\text{loc}}(\Omega)$. Then,

$$\int_\Omega F(x, u, z)\, dx \leq \liminf_{h\to\infty} \int_\Omega F(x, u_h, z_h)\, dx\,. \tag{4.4}$$

Proof. Setting

$$\mathcal{F}(u, z, A) = \int_A F(x, u, z)\, dx\,,$$

we can suppose that there exists the limit

$$\lim_{h\to\infty} \mathcal{F}(u_h, z_h, \Omega) =: \lambda$$

and that $u_h \to u$ almost everywhere in Ω.

Let $\tilde{\Omega} \subset\subset \Omega$. By the absolute continuity of the integral, for every $\epsilon > 0$ there exists a $\delta(\epsilon) > 0$ such that if $\Sigma \subset \tilde{\Omega}$ and $|\Sigma| < \delta$ we have[3] $\mathcal{F}(u, z, \Sigma) < \epsilon$.

From the preceding lemma and from the theorems of EGOROV and LUSIN we conclude that there exists a compact set $K \subset \tilde{\Omega}$ and a

[3]If $\mathcal{F}(u, z, \tilde{\Omega}) = +\infty$, $\mathcal{F}(u, z, \tilde{\Omega} - \Sigma) > \frac{1}{\epsilon}$.

number $R > 0$, such that $|\tilde{\Omega} - K| < \delta$ and

(α) $u_h, u \in C^0(K, M)$; $\sup_K |u| \leq R$, $\sup_K |u_h| \leq R$.
(β) $u_h \to u$ uniformly in K.
(γ) $F(x, u, z)$ is continuous in $K \times M \times \mathbf{R}^\nu$.

From Lemma 4.5 we get

$$\mathcal{F}(u, z, K) \leq \liminf_{h\to\infty} \mathcal{F}(u_h, z_h, K) \leq \lambda$$

and therefore[4]

$$\mathcal{F}(u, z, \tilde{\Omega}) < \lambda + \epsilon\,.$$

Since $\epsilon > 0$ is arbitrary, this implies $\mathcal{F}(u, z, \tilde{\Omega}) \leq \lambda$ for every $\tilde{\Omega} \subset\subset \Omega$, and hence the conclusion. □

We remark that assumption (i) of the theorem can be replaced by

(i′) $F(x, u, z)$ is measurable in x for every $(u, z) \in M \times \mathbf{R}^\nu$, and continuous in u for every $z \in \mathbf{R}^\nu$ and almost every $x \in \Omega$.

Actually it is not difficult to prove that if $g(u, z)$ is continuous in u and convex in z, then it is continuous in z uniformly with respect to u, and hence is continuous globally in (u, z).

Remark 4.1 Rereading the proofs of Theorems 4.4 and 4.5, and of Lemma 4.5, it is not difficult, keeping in mind Lemma 4.4, to realize that it is possible to substitute the condition $F(x, u, z) \geq 0$ with the more general assumption

$$F(x, u, z) \geq -c(|z| + |u| + g(x))$$

with $g \in L^1$. Moreover, if instead of the topology of $L^1_s \times L^1_w$ we use that of $L^k_s \times L^p_w$, with $k \geq 1$ and $p > 1$ (as we shall always do in what follows), it will be sufficient to assume

$$F(x, u, z) \geq -c(|z|^m + |u|^k + g)$$

with $g \in L^1$ and $m < p$. □

Example 4.1 Note that the above result does not hold for $m = p$, even in dimension one IOFFE [1]). To see that, consider the functional

---

[4] If $\mathcal{F}(u, z, \tilde{\Omega}) = +\infty$, $\lambda \geq \mathcal{F}(u, z, K) > \frac{1}{\epsilon}$ and hence $\lambda = +\infty$.

$$\mathcal{F}(u,z) = \int_0^1 \left(\left| \frac{u}{t} \right|^q \frac{1}{q} + \frac{u}{t} z \right) dt$$

$(\frac{1}{q} + \frac{1}{p} = 1)$, and the sequences

$$u_k = \begin{cases} tk^{\frac{1}{q}} & \text{if } 0 < t < k^{-1}\,, \\ 0 & \text{if } k^{-1} \leq t < 1\,, \end{cases}$$

$$z_k = \begin{cases} -k^{\frac{1}{p}} & \text{if } 0 < t < k^{-1}\,, \\ 0 & \text{if } k^{-1} \leq t < 1\,. \end{cases}$$

We have $F(t,u,z) \geq -\frac{|z|^p}{p}$. Moreover, $u_k \to 0$ uniformly, and $z_k \rightharpoonup 0$ in L^p. On the other hand, $\mathcal{F}(0,0) = 0$, whereas $\mathcal{F}(u_k, z_k) = -\frac{1}{p}$.

The semicontinuity theorem can be applied to functionals of type (4.1), where as usual Ω is an open set in $\mathbf{R}^n$, and u a mapping of Ω into a closed set $M \subset \mathbf{R}^N$. In this case $z = Du : \Omega \to \mathbf{R}^{nN}$ and recalling RELLICH's theorem we have the following:

Corollary 4.1 *Let F be as in the preceding theorem. The functional*

$$\mathcal{F}(u,\Omega) = \int_\Omega F(x,u,Du)\,dx$$

is LSC in the weak topology of $W^{1,1}_{\mathrm{loc}}(\Omega, M)$.

4.4 An Existence Theorem

Before proceeding further, we shall show how the results of this chapter, in particular Theorem 4.5 and its corollary, can be used to prove the existence of minima for the functional

$$\mathcal{F}(u) = \int_\Omega F(x,u,Du)\,dx\,.$$

By Theorem 4.1, it will be sufficient to show that there exists a minimizing sequence, convergent in the weak topology of $W^{1,1}_{\mathrm{loc}}$.

Generally speaking, the space $W^{1,1}_{\mathrm{loc}}$ is not the best for that purpose, since it is not reflexive. On the other hand, since a weakly convergent sequence in $W^{1,p}_{\mathrm{loc}}$ $(p > 1)$ is also weakly convergent in $W^{1,1}_{\mathrm{loc}}$, it will be sufficient to find a minimizing sequence which converges weakly in $W^{1,p}_{\mathrm{loc}}$ for some $p > 1$. The latter being a reflexive space, it will be enough to find a minimizing sequence, bounded in $W^{1,p}_{\mathrm{loc}}$.

The situation is very simple when *all* minimizing sequences are bounded. This happens for instance if the functional $\mathcal{F}$ is *coercive*, that is if

$$\lim_{\|u\|_{1,p}\to\infty} \mathcal{F}(u) = +\infty\,,$$

a condition that will be certainly satisfied if we assume that

$$F(x,u,z) \geq \nu(|u|^p + |z|^p) \tag{4.5}$$

with $\nu > 0$ and $p > 1$.

Until now, no mention as been made of boundary conditions, nor of other possible conditions imposed on the function $u(x)$ (except for the condition $u(x) \in M$ for almost every $x \in \Omega$, that we have already taken into account). In general, however, the problem consists of finding the minimum of the functional $\mathcal{F}$ among all the functions u satisfying suitable conditions, or in other words, that belong to a subset V of the space $W^{1,p}$. Such a subset V must be closed in the weak topology of $W^{1,p}$, since we want that the limit of a sequence of functions of V is itself in V.

We have the following:

Theorem 4.6 *Let Ω be an open set in $\mathbf{R}^n$, let M be a closed set in $\mathbf{R}^N$, and let $\mathcal{F}(u)$ be a lower semicontinuous functional in the weak topology of $W^{1,p}_{\mathrm{loc}}(\Omega, M)$, $p > 1$. Let V be a weakly closed subset of $W^{1,p}(\Omega, M)$, and assume that $\mathcal{F}$ is coercive in V:*

$$\lim_{\substack{\|u\|_{1,p}\to\infty \\ u\in V}} \mathcal{F}(u) = +\infty\,. \tag{4.6}$$

Then, $\mathcal{F}$ takes its minimum in V.

Of course, the coerciveness is guaranteed if the function F verifies (4.5); but it is possible, depending on the choice of V, that (4.6) holds under more general assumptions than (4.5).

The most usual condition consists in requesting that the function $u(x)$ assume given boundary values; in this case we must minimize the functional $\mathcal{F}$ among all the functions v taking prescribed values on $\partial\Omega$ (the DIRICHLET problem).

Formally, the DIRICHLET problem is posed by giving a function $U \in W^{1,p}(\Omega)$, and imposing the condition

$$u - U \in W^{1,p}_0(\Omega)\,.$$

In this case, the inequality

$$F(x,u,z) \geq |z|^p - b(x)|u|^\delta - a(x)\,, \tag{4.7}$$

with $\delta < p$, $a \in L^1(\Omega)$ and $b \in L^{\frac{p}{p-\delta}}$, is sufficient to guarantee the coerciveness when Ω has finite measure. In fact we have

$$|u|^\delta \le c(|U|^\delta + |u - U|^\delta)$$

and hence

$$b(x)|u|^\delta \le cb(x)|U|^\delta + \epsilon|u - U|^p + c(\epsilon)b^{\frac{p}{p-\delta}}\,.$$

On the other hand $u - U \in W_0^{1,p}(\Omega)$, and since Ω has finite measure:

$$\int_\Omega |u - U|^p\,dx \le c(\Omega)\int_\Omega |D(u - U)|^p\,dx\,. \tag{4.8}$$

But then

$$\begin{aligned}\mathcal{F}(u) &\ge \int_\Omega |Du|^p\,dx - \int_\Omega (b|u|^\delta + a)\,dx \\ &\ge \int_\Omega |Du|^p\,dx - \epsilon\,c(\Omega)\int_\Omega |D(u - U)|^p\,dx - c \\ &\ge \frac{1}{2}\int_\Omega |Du|^p\,dx - c\end{aligned}$$

provided we choose ϵ small enough.

We have in conclusion

$$\|u\|_{1,p}^p = \int_\Omega (|Du|^p + |u|^p)\,dx \le c\{\mathcal{F}(u) + 1\}$$

so that $\mathcal{F}$ is coercive.

We note that we cannot substitute $b(x)|u|^\delta$ in (4.7) with $A|u|^p$, even if A is a constant, unless A is small enough. For that, it will be sufficient to remark that the functional

$$\int_\Omega (|Du|^p - A|u|^p)\,dx$$

is not bounded below, unless $Ac(\Omega) < 1$, where $c(\Omega)$ is the best constant in the inequality (4.8). The constant $c(\Omega)$ depends on the geometry, rather than on the measure, of Ω; if $p = 2$ it coincides with the inverse of the first eigenvalue of the LAPLACE operator, that is, of the smallest constant λ_0 for which the equation

$$\Delta u + \lambda_0 u = 0$$

has a nonzero solution in $W_0^{1,2}(\Omega)$.

It is clear that if V is bounded, no supplementary condition on the function F is necessary. This happens for instance if we have conditions of the type

$$|Du(x)| \leq M \text{ a.e. in } \Omega$$

as it is sometimes the case in elasticity theory.

4.5 Notes and Comments

The role played by the convexity in the semicontinuity theorems was recognized since the beginnings of the direct methods in the calculus of variations, in particular in the works of TONELLI [2]. The main difficulty against a generalized use of direct methods was due essentially to the lack of suitable function spaces in which these minimum problems could be treated; a difficulty that was particularly sensitive in the case of multiple integrals.

In the case of a single independent variable, TONELLI recognized in the absolutely continuous functions, and even more so in those with bounded variation, two functional spaces that could be used profitably in the study of minimum problems, and showed that the use of lower semicontinuity instead of that of continuity could give to the classical WEIERSTRASS theorem the generality necessary for its application even to rather weak topologies.

The extension of TONELLI's methods to higher dimensions was attempted first in the direction of a generalization of these spaces to many dimensions. Real progress was achieved only with the introduction of SOBOLEV spaces, and later and in greater generality with the theory of *distributions*, in particular with the spaces BV of functions whose derivatives are measures, in which has been possible to approach with success the theory of minimal surfaces of codimension 1.

More recently, the *currents* have provided a natural ambient for treating geometrical problems, in the first place the problem of surfaces of least area (FEDERER [2], ALMGREN [1]), and have been used profitably in several problems, among which those arising from nonlinear elasticity (GIAQUINTA, MODICA and SOUČEK [2]).

The results of this chapter apply naturally to functionals of the form

$$\mathcal{F}(u, \Omega) = \int_\Omega F(x, u, Du)\, dx\,.$$

Taking into account RELLICH's theorem, if the function $F(x, u, z)$ is continuous in u and convex in z, the functional $\mathcal{F}$ is lower semicontinuous in the weak topology of $W^{1,1}_{\text{loc}}$. This result remains valid even if F is not continuous in u (AMBROSIO [1]).

In some cases this result can be ameliorated, and it holds for a topology weaker that that of $W^{1,1}$. In this context we quote the results of SERRIN [2], who proved that if F is continuous, non-negative and convex in z, and if one of the following conditions holds:

(i) F_x, F_z and F_{xz} are continuous,
(ii) F is strictly convex in z,
(iii)

$$\lim_{|z|\to+\infty} F(x,u,z) = +\infty$$

then for any $u_k, u \in W^{1,1}(\Omega)$, with $u_k \to u$ in $L^1(\Omega)$, one has

$$\mathcal{F}(u,\Omega) \leq \liminf_{k\to\infty} \mathcal{F}(u_k,\Omega)\,.$$

Condition (i) has been weakened in various ways by several authors. GORI and MARCELLINI [1] have shown that it can be replaced by

$$|F(x_1,u,z) - F(x_2,u,z)| \leq L|x_1 - x_2|$$

for every (x_1,u,z) and (x_2,u,z) in a compact set $K \subset \Omega \times \mathbf{R} \times \mathbf{R}^n$, with the constant L depending on K.

More recently, GORI and MARCELLINI [1] have proved that SERRIN's result holds if only one assumes, besides continuity, positivity, and convexity in z, that for every x and u the function $z \to F(x,u,z)$ is not constant on any straight line in $\mathbf{R}^n$. It is easily seen that the above condition is implied by either (ii) or (iii).

The same result, with F only continuous and convex in z, and without conditions at infinity, holds if F does not depend on x, even in the presence of discontinuities with respect to u (DE GIORGI, BUTTAZZO and DAL MASO [1]).

In all these theorems it is essential that u is a function with values in $\mathbf{R}$. If instead u takes values in $\mathbf{R}^N$, with $N > 1$, they do not hold any more, as EISEN [1] proved with an example.

Sometimes it happens that a functional $\mathcal{F} : V \to \mathbf{R}$ is not semicontinuous (or that V is not complete) in a given topology, otherwise particularly useful. In this case one can consider the *relaxed* functional $\bar{\mathcal{F}}$, defined in the completion $\bar{V}$ of V by the formula

$$\bar{\mathcal{F}}(u) = \inf \left\{ \liminf_{k\to\infty} \mathcal{F}(u_k);\ u_k \to u \right\}.$$

It is easily seen that $\mathcal{F}(u) \geq \bar{\mathcal{F}}(u)$, the equality holding if and only if $\mathcal{F}$ is lower semicontinuous in u. In this case, the relaxed functional $\bar{\mathcal{F}}$ is an extension of $\mathcal{F}$ to the space $\bar{V}$, and one has

$$\inf \mathcal{F} = \min \bar{\mathcal{F}}.$$

In general $\bar{\mathcal{F}}(u) < \mathcal{F}(u)$, and $\bar{\mathcal{F}}$ is the greatest lower semicontinuous functional which is less than or equal to $\mathcal{F}$.

A typical case is that of the area functional. Let V be the space of the functions of class $C^1(\bar{\Omega})$, taking given boundary values $u(x) = U(x)$ on $\partial\Omega$. The area $\mathcal{A}(u)$ of the graph of a function $u \in V$ is

$$\int_\Omega \sqrt{1+|Du|^2}\,dx = \sup\left\{ \int_\Omega [g_0 + uD_i g^i]\,dx;\ g_h \in C_0^1(\Omega), \sum_{h=0}^n g_h^2 \leq 1 \right\}. \qquad (4.9)$$

It is easily seen that the right-hand side is lower semicontinuous in the strong topology of L^1: if $u_k, u \in V$ and $u_k \to u$ strongly in L^1, then

$$\mathcal{A}(u) \leq \liminf_{k\to\infty} \mathcal{A}(u_k).$$

The completion of V in the topology of $L^1(\Omega)$ coincides with $L^1(\Omega)$; the relaxed functional is therefore defined in L^1, and has meaning also for functions that do not assume the given value U on the boundary. For these functions we have

$$\bar{\mathcal{A}}(u) = \int_\Omega \sqrt{1+|Du|^2}\,dx + \int_{\partial\Omega} |u-U|\,dH_{n-1},$$

where the first term on the right-hand side must be interpreted according to (4.9).[5]

[5] Of course, for many functions $v \in L^1$ we will have $\bar{\mathcal{A}}(v) = +\infty$. If we want to avoid that unpleasant feature, we should restrict ourselves to the functions for which the right-hand side of (4.9) is finite. These functions belong to the space $BV(\Omega)$ of the functions whose derivatives are measures in Ω; the right-hand side of (4.9) is then the total variation of the vector measure $(\mathcal{L}^n, Dv)$, where $\mathcal{L}^n$ is the LEBESGUE measure in $\mathbf{R}^n$.

In general the relaxed functional cannot be written as an integral; the cases in which that happens are of some interest, and have been studied by BUTTAZZO [1].

A generalization of the method of relaxation leads to the theory of Γ-convergence. Here, one considers a sequence of functionals $\mathcal{F}_k$, and a "limit" functional $\mathcal{F}$, and asks under which conditions the minima of the functionals $\mathcal{F}_k$ approximate the minima of $\mathcal{F}$.

This problem conducts to the definition of Γ-convergence. Limiting ourselves to the simpler case of sequential convergence, we shall say that the sequence $\mathcal{F}_k$ Γ-converges to $\mathcal{F}$ in a given topology, if
(i) for every sequence $u_k \to u$ one has $\mathcal{F}(u) \le \liminf_{k\to\infty} \mathcal{F}_k(u_k)$,
and
(ii) there exists a sequence $u_k \to u$ such that $\mathcal{F}(u) = \lim_{k\to\infty} \mathcal{F}_k(u_k)$.

It is easily seen that if u_k minimizes $\mathcal{F}_k$, and if $u_k \to u$, then u minimizes $\mathcal{F}$. Actually, if $v_k \to v$ and $\mathcal{F}_k(v_k) \to \mathcal{F}(v)$, since $\mathcal{F}_k(u_k) \le \mathcal{F}_k(v_k)$, we have

$$\mathcal{F}(u) \le \liminf_{k\to\infty} \mathcal{F}_k(u_k) \le \liminf_{k\to\infty} \mathcal{F}_k(v_k) = \mathcal{F}(v) .$$

Moreover, if $w_k \to u$ and $\mathcal{F}_k(w_k) \to \mathcal{F}(u)$, we have

$$\mathcal{F}(u) = \lim_{k\to\infty} \mathcal{F}_k(w_k) \ge \liminf_{k\to\infty} \mathcal{F}_k(u_k) \ge \mathcal{F}(u)$$

and since the same relation holds for any subsequence of u_k, we conclude that

$$\mathcal{F}_k(u_k) \to \mathcal{F}(u) .$$

The Γ-convergence is a very weak notion of convergence, and therefore it is possible to approximate particular functionals with others of a very different type. For instance (see MODICA and MORTOLA [1]), the sequence of functionals

$$\mathcal{F}_k(u) = \frac{3}{4}\int_\Omega \left(\frac{|Du|^2}{k} + k(1-u^2) \right) dx .$$

Γ-converges to the functional

$$\begin{cases} \displaystyle\int_\Omega |Du|\, dx , & \text{if } u \in BV(\Omega) \text{ and } |u| = 1 \text{ a.e.}, \\ +\infty & \text{otherwise}. \end{cases}$$

The literature on the Γ-convergence, and on the related problem of homogenization, is rather large, and we shall refer the reader to the works of DAL MASO [1] and DAL MASO and MODICA [1].

Finally, we mention that the study of general functionals

$$\mathcal{F}(u, z, \Omega) = \int_\Omega F(x, u(x), z(x))\, dx\,,$$

and in particular the semicontinuity Theorem 4.5, is useful in the theory of controls. In this case $z(x)$ is the control, $u(x)$ is the state of the system, and the functional $\mathcal{F}$ represents the total cost that must be minimized. Referring to the specialized literature (see for instance LIONS [2]) for a complete discussion, the following example can give an idea of the method used.

Suppose that we have a system governed by the differential equation

$$\begin{cases} \Delta u = z & \text{in } \Omega\,, \\ u = 0 & \text{on } \partial\Omega\,, \end{cases}$$

so that we can control the state u of the system by means of the control function z. Suppose that our goal is a state $U(x)$, and that the cost of the control z is given by the integral $\int_\Omega |z|^2\, dx$, whereas the price paid for not reaching the state U is $\int_\Omega g(u - U)\, dx$, with $g \geq 0$. The total cost will be then

$$\mathcal{F}(u, z) = \int_\Omega (g(u - U) + |z|^2)\, dx\,.$$

If we assume that g is continuous, the functional $\mathcal{F}$ is lower semicontinuous for the weak convergence in z and strong convergence in u in $L^2(\Omega)$.

Suppose now that z_k be a minimizing sequence of controls, and let u_k be the corresponding sequence of states. From the theory of elliptic equations (see Chapter 10) we deduce that $u_k \in W^{2,2}(\Omega)$, and that

$$\|u_k\|_{2,2} \leq c\|z_k\|_2\,.$$

Since z_k is a bounded sequence in L^2, u_k is bounded in $W^{2,2}$, and therefore, passing possibly to a subsequence, we can suppose that $u_k \to u$ and $z_k \rightharpoonup z$ in L^2. By Theorem 4.5, $z(x)$ is the optimal control.

Chapter 5

Quasi-Convex Functionals

5.1 Necessary Conditions

In the preceding chapter we have seen the central role played by the convexity in the proof of the semicontinuity, and hence in the theorems of existence of minima of functionals of the calculus of variations.

We shall begin this chapter by showing that this assumption is necessary for the lower semicontinuity of functionals of the type

$$\mathcal{F}(u, z) = \int_\Omega F(x, u(x), z(x))\, dx\,. \tag{5.1}$$

Theorem 5.1 *Let Ω be an open set of $\mathbf{R}^n$, and let $F(x, u, z)$ be a* CARATHEODORY *function in $\Omega \times \mathbf{R}^N \times \mathbf{R}^\nu$, with $F(\cdot, u, z)$ locally summable in Ω for every (u, z). Assume that for every $u \in \mathbf{R}^N$ the functional*

$$\mathcal{F}_u(z) = \int_\Omega F(x, u, z(x))\, dx$$

is lower semicontinuous in the weak topology of L^∞_{loc} of z.*

Then, for almost every $x \in \Omega$ and every $u \in \mathbf{R}^N$ the function $F(x, u, \cdot)$ is convex.

We note that the convergence in the weak* topology of L^∞_{loc} implies convergence in the weak topology of L^1_{loc}. Therefore, the above result, together with Theorem 4.5 of the preceding chapter, gives a necessary and sufficient condition for the lower semicontinuity.

Proof. Let $u \in \mathbf{R}^n$, $x_0 \in \Omega$ and let $Q \subset \Omega$ be a cube with center in x_0. For every $\lambda \in [0,1]$ there exists a sequence χ_h of characteristic functions of measurable sets $E_h \subset Q$ such that $\chi_h \overset{*}{\rightharpoonup} \lambda\chi_Q$.[1] Let a and b be two points of $\mathbf{R}^\nu$ and let $u \in \mathbf{R}^N$. We set

$$z_h = a\chi_h + b(1-\chi_h)\,; \quad z = a\lambda + b(1-\lambda)$$

in Q, and $z_h = z = 0$ in $\Omega - Q$. We have

$$\mathcal{F}_u(z_h) - \mathcal{F}_u(z) = \int_Q [F(x,u,a\chi_h + b(1-\chi_h)) - F(x,u,z)]\,dx \tag{5.2}$$

and since χ_h are characteristic functions:

$$\int_Q F(x,u,a\chi_h + b(1-\chi_h))\,dx$$
$$= \int_Q \chi_h F(x,u,a)\,dx + \int_Q (1-\chi_h)F(x,u,b)\,dx\,.$$

Introducing this relation in (5.2) and passing to the limit for $h \to \infty$ we obtain, taking into account the assumption of semicontinuity:

$$0 \le \liminf_{h\to\infty} \mathcal{F}_u(z_h) - \mathcal{F}_u(z)$$
$$= \lambda \int_Q F(x,u,a)\,dx + (1-\lambda)\int_Q F(x,u,b)\,dx - \int_Q F(x,u,z)\,dx\,.$$

Dividing by $|Q|$ and letting the side of Q go to zero, we get for almost every $x_0 \in \Omega$:

$$\lambda F(x_0,u,a) + (1-\lambda)F(x_0,u,b) \ge F(x_0,u,\lambda a + (1-\lambda)b) \tag{5.3}$$

that is the convexity of $F(x,u,\cdot)$.

The proof is not yet complete, since the set of zero measure for which (5.3) does not hold may depend on λ, u, a and b. But we can find countable dense sets $J \subset [0,1]$, $E \subset \mathbf{R}^N$, $A \subset \mathbf{R}^\nu$ and a set $Z \subset \Omega$ of zero measure such that (5.3) holds for every $\lambda \in J$, $u \in E$, $a, b \in A$ and for every $x \in \Omega - Z$. Moreover, we can suppose that $F(x,\cdot,\cdot)$ is continuous for every $x \in \Omega - Z$.

Every $\lambda \in [0,1]$, $u \in \mathbf{R}^N$ and $a, b \in \mathbf{R}^\nu$ is the limit of sequences $\lambda_h \in J$, $u_h \in E$ and $a_h, b_h \in A$. Writing (5.3) for these, and passing to the limit for $h \to \infty$ we obtain the required result. □

[1] We denote with χ_Q the characteristic function of Q.

In the theorem just proved the functions $u(x)$ ($\equiv u$) and $z(x)$ were completely independent, so that it does not apply to the situation we are most interesteding, namely that of functionals of the type

$$\mathcal{F}(u,\Omega)=\int_\Omega F(x,u,Du)\,dx\,, \tag{5.4}$$

that is when $z = Du$. And, in fact, the functions z_h and z involved in the proof of the theorem were not the gradients of the corresponding u_h and u, as it should be if we want to find necessary conditions for the semicontinuity of the functional (5.4).

Of course, this restricts considerably the choice of the possible sequences $z_h \rightharpoonup z$, so that the convexity in z, which we have shown to be necessary in the general case, might not be so essential in the new situation.

In this case, if we continue to assume that the functions $z_h = Du_h$ converge in the weak* topology of L^∞, so that $\|Du_h\|_\infty \le M$, it will not be restrictive to assume that the sequence u_h converges uniformly, since any other weaker convergence would reduce to that by the theorem of ASCOLI-ARZELÀ.

In conclusion, we may assume that the functional $\mathcal{F}$ is LSC with respect to convergence in the weak* topology of $W^{1,\infty}(\Omega) = \mathrm{Lip}\,(\Omega)$, or else, which is essentially equivalent, with respect to uniform convergence of bounded sequences of $\mathrm{Lip}\,(\Omega)$ (L-convergence):

Definition 5.1 *A sequence $u_h \in \mathrm{Lip}\,(\Omega)$ is said to be* L-convergent *to a function u if $u_h \to u$ uniformly in Ω and if there exists a constant M such that for every integer h*

$$[u_h]_{0,1}=\sup_\Omega|Du_h(x)|\le M\,. \tag{5.5}$$

It is clear that in this case the limit function u belongs to $\mathrm{Lip}\,(\Omega)$, and that its gradient is bounded by the same constant M.

Condition (5.5) tells us that the functions u_h are equi-Lipschitz-continuous in Ω. If Ω is bounded, it implies the weak convergence in $W^{1,s}(\Omega)$ for every $s \ge 1$.

This being established, we shall consider first the special case in which the function F depends only on z.

Theorem 5.2 *Let Ω be a bounded open set of $\mathbf{R}^n$, let $F(z)$ be a continuous function in $\mathbf{R}^{nN}$, and assume that the functional*

$$\mathcal{F}(u)=\int_\Omega F(Du(x))\,dx$$

is lower semicontinuous with respect to L-convergence. Then, for every function $\varphi \in C_0^1(\Omega)$ *and for every* $z_0 \in \mathbf{R}^{nN}$ *we have*

$$F(z_0)|\Omega| \leq \int_\Omega F(z_0 + D\varphi(x))\, dx\,. \tag{5.6}$$

Proof. Let Q be a cube containing Ω, and that modulo a homothety we can suppose to be the unit cube $[0,1]^n$. We can extend the function φ first setting it to be equal to zero in $Q - \Omega$, and then extending it periodically in $\mathbf{R}^n$:

$$\varphi(x_1, \ldots, x_n) = \varphi(\{x_1\}, \ldots, \{x_n\})\,,$$

where with $\{a\}$ we denote the *fractionary part* of a.

For any integer h we set

$$\varphi_h(x) = \frac{1}{h}\varphi(hx)\,,$$

$$u_h(x) = \langle z_0, x\rangle + \varphi_h(x)\,.$$

It is clear that $u_h \xrightarrow{L} \langle z_0, x\rangle$, and hence by the semicontinuity assumption:

$$|Q|F(z_0) \leq \liminf_{h\to\infty} \int_Q F(z_0 + D\varphi_h)\, dx\,.$$

Remarking that $D\varphi_h(x) = D\varphi(hx)$, we get, after a change of variables, $y = hx$:

$$|Q|F(z_0) \leq \liminf_{h\to\infty} h^{-n} \int_{hQ} F(z_0 + D\varphi(y))\, dy\,,$$

where $hQ = [0,h]^n$.

On the other hand φ has period one, and hence also $F(z_0 + D\varphi)$ has the same period; it follows that

$$\int_{hQ} F(z_0 + D\varphi(y))\, dy = h^n \int_Q F(z_0 + D\varphi(y))\, dy$$

and hence

$$|Q|F(z_0) \leq \int_Q F(z_0 + D\varphi(y))\, dy\,.$$

□

We can now deal with the general case.

Theorem 5.3 *Let Ω be a bounded open set of $\mathbf{R}^n$, let $F(x,u,z)$ be a continuous function in $\Omega \times \mathbf{R}^N \times \mathbf{R}^{nN}$, and assume that the functional*

$$\mathcal{F}(u,\Omega) = \int_\Omega F(x,u(x),Du(x))\,dx$$

is lower semicontinuous with respect to L-convergence. Then, for every function $\varphi \in C_0^1(\Omega)$ and for every $x_0 \in \Omega$, $u_0 \in \mathbf{R}^N$ and $z_0 \in \mathbf{R}^{nN}$ we have

$$F(x_0,u_0,z_0)|\Omega| \le \int_\Omega F(x_0,u_0,z_0 + D\varphi(x))\,dx\,. \tag{5.7}$$

Proof. As above, we can suppose that Ω be contained in the unit cube Q, and that φ is extended periodically in $\mathbf{R}^n$. Let $x_0 \in \Omega$, and let W be a cube contained in Ω, with center x_0 and side t. We define

$$\begin{aligned} u(x) &= u_0 + \langle z_0, x - x_0\rangle\,,\\ \varphi_h(x) &= \frac{t}{h}\varphi\left(\frac{h(x-x_0)}{t}\right),\\ u_h(x) &= u(x) + \tilde{\varphi}_h(x)\,, \end{aligned}$$

where

$$\tilde{\varphi}_h(x) = \begin{cases} \varphi_h(x) & \text{if } x \in W\,,\\ 0 & \text{if } x \in \Omega - W\,. \end{cases}$$

Since $\varphi_h(x) = 0$ in a neighborhood of ∂W, the function $\tilde{\varphi}_h$ belongs to $C_0^1(W)$.

To get an estimate for $\mathcal{F}(u_h, W)$, we decompose W into h^n equal cubes W_i of centers x_i. We have

$$\begin{aligned} \mathcal{F}(u_h,W) &= \int_W F(x, u+\varphi_h, z_0 + D\varphi_h)\,dx\\ &= \sum_{i=1}^{h^n}\int_{W_i} F\left(x_i, u(x_i), z_0 + D\varphi\left(\frac{h(x-x_0)}{t}\right)\right)dx\\ &\quad + \sum_{i=1}^{h^n}\int_{W_i}[F(x,u+\varphi_h,z_0+D\varphi_h)\\ &\quad - F(x_i,u(x_i),z_0+D\varphi_h)]\,dx\\ &= A_h + B_h\,. \end{aligned}$$

For what concerns the term B_h we remark that the derivatives of φ_h are equibounded, and therefore the arguments of the function F remain confined in a compact set $\Theta \subset \Omega \times \mathbf{R}^N \times \mathbf{R}^{nN}$. The function F is then uniformly continuous in Θ, and hence $B_h \to 0$ when $h \to \infty$.

On the other hand $h(x_i - x_0)/t$ is a vector with integral components, and therefore by the periodicity of φ we have

$$A_h = \sum_{i=1}^{h^n} \int_{W_i} F\left(x_i, u(x_i), z_0 + D\varphi\left(\frac{h(x - x_i)}{t}\right)\right) dx\,,$$

which, after a change of variables $y = h(x - x_i)/t$, becomes

$$A_h = \sum_{i=1}^{h^n} \left(\frac{t}{h}\right)^n \int_Q F(x_i, u(x_i), z_0 + D\varphi(y))\, dy\,.$$

We write for simplicity

$$g(x) = \int_Q F(x, u(x), z_0 + D\varphi(y))\, dy\,.$$

The function g is continuous, and we have

$$A_h = \sum_{i=1}^{h^n} g(x_i)|W_i|$$

and hence

$$\lim_{h\to\infty} A_h = \int_W dx \int_Q F(x, u(x), z_0 + D\varphi(y))\, dy\,.$$

On the other hand the sequence u_h L-converges to u, so that the lower semicontinuity of $\mathcal{F}$ gives

$$\lim_{h\to\infty} A_h = \lim_{h\to\infty} \mathcal{F}(u_h, W) \geq \mathcal{F}(u, W) = \int_W F(x, u(x), z_0)\, dx$$

and the conclusion is obtained dividing by $|W|$ and letting the side t of W go to zero. □

As we shall see, condition (5.7) will play a major role not only in the semicontinuity theory, but also in the regularity of minima. We are then induced to formulate the following definition.

Definition 5.2 *The functional*

$$\mathcal{F}(u, \Omega) = \int_\Omega F(x, u, Du)\, dx$$

is called quasi-convex *if for every* $(x_0, u_0, z_0) \in \Omega \times \mathbf{R}^N \times \mathbf{R}^{nN}$ *and for every* $\varphi \in C_0^1(\Omega, \mathbf{R}^N)$ *we have*

$$F(x_0, u_0, z_0) \leq \fint_\Omega F(x_0, u_0, z_0 + D\varphi(x))\, dx\,. \tag{5.8}$$

Geometrically, quasi-convexityindexquasi-convexity means that the linear functions

$$u(x) = a + \langle z_0, x\rangle$$

minimize the "frozen" functional

$$\mathcal{F}^0(u, \Omega) = \int_\Omega F(x_0, u_0, Du(x))\, dx\,.$$

Remark 5.1 We remark that if (5.8) holds for an open set Ω, it holds also for every open set $\Lambda \subset \mathbf{R}^n$. Actually, we can choose t and $\bar{x}$ in such a way that

$$\Lambda_1 = \{x \in \mathbf{R}^n : tx + \bar{x} \in \Lambda\} \subset \Omega\,.$$

If $\operatorname{supp}(\eta) \subset \Lambda$, the function $\varphi(x) = t^{-1}\eta(y) =: t^{-1}\eta(tx + \bar{x})$ belongs to $C_0^1(\Omega)$, and hence

$$\begin{aligned}\fint_\Lambda F(x_0, u_0, z_0 + D\eta(y))\, dy &= \fint_{\Lambda_1} F(x_0, u_0, z_0 + D\varphi(x))\, dx \\ &= \frac{1}{|\Lambda_1|}\Big\{ \int_\Omega F(x_0, u_0, z_0 + D\varphi(x))\, dx \\ &\quad - F(x_0, u_0, z_0)|\Omega - \Lambda_1|\Big\} \\ &\geq F(x_0, u_0, z_0)\,.\end{aligned}$$

As a consequence we can speak of quasi-convex functions: a function F being *quasi-convex* if (5.8) holds in some open set $\Omega \subset \mathbf{R}^n$. □

It is easily seen that quasi-convexity is strictly weaker than convexity. In fact, if $F(x, u, z)$ is a continuous function, convex in z, there holds

$$F(x_0, u_0, z_0 + D\varphi) \geq F(x_0, u_0, z_0) + \langle \lambda_0, D\varphi\rangle$$

for some $\lambda_0 \in \mathbf{R}^{nN}$. Integrating over Ω,

$$\int_\Omega F(x_0, u_0, z_0 + D\varphi)\, dx \geq F(x_0, u_0, z_0)|\Omega|$$

since

$$\int_\Omega \langle \lambda_0, D\varphi \rangle \, dx = 0 .$$

It follows that any convex function F is quasi-convex.

The converse is not true, since the function $F(z) = \det(z)$ $(N = n)$ is quasi-convex but it is not convex.

To see that, let $g(x)$ be a function of class $C^2(\mathbf{R}^n, \mathbf{R}^n)$, and let ω be the differential form

$$\det(Dg) dx_1 \wedge dx_2 \wedge \cdots \wedge dx_n .$$

We have

$$\omega = dg_1 \wedge dg_2 \wedge \cdots \wedge dg_n = d(g_1 \wedge dg_2 \wedge \cdots \wedge dg_n) \tag{5.9}$$

since $dd = 0$, and therefore

$$\int_\Omega \det(Dg) \, dx_1 \wedge dx_2 \wedge \cdots \wedge dx_n = \int_{\partial\Omega} g_1 \wedge dg_2 \wedge \cdots \wedge dg_n .$$

In particular, the integral on the left-hand side is zero whenever one among the functions $g_1, \ldots, g_n$ is zero on $\partial\Omega$. Moreover, if two functions g and h coincide in a neighborhood of $\partial\Omega$, we have

$$\int_\Omega \det(Dg) \, dx = \int_\Omega \det(Dh) \, dx .$$

The same result (without assuming $N = n$) holds if we substitute if the determinant any minor of the matrix Dg.

In particular, if φ is a function of class $C_0^2(\Omega)$, the functions $u(x) = \langle z_0, x \rangle$ and $u + \varphi$ have the some value in a neighborhood of $\partial\Omega$, and hence

$$\int_\Omega \det(Du + D\varphi) \, dx = \int_\Omega \det(z_0 + D\varphi) \, dx = |\Omega| \det(z_0) .$$

By approximation with C^2 functions we can easily see that the above relation holds for an arbitrary function $\varphi \in W_0^{1,n}(\Omega)$, so that the function $\det(z)$ is quasi-convex.[2]

More generally, let z be a $n \times m$ matrix, and let $M(z)$ be the vector whose components are the minors of the matrix z. If $g(M)$ is a convex

[2] Actually, the value of the functional $\mathcal{F}(u) = \int \det(Du) \, dx$ depends only on the boundary values of u, and therefore it is the same for functions having the same values on $\partial\Omega$. Consequently, the EULER equation for $\mathcal{F}$ is identically satisfied. Functionals of this kind are called *null Lagrangian*.

function of M, the function

$$F(z) = g(M(z))$$

is called *polyconvex*. We have

Proposition 5.1 *A polyconvex function is quasi-convex.*

Proof. If $g(w)$ is a convex function and $w(x)$ a summable function in Ω, we have the JENSEN inequality

$$\fint_{\Omega} g(w(x))\,dx \geq g\left(\fint_{\Omega} w(x)\,dx\right), \tag{5.10}$$

Setting $u = \langle z_0, x\rangle$, we have

$$\begin{aligned}\fint_{\Omega} g(M(z_0 + D\varphi))\,dx &\geq g\left(\fint_{\Omega} M(z_0 + D\varphi)\,dx\right)\\ &= g\left(\fint_{\Omega} M(z_0)\,dx\right) = g(M(z_0))\end{aligned}$$

from which the result follows at once. □

A last and weakest form of convexity is the so-called *rank-one convexity.*[3]

Definition 5.3 *A function $F(x, u, z)$ defined in $\Omega \times \mathbf{R}^N \times \mathbf{R}^{nN}$ is* rank-one convex *if for every $(x_0, u_0, z_0) \in \Omega \times \mathbf{R}^N \times \mathbf{R}^{nN}$ the function*

$$g(\xi, \eta) = F(x_0, u_0, z_0 + \xi \otimes \eta)$$

is separately convex in $\xi \in \mathbf{R}^n$ and $\eta \in \mathbf{R}^N$.

Proposition 5.2 *A continuous quasi-convex function is rank-one convex.*

Proof. We can suppose that F depends only on z. Assume first that F is of class C^2. Since F is quasi-convex, the affine functions minimize the functional $\mathcal{F}$, and therefore if $u(x) = z_0\, x$ is an affine function and if

[3]In his book [3], modifying a terminology that he himself had previously introduced in [2], MORREY calls quasi-convexity the rank-one convexity, and strong quasi-convexity what, according to current usage, we have called quasi-convexity.

$\varphi \in C_0^1(\Omega, \mathbf{R}^N)$, the function

$$G(t) = \mathcal{F}(u + t\varphi)$$

has a minimum for $t = 0$. We have therefore $G'(0) = 0$ and $G''(0) \geq 0$, and hence

$$\int_\Omega \frac{\partial^2 F}{\partial z_k^\alpha \partial z_j^\beta}(z_0) D_k \varphi^\alpha D_j \varphi^\beta \, dx \geq 0 \tag{5.11}$$

for every $\varphi \in C_0^1(\Omega, \mathbf{R}^N)$.

Setting $\varphi = \lambda + i\mu$, writing (5.11) for λ and μ and summing, we get

$$\operatorname{Re} \int_\Omega \frac{\partial^2 F}{\partial z_k^\alpha \partial z_j^\beta}(z_0) D_k \varphi^\alpha D_j \bar{\varphi}^\beta \, dx \geq 0\,. \tag{5.12}$$

We choose now $\varphi = \eta e^{i\tau\langle \xi, x\rangle} \gamma(x)$, where $\xi \in \mathbf{R}^n$, $\eta \in \mathbf{R}^N$, and γ is a function in $C_0^\infty(\Omega, \mathbf{R})$. From (5.12) we obtain

$$\int_\Omega \frac{\partial^2 F}{\partial z_k^\alpha \partial z_j^\beta}(z_0) \eta^\alpha \eta^\beta [\tau^2 \xi_k \xi_j \gamma^2 + D_k \gamma D_j \gamma] \, dx \geq 0\,.$$

Dividing by τ^2 and letting τ go to infinity we get

$$\int_\Omega \frac{\partial^2 F}{\partial z_k^\alpha \partial z_j^\beta}(z_0) \xi_k \xi_j \eta^\alpha \eta^\beta \gamma^2 \, dx \geq 0$$

for every $\gamma \in C_0^\infty(\Omega)$, and in conclusion

$$\frac{\partial^2 F}{\partial z_k^\alpha \partial z_j^\beta}(z_0) \xi_k \xi_j \eta^\alpha \eta^\beta \geq 0\,. \tag{5.13}$$

The last inequality, that carries the name of the LEGENDRE–HADAMARD *condition*, implies the separate convexity in ξ and η.

Suppose now that only F is continuous, and let

$$F_\epsilon(z) = \int F(z - w) \varphi_\epsilon(w) \, dw$$

be the mollified function of F. If ϑ is a function in $C_0^1(\Omega)$, we have

$$\begin{aligned} \int_\Omega F_\epsilon(z_0 + D\vartheta) \, dx &= \int \varphi_\epsilon(w) \int_\Omega F(z_0 + D\vartheta - w) \, dx \\ &\geq \int \varphi_\epsilon(w) F(z_0 - w) |\Omega| \, dw = |\Omega| F_\epsilon(z_0) \end{aligned}$$

and hence F_ϵ is quasi-convex.

It follows from the above that F_ϵ is rank-one convex, and passing to the limit as $\epsilon \to 0$ the same is true for F. □

We have then a series of conditions of increasing generality, since convexity implies policonvexity (we can consider the single components of the gradient as rank-one minors), the latter in turn implies quasi-convexity, which finally implies rank-one convexity.

Generally speaking, the opposite implications do not hold,[4] except in special situations.

A first case, in which all the definitions above are equivalent, is when u is a *scalar* function, that is when the codomain has dimension $N = 1$. In fact in this case the LEGENDRE–HADAMARD condition reduces to

$$\frac{\partial^2 F}{\partial z_i \partial z_j}(z_0)\xi_i\xi_j \geq 0$$

and is equivalent to the convexity of the function F.[5]

A second case of some importance, in which rank-one convexity implies quasi-convexity, is when the function $F(x, u, z)$ is quadratic in z:

$$F(x, u, z) = A^{ij}_{\alpha\beta}(x, u) z^\alpha_i z^\beta_j \,.$$

When F has the above form, setting for simplicity $A = A(x_0, u_0)$, we have

$$\int_\Omega F(x_0, u_0, z_0 + D\varphi)\, dx = \int_\Omega [\langle Az_0, z_0\rangle + \langle AD\varphi, D\varphi\rangle]\, dx$$

since the remaining integrals are zero because φ has compact support. To prove the quasi-convexity it will therefore suffice to show that

$$\int_\Omega \langle AD\varphi, D\varphi\rangle\, dx \geq 0$$

for every $\varphi \in C^\infty_0(\Omega)$.

More generally, assume that instead of (5.13) we have

$$A^{ij}_{\alpha\beta}\xi_i\xi_j\eta^\alpha\eta^\beta \geq \nu|\xi|^2|\eta|^2 \tag{5.14}$$

with $\nu \geq 0$.

[4]For a more detailed discussion we refer to the final section of this chapter.

[5]If F is not regular, we can proceed as above, approximating with regular functions. We note that the same conclusion also holds when the domain has dimension $n = 1$.

In this case we can prove the following:

Lemma 5.1 *Let A be a constant matrix, satisfying* (5.14). *For every $\zeta \in W_0^{1,2}$ we have*

$$\int A_{\alpha\beta}^{ij} D_j \zeta^\beta D_i \zeta^\alpha \, dx \geq \nu \int |D\zeta|^2 \, dx \, .$$

Proof. Denoting by $\hat{f}(\xi)$ the FOURIER transform of a function f:

$$\hat{f}(\xi) = \frac{1}{(2\pi i)^n} \int f(x) e^{i\langle x, \xi \rangle} \, dx \, ,$$

we have $\widehat{D_h f}(\xi) = i\xi_h \hat{f}(\xi)$, and

$$\int D_j \zeta^\beta D_i \zeta^\alpha \, dx = \int D_j \zeta^\beta \overline{D_i \zeta^\alpha} \, dx = \int \xi_i \xi_j \widehat{\zeta^\alpha} \overline{\widehat{\zeta^\beta}} \, d\xi \, .$$

On the other hand, if in (5.14) we allow the vector η to take complex values: $\eta^\alpha = \varrho^\alpha + i\pi^\alpha$, we get

$$A_{\alpha\beta}^{ij} \xi_i \xi_j \eta^\alpha \overline{\eta^\beta} = A_{\alpha\beta}^{ij} \xi_i \xi_j [\varrho^\alpha \varrho^\beta + \pi^\alpha \pi^\beta + i(\varrho^\alpha \pi^\beta - \varrho^\beta \pi^\alpha)]$$

and hence

$$\operatorname{Re} A_{\alpha\beta}^{ij} \xi_i \xi_j \eta^\alpha \overline{\eta^\beta} \geq \nu |\xi|^2 |\eta|^2 \, .$$

As a consequence, recalling PARSEVAL's identity:

$$\begin{aligned} \int A_{\alpha\beta}^{ij} D_j \zeta^\beta D_i \zeta^\alpha \, dx &= \operatorname{Re} \int A_{\alpha\beta}^{ij} \xi_i \xi_j \widehat{\zeta^\alpha} \overline{\widehat{\zeta^\beta}} \, d\xi \\ &\geq \nu \int |\xi|^2 |\hat{\zeta}|^2 \, d\xi = \nu \int |D\zeta|^2 \, dx \end{aligned}$$

and the proof is concluded. □

5.2 First Semicontinuity Results

The results of the preceding section force us to ask whether the quasi-convexity is also sufficient for the lower semicontinuity in a topology weak enough to guarantee the existence of minima. Results in this direction are rather recent, and have been proved under increasingly less restrictive conditions.

In this section we shall consider the relatively simpler case in which the integrand F depends only on z; the general case will be treated later.

To begin, we need the following lemma:

Lemma 5.2 *Let $F(z)$ be a rank-one convex function, such that*

$$|F(z)| \leq c(\lambda + |z|)^p, \quad \lambda \geq 0. \tag{5.15}$$

with $p \geq 1$. Then,

$$|F(z) - F(w)| \leq c(\lambda + |z| + |w|)^{p-1}|z - w|. \tag{5.16}$$

Moreover, if F has first derivatives, we have

$$|F_z| \leq c(\lambda + |z|)^{p-1}. \tag{5.17}$$

Proof. A matrix π having only one element different from zero is obviously of rank 1, and hence if $F(\zeta)$ is rank-one convex, the function $g(t) = F(\zeta + t\pi)$ is convex, and hence

$$G(t) = \frac{g(t) - g(0)}{t}$$

is increasing.

Consider now the matrix z, of dimension $n \times N$, as being a vector in $\mathbf{R}^{nN}$ with components z_k; let $w \in \mathbf{R}^{nN}$, and for $k = 0, 1, 2, \ldots, nN$ define

$$z^{(k)} = (w_1, \ldots, w_k, z_{k+1}, \ldots, z_{nN}).$$

We have $z^{(0)} = z$ and $z^{(nN)} = w$; moreover, taking $\zeta = z^{(k)}$, $\pi = z^{(k+1)} - z^{(k)}$ and $t = \frac{\lambda+|z|+|w|}{|z-w|} > 1$ we obtain

$$\begin{aligned} F(z^{(k+1)}) - F(z^{(k)}) &= G(1) \leq G(t) \\ &= \frac{F(z^{(k)} + t(z^{(k+1)} - z^{(k)})) - F(z^{(k)})}{\lambda + |z| + |w|}|z - w|. \end{aligned}$$

Using (5.15) and remarking that $|z^{(k)}| \leq |z| + |w|$ and $|z^{(k)} + t(z^{(k+1)} - z^{(k)})| \leq c(\lambda + |z| + |w|)$, we get

$$F(z^{(k+1)}) - F(z^{(k)}) \leq c(\lambda + |z| + |w|)^{p-1}|z - w|,$$

from which, summing over k and then exchanging z with w, we deduce immediately (5.16).

Now let v be a vector in $\mathbf{R}^{nN}$, with $|v| = 1$. From (5.16) we get

$$\left| \frac{F(z + tv) - F(z)}{t} \right| \leq c(\lambda + |z + tv| + |z|)^{p-1},$$

and (5.17) follows immediately letting t go to zero. □

We can prove now the first semicontinuity theorem.

Theorem 5.4 (MARCELLINI [1]) *Let $F(z)$ be a quasi-convex function, such that*

$$0 \leq F(z) \leq c(1 + |z|^p)$$

with $p \geq 1$. Then, the functional

$$\mathcal{F}(u, \Omega) = \int_\Omega F(Du)\, dx$$

is lower semicontinuous in the weak topology of $W^{1,p}_{\text{loc}}(\Omega, \mathbf{R}^N)$.

Proof. Let $u_h \rightharpoonup u$ in $W^{1,p}(\Omega)$, and assume first that u is an affine function, so that $Du = z_0 =$ constant. If we had $u_h = u$ on $\partial\Omega$ (that is $u_h = u + \varphi_h$ with $\varphi_h \in W^{1,p}_0(\Omega)$), then from the assumption of quasi-convexity we would get

$$\int_\Omega F(Du_h)\, dx \geq F(z_0)|\Omega| = \int_\Omega F(Du)\, dx$$

and the semicontinuity would follow immediately. The problem therefore consists of modifying suitably the functions u_h near $\partial\Omega$, in such a way that they take the value u on $\partial\Omega$ without changing excessively the value of the functional. For that, let $\Omega_0 \subset\subset \Omega$ and let L be an integer. We set $R = \frac{1}{2}\text{dist}\,(\Omega_0, \partial\Omega)$, and for $i = 1, 2, \ldots, L$ we define

$$\Omega_i = \left\{x \in \Omega : \text{dist}\,(x, \Omega_0) < \frac{i}{L}R\right\}.$$

Now choose functions $\psi_i \in C^1_0(\Omega_i)$ such that

$$0 \leq \psi_i \leq 1\,; \quad \psi_i = 1 \quad \text{in } \Omega_{i-1}\,; \quad |D\psi_i| \leq \frac{2(L+1)}{R}$$

and set

$$v_{ih} = u + \psi_i(u_h - u)\,.$$

The functions $v_{ih} - u$ belong to $W^{1,p}_0(\Omega)$. From what we have said above it follows that

$$\int_\Omega F(Dv_{ih})\, dx \geq F(z_0)|\Omega| = \int_\Omega F(Du)\, dx\,,$$

and hence

$$\int_\Omega F(Du)\,dx \le \int_{\Omega-\Omega_i} F(Du)\,dx + \int_{\Omega_i-\Omega_{i-1}} F(Dv_{ih})\,dx + \int_{\Omega_{i-1}} F(Du_h)\,dx$$
$$\le \int_{\Omega-\Omega_0} F(Du)\,dx + \int_{\Omega_i-\Omega_{i-1}} F(Dv_{ih})\,dx + \int_\Omega F(Du_h)\,dx\,. \tag{5.18}$$

On the other hand

$$Dv_{ih} = (1-\psi_i)Du + \psi_i Du_h + (u_h - u)D\psi_i\,,$$

and therefore

$$|Dv_{ih}|^p \le c\left(|Du|^p + |Du_h|^p + \frac{(L+1)^p}{R^p}|u_h - u|^p\right).$$

We can use this inequality to estimate the penultimate integral in (5.18). Moreover, if we sum over i, and divide by L, we easily get

$$\int_{\Omega_0} F(Du)\,dx \le \int_\Omega F(Du_h)\,dx$$
$$+\frac{c}{L}\int_\Omega \left(1 + |Du|^p + |Du_h|^p + \frac{(L+1)^p}{R^p}|u_h - u|^p\right)dx\,. \tag{5.19}$$

Since $u_h \rightharpoonup u$ in $W^{1,p}$, by RELLICH's theorem we deduce that $u_h \to u$ in L^p, and the derivatives Du_h are bounded in L^p. If we let h go to infinity in (5.19), the last term tends to zero, and therefore:

$$\int_{\Omega_0} F(Du)\,dx \le \liminf_{h\to\infty} \int_\Omega F(Du_h)\,dx + \frac{c}{L}\,.$$

Passing to the limit as $L \to \infty$ and letting Ω_0 tend to Ω, we get

$$\int_\Omega F(Du)\,dx \le \liminf_{h\to\infty} \int_\Omega F(Du_h)\,dx \tag{5.20}$$

for every affine function $u(x)$.

Now let u be a generic function in $W^{1,p}$. Consider a countable family of pairwise disjoint open cubes $Q_i \subset \Omega$, such that $|\Omega - \cup Q_i| = 0$. In any of these cubes, we set

$$z_i = \fint_{Q_i} Du\;dx$$

and let $\bar{Z}(x)$ be the function that in every cube Q_i takes the constant value z_i.

When the maximum diameter of the cubes Q_i tends to zero, the function $\bar{Z}$ tends to Du in $L^p(\Omega)$; it follows that for every $\epsilon > 0$ there exists a family Q_i such that

$$\int_\Omega |Du - \bar{Z}|^p \, dx = \sum_{i=1}^{\infty} \int_{Q_i} |Du - z_i|^p \, dx < \epsilon^p .$$

Let now u_h be a sequence converging weakly to u in $W^{1,p}(\Omega)$. For $x \in Q_i$, let $v_h^{(i)}(x) = u_h(x) - u(x) + \langle z_i, x \rangle$. When h tends to infinity, the sequence $v_h^{(i)}$ tends to $\langle z_i, x \rangle$ in the weak topology of $W^{1,p}(Q_i)$, and therefore, according to what we have just proved, we have

$$\liminf_{h \to \infty} \int_{Q_i} F(Dv_h^{(i)}) \, dx \geq \int_{Q_i} F(z_i) \, dx$$

and the same inequality holds if we sum over i.

On the other hand, using the preceding lemma and the HÖLDER inequality, and summing over i, we obtain

$$\begin{aligned} &\left| \int_\Omega F(Du_h) \, dx - \sum_i \int_{Q_i} F(Dv_h^{(i)}) \, dx \right| \\ &\leq c \sum_i \int_{Q_i} \left(1 + |Du_h|^{p-1} + |Dv_h^{(i)}|^{p-1}\right) |Du - z_i| \, dx \\ &\leq c \left\{ \sum_i \int_{Q_i} (1 + |Du_h|^p + |Dv_h^{(i)}|^p) \, dx \right\}^{\frac{p-1}{p}} \left\{ \sum_i \int_{Q_i} |Du - z_i|^p \, dx \right\}^{\frac{1}{p}} \\ &\leq c\epsilon . \end{aligned}$$

In a similar way we prove that

$$\left| \int_\Omega F(Du) \, dx - \sum_i \int_{Q_i} F(z_i) \, dx \right| < c\epsilon$$

and in conclusion:

$$\int_\Omega F(Du) \, dx \leq \liminf_{h \to \infty} \int_\Omega F(Du_h) \, dx + c\epsilon ,$$

from which we get the semicontinuity under weak convergence in $W^{1,p}(\Omega)$.

Finally, if $u_h \rightharpoonup u$ in $W^{1,p}_{\text{loc}}(\Omega)$, we have from any $\Sigma \subset\subset \Omega$:

$$\mathcal{F}(u, \Sigma) \leq \liminf_{h\to\infty} \mathcal{F}(u_h, \Sigma) \leq \liminf_{h\to\infty} \mathcal{F}(u_h, \Omega)$$

and the theorem follows letting Σ tend to Ω. □

Remark 5.2 The conclusion of the theorem remains valid if we only assume that $-c(1+|z|^r) \leq F(z) \leq c(1+|z|^p)$ with $r < p$. In this case we must add to the right-hand side of (5.18) the quantity

$$-\int_{\Omega_i - \Omega_0} F(Du)\,dx - \int_{\Omega - \Omega_{i-1}} F(Du_h)\,dx$$

$$\leq c \int_{\Omega - \Omega_0} (1 + |Du| + |Du_h|)^r\,dx\,.$$

The last integral can be estimated by

$$|\Omega - \Omega_0|^{\frac{p-r}{p}} \left(\int_\Omega (1 + |Du| + |Du_h|)^p\,dx \right)^{\frac{r}{p}} \leq c|\Omega - \Omega_0|^{\frac{p-r}{p}}\,,$$

that tends to zero when $\Omega_0 \to \Omega$.

Everything else remains unchanged. □

The preceding result continues to hold when the function F depends also on x and u. In this case the proof is considerably more complex, and we shall need some preliminary results, many of which are interesting in themselves, independently of their application to the semicontinuity.

5.3 The Quasi-Convex Envelope

Definition 5.4 Let $G(z) \geq 0$ be a function defined in $\mathbf{R}^{nN}$, and let Ω be an open set in $\mathbf{R}^n$. We set

$$\gamma_\Omega(z) = \inf \left\{ \fint_\Omega G(z + D\varphi(x))\,dx;\ \varphi \in C_0^\infty(\Omega) \right\}. \tag{5.21}$$

It is easily seen that γ_Ω is invariant under homothety. Actually, if Ω_1 is an open set, homothetic to $\Omega : \Omega_1 = x_0 + \lambda\Omega$, and if $\varphi_1 \in C_0^\infty(\Omega_1)$, setting $\varphi(x) = \lambda^{-1}\varphi_1(x_0 + \lambda x)$, we have $\varphi \in C_0^\infty(\Omega)$, and

$$\fint_\Omega G(z + D\varphi(y))\,dy = \fint_{\Omega_1} G(z + D\varphi_1(x))\,dx\,,$$

from which immediately follows $\gamma_\Omega = \gamma_{\Omega_1}$.

We shall actually prove that the function γ does not depends on Ω. For that, we need the following:

Lemma 5.3 *Let A and Ω be two open sets in $\mathbf{R}^n$, with $|\partial\Omega| = 0$. For every $\epsilon > 0$ there exists a finite number of open sets Ω_i, $i = 1, 2, \ldots, N$, homothetic to Ω and pairwise disjoint, all contained in A and such that*

$$|A - \cup\Omega_i| < \epsilon\,.$$

Proof. Let Q be the unit cube, and let Ω_0 be an open set homothetic to Ω and contained in Q. Let $2\vartheta = |\Omega_0|$. The open set A is the countable union of cubes, so that there exists a finite family of pairwise disjoint cubes contained in A, whose union has measure not less than $\frac{1}{2}|A|$. In any of these cubes Q_i we can put an open set Ω_i homothetic to Ω, with $|\Omega_i| = 2\vartheta|Q_i|$. The union Z_1 of these open sets has measure not less than $\vartheta|A|$, and therefore the open set $A - \overline{Z_1}$ has measure not greater than $(1-\vartheta)|A|$.

We can repeat the above argument with $A - \overline{Z_1}$ instead of A. We will find a finite family of open sets homothetic to Ω, whose union Z_2 is such that $|A - \overline{Z_1} - \overline{Z_2}| \le (1-\vartheta)^2|A|$. Continuing in this way, we obtain k finite families of open sets homothetic to Ω and pairwise disjoint, with

$$|A - \overline{Z_1} - \overline{Z_2} - \cdots - \overline{Z_k}| \le (1-\vartheta)^k|A|\,.$$

The conclusion follows at once taking k in such a way that $(1-\vartheta)^k < \epsilon$, and recalling that $\partial\Omega$, and therefore ∂Z_i, has zero measure for every i. □

Proposition 5.3 *If A and Ω are two bounded open sets in $\mathbf{R}^n$, we have $\gamma_A = \gamma_\Omega$.*

Proof. Suppose first that $|\partial\Omega| = 0$, and let us begin by showing that if $\{\Omega_i\}$ is a finite family of pairwise disjoint open sets, all homothetic to Ω, and if $\Sigma = \cup\Omega_i$, we have $\gamma_\Sigma = \gamma_\Omega$.

Let $\varphi \in C_0^\infty(\Omega)$ be such that

$$\gamma_\Omega(z) \ge \fint_\Omega G(z + D\varphi)\,dx - \epsilon\,. \tag{5.22}$$

If φ_i are the corresponding functions in $C_0^\infty(\Omega_i)$, and if we set $\psi = \sum \varphi_i$, we have $\psi \in C_0^\infty(\Sigma)$, and

$$\fint_\Sigma G(z + D\psi)\,dx = \frac{1}{|\Sigma|}\sum_i \int_{\Omega_i} G(z + D\varphi_i)\,dx \le \frac{1}{|\Sigma|}\sum_i |\Omega_i|(\gamma_\Omega(z) + \epsilon)\,,$$

from which $\gamma_\Sigma \le \gamma_\Omega$ follows at once.

On the other hand, since the sets Ω_i are disjoint, if $\psi \in C_0^\infty(\Sigma)$, its restriction ψ_i to Ω_i belongs to $C_0^\infty(\Omega_i)$. Consequently, if ψ is such that

$$\gamma_\Sigma(z) \geq \fint_\Sigma G(z + D\psi)\, dx - \epsilon\,,$$

we have

$$\begin{aligned}\gamma_\Sigma(z) &\geq \frac{1}{|\Sigma|}\sum_i \int_{\Omega_i} G(z + D\psi_i)\, dx - \epsilon \geq \frac{1}{|\Sigma|}\sum_i \gamma_\Omega(z)|\Omega_i| - \epsilon \\ &= \gamma_\Omega(z) - \epsilon\end{aligned}$$

so that $\gamma_\Sigma \geq \gamma_\Omega$.

Let now A be a second open set in $\mathbf{R}^n$, and choose the Ω_i as in Lemma 5.3. If φ satisfies (5.22), let φ_i be the corresponding functions in Ω_i, and let $\psi = \sum_i \varphi_i$. We have $\psi \in C_0^\infty(A)$, and hence

$$\gamma_A(z) \leq \fint_A G(z + D\psi)\, dx \leq \fint_\Sigma G(z + D\psi)\, dx + G(z)|A - \Sigma|\,.$$

It follows immediately that $\gamma_A(z) \leq \gamma_\Omega(z) + \epsilon + G(z)\epsilon$, so that, if $|\partial\Omega| = 0$, we have

$$\gamma_A \leq \gamma_\Omega\,.$$

Assume now that $|\partial\Omega| > 0$. Let $\varphi \in C_0^\infty(\Omega)$ be a function satisfying (5.22), and let $\Lambda \subset \Omega$ be an open set with boundary of zero measure, containing the support of φ and such that $|\Omega - \Lambda| < \epsilon < \frac{1}{2}|\Omega|$. We have

$$\gamma_\Lambda(z) \leq \fint_\Lambda G(z + D\varphi)\, dx \leq \frac{|\Omega|}{|\Lambda|}(\gamma_\Omega(z) + \epsilon) < (\gamma_\Omega(z) + \epsilon)\left(1 + \frac{\epsilon}{|\Lambda|}\right).$$

On the other hand, since $|\partial\Lambda| = 0$, we have $\gamma_A(z) \leq \gamma_\Lambda(z)$, and therefore

$$\gamma_A(z) < (\gamma_\Omega(z) + \epsilon)\left(1 + \frac{2\epsilon}{|\Omega|}\right)$$

so that $\gamma_A \leq \gamma_\Omega$ in this case too.

Exchanging the roles of A and Ω we get the opposite inequality, and the conclusion of the proposition. □

In particular, we can write simply $\gamma(z)$ instead of $\gamma_A(z)$.

Remark 5.3 If G is a continuous function, satisfying

$$0 \leq G(z) \leq c(1 + |z|^p)\,, \quad p \geq 1\,, \tag{5.23}$$

the functional

$$\mathcal{G}(u) = \int_\Omega G(z + Du)\,dx$$

is continuous in the strong topology of $W^{1,p}(\Omega)$.[6]

As a consequence, we have

$$\gamma(z) = \inf\left\{ \fint_\Omega G(z + D\varphi(x))\,dx;\ \varphi \in W_0^{1,p}(\Omega)\right\}.$$

In fact in our hypotheses the infimum in $W_0^{1,p}$ coincides with the infimum taken on any dense set, such as for instance C_0^∞ or else the set of piecewise affine functions[7] with compact support in Ω. □

We shall now prove that under suitable assumptions for G, the function γ is quasi-convex. We shall assume that

$$\nu|z|^p \le G(z) \le c(1 + |z|^p) \tag{5.24}$$

with $\nu > 0$, and moreover that G is continuous in z; more precisely, that there exists a continuous function $\omega(t)$, with $\omega(0) = 0$ and such that

$$|G(z) - G(w)| \le (1 + |z|^p + |w|^p)\omega(|z - w|)\,. \tag{5.25}$$

Lemma 5.4 *With the assumptions* (5.24) *and* (5.25), *the function* $\gamma(z)$ *is continuous in* $\mathbf{R}^{nN}$.

Proof. By (5.24) we can assume that the functions φ in (5.21) satisfy $\|D\varphi\|_p \le K$. We then have

$$\left|\fint_\Omega G(z + D\varphi)\,dx - \fint_\Omega G(w + D\varphi)\,dx\right|$$
$$\le c\omega(|z - w|)(1 + |z|^p + |w|^p + K^p)$$

and the conclusion follows at once. □

Remark 5.4 The preceding lemma continues to hold if G (and hence γ) depends continuously on a parameter u, and satisfies (5.24) with ν and c

---

[6] It will be sufficient to consider a sequence u_k converging to u strongly in $W^{1,p}$ and such that $u_k \to u$ and $Du_k \to Du$ almost everywhere, and to apply FATOU's lemma to the sequences $\mathcal{G}(u_k)$ and $c\int_\Omega (1 + |Du_k|^p)\,dx - \mathcal{G}(u_k)$.

[7] We recall that a function w is piecewise affine in Ω if it is continuous and if there exists a finite number of open sets $A_1, \ldots, A_n$, with $\cup \overline{A_i} = \overline{\Omega}$ and such that the restriction of w to any one of them is an affine function.

independent of u, and with (5.25) replaced by

$$|G(u,z) - G(v,w)| \leq (1 + |z|^p + |w|^p)\omega(|u-v| + |z-w|) . \qquad (5.26)$$

Of course, in this case $\gamma(u,z)$ will be continuous in the pair (u,z). □

Theorem 5.5 (DACOROGNA [1]) *Always assuming* (5.24) *and* (5.25), *the function* $\gamma(z)$ *is the* quasi-convex envelope *of* G, *that is the greatest quasi-convex function less than or equal to* G.

Proof. Let us begin by showing that γ is quasi-convex. Let Q be a cube in $\mathbf{R}^n$, and let ψ be a piecewise affine function with compact support in Q. Let $A_1, \ldots, A_k$ be the open sets such that $D\psi = w_i =$ constant in A_i. We have

$$\int_Q \gamma(z + D\psi)\, dx = \sum_{i=1}^{k} |A_i|\gamma(z + w_i).$$

For $\epsilon > 0$, let $\varphi_i \in C_0^\infty(A_i)$ be a function such that

$$\gamma(z + w_i) \geq \fint_{A_i} G(z + w_i + D\varphi_i)\, dx - 2\epsilon ,$$

and define $\eta = \psi + \sum \varphi_i$.

We then have

$$\sum_{i=1}^{k} |A_i|\gamma(z + w_i) \geq \int_Q G(z + D\eta)\, dx - \epsilon|Q| \geq (\gamma(z) - \epsilon)|Q| ,$$

and therefore

$$\int_Q \gamma(z + D\psi)\, dx \geq |Q|\gamma(z) .$$

Since the piecewise affine functions are dense in $W_0^{1,p}$, the preceding inequality holds for every $\psi \in W_0^{1,p}(Q)$.

Let now Ω be a bounded open set of $\mathbf{R}^n$, and let Q be a cube containing Ω. We have for every $\psi \in W_0^{1,p}(\Omega)$:

$$\int_\Omega \gamma(z + D\psi)\, dx = \int_Q \gamma(z + D\psi)\, dx - \gamma(z)|Q - \Omega| \geq |\Omega|\gamma(z) .$$

It is evident that $\gamma(z) \leq G(z)$. Now let $F(z) \leq G(z)$ be a quasi-convex function. For $\varphi \in C_0^\infty(\Omega)$ we have

$$F(z) \leq \fint_\Omega F(z + D\varphi)\, dx \leq \fint_\Omega G(z + D\varphi)\, dx$$

from which follows immediately that $F(z) \leq \gamma(z)$. □

5.4 The Ekeland Variational Principle

In this section we shall prove a result, known as the EKELAND *variational principle* [1–3], that we shall use quite often later.

Theorem 5.6 *Let (V, d) be a complete metric space, and let $\mathcal{F} : V \to \bar{\mathbf{R}}$ be a lower semicontinuous function (in the metric topology), bounded from below and taking a finite value at some point.*

Assume that for some $u \in V$ and some $\epsilon > 0$ we have

$$\mathcal{F}(u) \leq \inf_V \mathcal{F} + \epsilon\,.$$

Then, there exists a point $v \in V$ such that

(i) $d(u, v) \leq 1$;
(ii) $\mathcal{F}(v) \leq \mathcal{F}(u)$;
(iii) $\mathcal{F}(v) \leq \mathcal{F}(w) + \epsilon d(v, w) \quad \forall\, w \in V$.

Proof. Let us define by induction a sequence $u_k \in V$ in the following way. Set first $u_1 = u$. Suppose now that we have defined $u_1, u_2, \ldots, u_k$. The set

$$S_k = \{w \in V : \mathcal{F}(w) \leq \mathcal{F}(u_k) - \epsilon d(u_k, w)\}$$

is non-empty, since it contains u_k. There exists therefore a point $u_{k+1} \in S_k$ such that

$$\mathcal{F}(u_{k+1}) \leq \frac{1}{2}\left\{\mathcal{F}(u_k) + \inf_{S_k} \mathcal{F}\right\}. \tag{5.27}$$

We will show that u_k is a CAUCHY sequence. Since $u_{k+1} \in S_k$, we have

$$\epsilon d(u_{k+1}, u_k) \leq \mathcal{F}(u_k) - \mathcal{F}(u_{k+1}) \tag{5.28}$$

and hence

$$\epsilon d(u_{k+m}, u_k) \leq \epsilon \sum_{i=1}^{m} d(u_{k+i}, u_{k+i-1}) \leq \mathcal{F}(u_k) - \mathcal{F}(u_{k+m})\,. \tag{5.29}$$

On the other hand (5.28) implies that the sequence $\mathcal{F}(u_k)$ decreases, and therefore, $\mathcal{F}$ being bounded below in V, it will converge to some real number α. By inequality (5.29), u_k is then a CAUCHY sequence.

Let $v = \lim_{k\to\infty} u_k$. From the semicontinuity of $\mathcal{F}$ it follows that

$$\mathcal{F}(v) \leq \liminf_{m\to\infty} \mathcal{F}(u_{k+m}) = \alpha\,,$$

and therefore letting $m \to \infty$ in (5.29),

$$\epsilon d(u_k, v) \le \mathcal{F}(u_k) - \mathcal{F}(v). \tag{5.30}$$

In particular, taking $k = 1$, we have

$$0 \le \epsilon d(u, v) \le \mathcal{F}(u) - \mathcal{F}(v) \le \mathcal{F}(u) - \inf_V \mathcal{F} \le \epsilon$$

so that (i) and (ii) are satisfied.

Suppose now that (iii) does not hold. In this case there would exist $w \in V$ such that

$$\mathcal{F}(w) < \mathcal{F}(v) - \epsilon d(w, v). \tag{5.31}$$

Taking into account (5.30), we have in this case

$$\mathcal{F}(w) < \mathcal{F}(u_k) - \epsilon d(u_k, w)$$

for every k. From the definition of S_k, it would follow that $w \in S_k$ for every k, and hence

$$\inf_{S_k} \mathcal{F} \le \mathcal{F}(w).$$

On the other hand, we get from (5.27),

$$2\mathcal{F}(u_{k+1}) - \mathcal{F}(u_k) \le \mathcal{F}(w) < \mathcal{F}(v) - \epsilon d(v, w)$$

and passing to the limit as $k \to \infty$:

$$\mathcal{F}(v) \le \mathcal{F}(w) < \mathcal{F}(v) - \epsilon d(v, w).$$

But this cannot hold, and hence (iii) is proved. □

Remark 5.5 If we introduce in V the new distance $d_1 = \epsilon^{-\frac{1}{2}} d$, the topology of V remains the same. In particular (V, d_1) is a complete metric space, and $\mathcal{F}$ is lower semicontinuous. From the preceding theorem it follows that if $\mathcal{F}(u) \le \inf_V \mathcal{F} + \epsilon$, there exists $v \in V$ (of course different from that of the preceding theorem) such that

(i′) $d(u, v) \le \epsilon^{\frac{1}{2}}$;
(ii′) $\mathcal{F}(v) \le \mathcal{F}(u)$;
(iii′) $\mathcal{F}(v) \le \mathcal{F}(w) + \epsilon^{\frac{1}{2}} d(v, w) \quad \forall\, w \in V$.

In particular, if u_k is a minimizing sequence, that is if $\epsilon_k = \mathcal{F}(u_k) - \inf \mathcal{F}$ tends to zero, the corresponding sequence v_k is itself a minimizing sequence. Moreover, we have

$$\mathcal{F}(v_k) \le \mathcal{F}(w) + \epsilon_k^{\frac{1}{2}} d(v_k, w) \tag{5.32}$$

for every $w \in V$. □

5.5 Semicontinuity

We can now prove the main result of this chapter. For that, we shall consider a CARATHEODORY function $F(x, u, z)$, defined in $\Omega \times \mathbf{R}^N \times \mathbf{R}^{nN}$, and satisfying the conditions:

$$-a(|z|^r + |u|^t) - h(x) \le F(x, u, z) \le g(x, u)(1 + |z|^p) \tag{5.33}$$

with $p > 1$, $1 \le r < p$, $1 \le t < p^* = \frac{pn}{n-p}$ ($t \ge 1$ if $p \ge n$). As for the functions h and g, we shall assume that $h \in L^1(\Omega)$ and that $g \ge 0$ is a CARATHEODORY function in $\Omega \times \mathbf{R}^N$.

Under these assumptions we shall prove the following semicontinuity result:

Theorem 5.7 (ACERBI and FUSCO [AF1]) *Let $F(x, u, z)$ be a quasi-convex function, satisfying the conditions* (5.33). *Then, the functional*

$$\mathcal{F}(u) = \int_\Omega F(x, u(x), Du(x))\, dx$$

is lower semicontinuous in the weak topology of $W^{1,p}_{\mathrm{loc}}(\Omega, \mathbf{R}^N)$.

The proof of the theorem will be made in a series of steps. In the first place, we shall prove the theorem when the lower bound in (5.33) is replaced by

$$F(x, u, z) \ge \nu |z|^p, \quad \nu > 0. \tag{5.34}$$

For $i \in \mathbf{N}$ we call $\vartheta_i(t)$ a continuous function taking the values one for $t \le i - 1$ and zero for $t \ge i$, and we set

$$\eta_i(x, u) = \begin{cases} \vartheta_i(|u|) & \text{if } g(x, u) \le i\,, \\ \dfrac{i\vartheta_i(|u|)}{g(x, u)} & \text{if } g(x, u) > i\,. \end{cases}$$

Moreover, we set

$$F_i(x, u, z) = \eta_i(x, u) F(x, u, z) + (1 - \eta_i(x, u))\nu |z|^p\,.$$

The functions F_i form an increasing sequence of quasi-convex CARATHEODORY functions, and satisfy

$$F_i(x,u,z) \geq \nu|z|^p\,; \quad F_i(x,u,z) = \nu|z|^p\,; \quad \text{if } |u| \geq i\,;$$

$$F_i(x,u,z) \leq (i+\nu)(1+|z|^p)\,;$$

$$F_i(x,u,z) = F(x,u,z) \text{ if } i > g(x,u) + |u| + 1\,;$$

$$\lim_{i\to\infty} F_i(x,u,z) = \sup_i F_i(x,u,z) = F(x,u,z)\,.$$

Denoting with $\mathcal{F}_i(u)$ the corresponding functional, we have $\mathcal{F}(u) = \sup_i \mathcal{F}_i(u)$, and therefore, since the supremum of a family of lower semicontinuous functionals is l.s.c., it will suffice to prove the theorem for each of the functionals $\mathcal{F}_i(u)$. In other words, we can assume that

$$\nu|z|^p \leq F(x,u,z) \leq A(1+|z|^p) \tag{5.35}$$

and moreover that $F(x,u,z) \equiv \nu|z|^p$ for $|u| \geq \mu$.

We want now to perform a similar operation with respect to z. We set

$$G_i(x,u,z) = \vartheta_i(|z|)F(x,u,z) + (1-\vartheta_i(|z|))\nu|z|^p\,.$$

The functions G_i satisfy (5.35), and coincide with $\nu|z|^p$ for $|z| > i$ (and of course for $|u| > \mu$). On the other hand, in the above operation we lose the quasi-convexity, so that it will be necessary to replace the functions G_i with other quasi-convex functions. Recalling the results of Sec. 5.3, the natural candidates are the quasi-convex envelopes

$$g_i(x,u,z) = \inf\left\{ \fint_\Omega G_i(x,u,z+D\varphi(y))\,dy,\ \varphi \in C_0^\infty(\Omega)\right\}.$$

Actually the functions G_i satisfy all the assumptions stated in Sec. 5.3; in particular they are continuous in (u,z) for almost every $x \in \Omega$, and verify (5.26). In fact, for almost every $x \in \Omega$, G_i are uniformly continuous in the compact set $|u| \leq \mu+1$, $|z| \leq i+1$, whereas if either $|u| \geq \mu$ or $|z| \geq i$ they coincide with $\nu|z|^p$. We can therefore apply the Lemma 5.16, from which it follows that the functions g_i are continuous in (u,z) for almost every $x \in \Omega$. Moreover, we have obviously $g_i \leq G_i$, and from the inequality $\nu|z|^p \leq G_i(x,u,z)$ (the function $\nu|z|^p$ are convex, and hence quasi-convex), it follows that $g_i(x,u,z) \geq \nu|z|^p$. In particular, we have $g_i(x,u,z) = \nu|z|^p$ whenever either $|u| \geq \mu$ or $|z| \geq i$. Finally, like G_i, the sequence g_i is increasing.

Lemma 5.5 *Let $p > 1$. For $i \to \infty$ the sequence $g_i(x, u, z)$ converges to $F(x, u, z)$.*

Proof. For every fixed i, let $w_i \in C_0^\infty(\Omega)$ be a function such that

$$g_i(x, u, z) > \fint_\Omega G_i(x, u, z + Dw_i(y))\, dy - \frac{1}{i}.$$

Let us consider now the metric space $W_0^{1,1}(\Omega)$, with the distance

$$d(v, w) = \int_\Omega |Dv - Dw|\, dx.$$

Applying EKELAND's theorem (Theorem 5.6) to the functional

$$\Theta(w) = \fint_\Omega G_i(x, u, z + Dw(x))\, dx$$

we obtain a sequence v_i, with

$$\fint_\Omega G_i(x, u, z + Dv_i(y))\, dy < g_i(x, u, z) + \frac{1}{i}$$

and such that every v_i minimizes in $W_0^{1,1}(\Omega)$ the functional

$$\Gamma_i(\varphi) = \fint_\Omega G_i(x, u, z + D\varphi(y))\, dy + \frac{1}{i} \int_\Omega |D\varphi(y) - Dv_i(y)|\, dy.$$

Denoting by Σ the support of $v_i - \varphi$, we get

$$\begin{aligned}\int_\Omega |D\varphi(y) - Dv_i(y)|\, dy &\leq \int_\Sigma (|z + Dv_i| + |z + D\varphi|)\, dy \\ &\leq \int_\Sigma (c + \epsilon|z + Dv_i|^p + |z + D\varphi|^p)\, dy\end{aligned}$$

and therefore, taking ϵ small enough:

$$\int_\Sigma (1 + |Dv_i(y)|)^p\, dy \leq Q \int_\Sigma (1 + |D\varphi(y)|)^p\, dy.$$

In conclusion, the function v_i is a Q-minimum of the functional

$$\int_\Omega (1 + |Dv|)^p\, dx$$

(see Definition 6.1), and hence the conclusion of Theorem 6.7 holds. Since the sequence v_i is obviously equibounded in $W^{1,p}(\Omega)$, by the above theorem it will be bounded in $W^{1,p+\tau}_{\text{loc}}(\Omega)$, for some $\tau > 0$, independent of i.

Now let $\Omega_0 \subset\subset \Omega$, and let

$$\Omega_i = \{y \in \Omega_0 : |z + Dv_i(y)| \geq i - 1\}.$$

Since

$$(i-1)^p |\Omega_i| \leq \int_{\Omega_0} |z + Dv_i|^p \, dx \leq c,$$

we have $|\Omega_i| \to 0$, and therefore the sequence

$$\int_{\Omega_i} |Dv_i|^p \, dy \leq |\Omega_i|^{\frac{\tau}{p+\tau}} \left(\int_{\Omega_0} |Dv_i|^{p+\tau} \, dy \right)^{\frac{p}{p+\tau}}$$

will converge to zero.

We have therefore

$$\begin{aligned} g_i + \frac{1}{i} &> \frac{1}{|\Omega|} \int_{\Omega_0} G_i(x, u, z + Dv_i(y)) \, dy \\ &\geq \frac{1}{|\Omega|} \int_{\Omega_0 - \Omega_i} F(x, u, z + Dv_i(y)) \, dy \\ &\geq \frac{1}{|\Omega|} \int_{\Omega_0} F(x, u, z + Dv_i(y)) \, dy \\ &\quad - \frac{A}{|\Omega|} \int_{\Omega_i} (1 + |z + Dv_i|^p) \, dy. \end{aligned}$$

When $i \to \infty$, a subsequence of v_i will converge weakly in $W^{1,p}(\Omega)$ to a function v, whereas the last integral on the right-hand side will tend to zero. We have therefore, taking into account Theorem 5.4:

$$\begin{aligned} \lim_{i \to \infty} g_i(x, u, z) &\geq \fint_{\Omega} F(x, u, z + Dv(y)) \, dy \\ &\quad - \frac{A}{|\Omega|} \int_{\Omega - \Omega_0} (1|z + Dv|^p) \, dy \\ &\geq F(x, u, z) - \frac{A}{|\Omega|} \int_{\Omega - \Omega_0} (1 + |z + Dv|^p) \, dy \end{aligned}$$

and letting Ω_0 tend to Ω:

$$\lim_{i \to \infty} g_i(x, u, z) \geq F(x, u, z).$$

The opposite inequality is trivial, since $g_i \leq G_i \leq F$. □

Recalling again that the supremum of a family of semicontinuous functionals is a semicontinuous functional itself, we can conclude from the lemma just proved that it is sufficient to prove the semicontinuity of the functionals

$$\mathcal{G}_i(u) = \int_\Omega g_i(x, u, Du)\, dx\,;$$

in other words we can suppose that the function F satisfies

$$\nu |z|^p \leq F(x, u, z) \leq A(1 + |z|^p)$$

and moreover

$$F(x, u, z) \equiv \nu |z|^p \text{ for } |u| \geq \mu \text{ and } |z| \geq \mu\,.$$

Since $F(x, u, z)$ is a CARATHEODORY function, we can apply the lemma of SCORZA DRAGONI (Lemma 4.6) and conclude that for every $\epsilon > 0$ there exists a compact set $K \subset \Omega$, with $|\Omega - K| < \epsilon$, such that F is continuous in $K \times \mathbf{R}^N \times \mathbf{R}^{nN}$.

Lemma 5.6 *There exists a bounded continuous function $\omega(t)$, with $\omega(0) = 0$ and such that*

$$|F(x, u, z) - F(y, v, z)| \leq \omega(|x - y| + |u - v|)$$

for every $x, y \in K$, $u, v \in \mathbf{R}^N$ and for every $z \in \mathbf{R}^{nN}$.

Proof. It follows immediately from the continuity of F and from the fact that $F \equiv \nu |z|^p$ if either $|u| \geq \mu$ or $|z| \geq \mu$. □

We can now conclude the proof of Theorem 5.7.

Consider a sequence u_j weakly convergent in $W^{1,p}$ to a function u.

As before, let $\mathcal{R}$ be a countable family of pairwise disjoint cubes Q_i, such that $|\Omega - \cup_i Q_i| = 0$. Let $\bar{x}_i$ be the center of the cube Q_i, and let

$$\bar{u}_i = \fint_{Q_i} u\, dx\,.$$

Moreover, let $\bar{X}$ and $\bar{U}$ be the functions taking, respectively, the values $\bar{x}_i$ and $\bar{u}_i$ in the cube Q_i.

Starting from a given partition $\mathcal{R}$, let us consider the sequence $\mathcal{R}_h$, where $\mathcal{R}_h$ is obtained from $\mathcal{R}_{h-1}$ by dividing every cube in 2^n equal cubes, and let $\overline{X_h}$ and $\overline{U_h}$ be the corresponding functions.

The sequences $\bar{X}_h$ and $\bar{U}_h$ converge almost everywhere to x and u, respectively; by the dominated convergence theorem (remember that ω is bounded) it is possible to choose a partition $\mathcal{R}_h$ in such a way that

$$\int_\Omega \omega(|x - \bar{X}_h| + |u - \bar{U}_h|)\, dx < \epsilon\,.$$

We have therefore

$$\begin{aligned}\int_\Omega F(x, u_j, Du_j)\, dx &= \int_{\Omega - K} \{F(x, u_j, Du_j) - F(\bar{X}_h, \bar{U}_h, Du_j)\}\, dx \\ &\quad + \int_K \{F(x, u_j, Du_j) - F(x, u, Du_j)\}\, dx \\ &\quad + \int_K \{F(x, u, Du_j) - F(\bar{X}_h, \bar{U}_h, Du_j)\}\, dx \\ &\quad + \int_\Omega F(\bar{X}_h, \bar{U}_h, Du_j)\, dx\,.\end{aligned}$$

It follows that

$$\begin{aligned}\int_\Omega F(x, u_j, Du_j)\, dx &\geq -c|\Omega - K| - \int_K \omega(|u_j - u|)\, dx - \epsilon \\ &\quad + \int_\Omega F(\bar{X}_h, \bar{U}_h, Du_j)\, dx\,.\end{aligned}$$

Let now $j \to \infty$; if we remark that the last integral is the sum of integrals over the cubes of the partition, in each of which the integrand depends only on z, we can apply Theorem 5.4, obtaining

$$\liminf_{j \to \infty} \int_\Omega F(x, u_j, Du_j)\, dx \geq -c\epsilon + \int_\Omega F(\bar{X}_h, \bar{U}_h, Du)\, dx\,,$$

form which we arrive to the conclusion by letting first $h \to \infty$ and then $\epsilon \to 0$.

Our Theorem 5.7 is thus proved when F satisfies the estimate $F(x, u, z) \geq \nu|z|^p$. To conclude the proof, it only remains to eliminate that assumption.

For $\epsilon > 0$ define

$$F_\epsilon(x, u, z) = F(x, u, z) + a|u|^t + h(x) + 2\epsilon|z|^p + M\,.$$

Since $p > r$, we can choose the constant M in such a way that $F_\epsilon(x, u, z) \geq \epsilon|z|^p$; for what we have proved above, the function F_ϵ will be lower semicontinuous.

Let now $u_i \rightharpoonup u$ in $W^{1,p}_{\mathrm{loc}}(\Omega)$, and let $\Omega_0 \subset\subset \Omega$ be an open set with regular boundary. Since $t < p^*$, the sequence u_i will converge to u in the strong topology of $L^t(\Omega_0)$. We have therefore:

$$\begin{aligned}
&\liminf_{i\to\infty} \int_\Omega F(x, u_i, Du_i)\, dx \\
&\geq \liminf_{i\to\infty} \int_{\Omega_0} F_\epsilon(x, u_i, Du_i)\, dx - \lim_{i\to\infty} \int_{\Omega_0} \{a|u_i|^t + h(x) + M\}\, dx - 2c\epsilon \\
&\geq \int_{\Omega_0} F_\epsilon(x, u, Du)\, dx - \int_{\Omega_0} \{a|u|^t + h(x) + M\}\, dx - 2c\epsilon \\
&\geq \int_{\Omega_0} F(x, u, Du)\, dx - 2c\epsilon
\end{aligned}$$

and the conclusion follows letting first $\epsilon \to 0$, and then letting Ω_0 tend to Ω.

Remark 5.6 The preceding theorem does not hold if either $r = p$, or $t = p^*$. In the first case, MURAT and TARTAR [1] have shown that the functional

$$\int_\Omega \det(Du)\, dx$$

$(n = N = 2)$ is not continuous in the weak topology of $W^{1,2}$ (note that the functions $\det(z)$ and $-\det(z)$ are both quasi-convex). In the second, it will be sufficient to consider the functional

$$\int_\Omega (|Du|^p - |u|^{p^*})\, dx$$

and a sequence u_k converging *strongly* to u in $W^{1,p}$, but not in L^{p^*}. □

5.6 Coerciveness and Existence

The above results can be applied to prove the existence of minima of functionals. As we said in the preceding chapter, once the semicontinuity has been proved, the existence of minima under suitable conditions will depend essentially on the coerciveness of the functional under discussion. In the preceding chapter we have seen that for the DIRICHLET problem the coerciveness follows from estimates of the type

$$|z|^p - \beta(x)|u|^\delta - g(x) \leq F(x, u, z) \leq L|z|^p + b(x)|u|^\gamma + g(x)\,. \tag{5.36}$$

Now these assumptions are quite natural for functions convex in z, much less when $F(x,u,z)$ is only quasi-convex, or even polyconvex. Actually, if we consider only the DIRICHLET problem, or more precisely if we look for coerciveness in the space

$$V = U + W_0^{1,p}(\Omega) = \{v \in W^{1,p}(\Omega) : v - U \in W_0^{1,p}\},$$

(5.36) can be replaced by

$$|F(x,u,z)| \leq L|z|^p + b(x)|u|^\delta + a(x), \tag{5.37}$$

$$F(x,u,z) \geq \tilde{F}(z) - b(x)|u|^\delta - a(x), \tag{5.38}$$

where, as above, $\delta < p$, $b \in L^{\frac{p}{p-\delta}}$, $a \in L^1$, and where $\tilde{F}(z)$ is a function *strictly* quasi-convex in 0, that is, such that

$$\nu \int_\Omega |D\varphi|^p \, dx \leq \int_\Omega [\tilde{F}(D\varphi) - \tilde{F}(0)] \, dx \tag{5.39}$$

for every $\varphi \in W_0^{1,p}(\Omega)$.[8] Adding possibly a constant to the function a, we can suppose $\tilde{F}(0) = 0$.

With these assumptions, for every $u \in V$, setting $\varphi = u - U$, we have

$$\begin{aligned}
\int_\Omega |Du|^p \, dx &\leq c \int_\Omega |D\varphi|^p \, dx + c \int_\Omega |DU|^p \, dx \\
&\leq c \int_\Omega \tilde{F}(D\varphi) \, dx + c \int_\Omega |DU|^p \, dx \\
&\leq c \int_\Omega F(x,u,D\varphi) \, dx \\
&\quad + c \int_\Omega (|DU|^p + b(x)|u|^\delta + a(x)) \, dx .
\end{aligned}$$

On the other hand, we get from Lemma 5.2:

$$\begin{aligned}
\int_\Omega F(x,u,D\varphi) \, dx &\leq \int_\Omega F(x,u,Du) \, dx + \int_\Omega [F(x,u,D\varphi) - F(x,u,Du)] \, dx \\
&\leq \int_\Omega F(x,u,Du) \, dx \\
&\quad + c \int_\Omega (|Du| + |DU| + \vartheta(x,u))^{p-1} |DU| \, dx ,
\end{aligned}$$

where $\vartheta(x,u) =: [b|u|^\delta + a(x)]^{1/p}$.

[8] This last assumption will be discussed in more detail in the next chapter.

Using the standard inequality $A^{p-1}B \leq \epsilon A^p + c(\epsilon)B^p$, we obtain

$$\int_\Omega |Du|^p\,dx \leq c\int_\Omega F(x,u,Du)\,dx + c\int_\Omega (|DU|^p + a)\,dx$$
$$+ c\int_\Omega b|u|^\delta\,dx + \epsilon\int_\Omega |Du|^p\,dx\,.$$

The last integral can be subtracted from the left-hand side; the penultimate can be estimated by

$$c\int_\Omega b(|\varphi|^\delta + |U|^\delta)\,dx \leq \epsilon\int_\Omega |\varphi|^p\,dx + c\int_\Omega (b^{\frac{p}{p-\delta}} + b|U|^\delta)\,dx$$
$$\leq c\epsilon\int_\Omega |D\varphi|^p\,dx + c\int_\Omega (b^{\frac{p}{p-\delta}} + b|U|^\delta)\,dx$$
$$\leq c\epsilon\int_\Omega |Du|^p\,dx + c\int_\Omega (b^{\frac{p}{p-\delta}} + |U|^p + |DU|^p)\,dx\,.$$

In conclusion:

$$\int_\Omega |Du|^p\,dx \leq c\int_\Omega F(x,u,Du)\,dx + c\int_\Omega (|DU|^p + |U|^p + b^{\frac{p}{p-\delta}} + a)\,dx$$

from which the coerciveness of $\mathcal{F}$ follows at once.

In any case, independently of the adequacy of the assumptions, it is not difficult to prove existence theorems starting from the semicontinuity Theorem 5.7, and from coerciveness assumptions of the type (5.36) or (5.37), (5.38). The proof proceeds in the same way as for the "scalar" functionals discussed in the preceding chapter, to which we refer for details.

5.7 Notes and Comments

We have already said that the assumptions of convexity, policonvexity, quasi-convexity and rank-one convexity are successively more general, being all equivalent if either $N = 1$ or $n = 1$. The existence of non-convex polyconvex functions is easily proved; it will suffice to consider, in the case $N = n$, the determinant $\det(z)$, or else any minor of the matrix z.

Examples of rank-one convex functions which are not polyconvex have been given by TERPSTRA [1] in the case $N = n \geq 3$. The case $N = n = 2$ was discussed by DACOROGNA and MARCELLINI [1], and later by ALIBERT and DACOROGNA [1], who proved the following result:

The function $|z|^2(|z|^2 - 2\gamma \det(z))$, $\gamma \in \mathbf{R}$, is:

(i) convex, if and only if $|\gamma| \leq \frac{2}{3}\sqrt{2}$,
(ii) polyconvex if and only if $|\gamma| \leq 1$,
(iii) quasi-convex if and only if $|\gamma| \leq 1 + \epsilon$, for a suitable $\epsilon > 0$,
(iv) rank-one convex if and only if $|\gamma| \leq \frac{2}{\sqrt{3}}$.

It is not known whether $1 + \epsilon$ in (iii) is strictly less than $\frac{2}{\sqrt{3}}$. Later, ŠVERÁK [1] gave, for $n = N \geq 3$, an example of a rank-one convex function which is not quasi-convex, thus concluding the proof that the four concepts introduced are all different. The case $n = N = 2$ is still open.

The semicontinuity Theorem 5.7 is due essentially to ACERBI and FUSCO [1], who obtained it under slightly more stringent assumptions. Our proof follows that by MARCELLINI [1]. As the reader will recognize, the main point of the proof consists in the estimate of integrals of the type

$$\int_{A_i} |Du_i|^p \, dx \,,$$

where A_i is the set in which $|Du_i| > i$. The proof simplifies essentially if we only require the lower semicontinuity in the weak topology of $W^{1,p+\epsilon}$, with $\epsilon > 0$, or else if $p = 1$, since in both these cases the integrals in question are equi-absolutely continuous (see FUSCO [1]).

When the integrand function $F = F(x, z)$ is independent of u and continuous in $\Omega \times \mathbf{R}^{nN}$, the semicontinuity theorem continues to hold, even for weak convergence in $W^{1,q}$ with $q > \frac{pn}{n+1}$,[9] provided the function F satisfies the technical condition $F(x, tz) \leq c(1 + F(x, z))$ for $t \in [0, 1]$ (MARCELLINI [2]).

When $F = F(z)$ is polyconvex, and $n = N$, we have the semicontinuity under weak convergence in $W^{1,q}$, with $q > n - 1$ (MARCELLINI [2]). This result does not hold for $q < n - 1$ (MALY [2]).

Polyconvex integrands are of some importance in finite elasticity. If $u : \Omega \to \mathbf{R}^n$ indicates the position of the point x after the deformation, $\det(Du)$ gives at every point the ratio between the volume elements after and before the deformation. It is reasonable to expect that the deformation energy tends to infinity when $\det(Du)$ tends either to zero or to infinity. For instance, a reasonable functional describing the elastic energy could be

---

[9] More precisely, if $u_k, u \in W^{1,p}$ and $u_k \rightharpoonup u$ in $W^{1,q}$, then $\mathcal{F}(u) \leq \liminf \mathcal{F}(u_k)$.

$$\mathcal{F}(u) = \int_\Omega \{|Du|^2 + g(\det(Du))\}\, dx\,,$$

where $g(t)$ is a convex function, tending to infinity for $t \to 0$ and for $t \to +\infty$.

The study of functionals of this type encounters several difficulties, even in the relatively simple case in which $g(t)$ remains bounded for $t \to 0$. Actually, with the exception of a few special cases, the function

$$F(z) = |z|^2 + g(\det(z))$$

does not have the same behavior from above and from below (think only about functions like $|z|^2 + |\det(z)|^2$). In these cases, (5.36) does not hold, and will be substituted by a weaker condition such as

$$|z|^p \leq F(z) \leq c(|z|^k + 1)$$

with $p < k$. This has led to a number of studies of functionals bounded from above and from below by different powers of $|z|$ (see for instance MARCELLINI [4, 6]); some positive results have been obtained when the two exponents are rather close to each other, an assumption that is not easily satisfied in the applications. If, on the contrary, the exponents are quite different, it is possible to find counterexamples to the regularity of the minima, even in the scalar case (GIAQUINTA [2], MARCELLINI [3]).

Polyconvex functionals have already been described already in the book by MORREY [3], and have been studied in detail by BALL [1], even in connection with problems in finite elasticity. More recently, in a series of papers GIAQUINTA, G. MODICA and J. SOUČEK [1, 2] have extended these functionals to the parametric case, obtaining important results of existence of minima.

Finally, we note that it is possible to define quasi-convexity even for functionals depending on the derivatives of higher order (MEYERS [3], FUSCO [1], MU and LI [1]). It is possible to extend some of the results of this chapter to these functionals.

Chapter 6

Quasi-Minima

6.1 Preliminaries

In the preceding chapters we have discussed the existence of minima of regular functionals of the calculus of variations, and we have proved that under suitable assumptions of convexity (or quasi-convexity in the case of vector-valued functions) of the function $F(x, u, z)$ with respect to the variable z, the functional

$$\mathcal{F}(u, \Omega) =: \int_{\Omega} F(x, u, Du)\, dx \tag{6.1}$$

attains its minimum value among the functions of $W^{1,p}(\Omega, \mathbf{R}^N)$ which take given values at the boundary of Ω (DIRICHLET problem).

The problem remains of determining if and under what conditions the minimizing function $u(x)$ has additional regularity properties, beyond those deriving from its belonging to the class $W^{1,p}$. To this problem of regularity we shall devote the remainder of the present volume.

In order to make the results independent of the boundary value problem (or better in order to separate the interior regularity, which does not depend on the boundary data, from the boundary regularity), we remark in the first place that it is possible to consider minimizing functions that are summable only locally. More precisely, let us assume that the function $F(x, u, z)$ satisfies the inequality

$$|F(x, u, z)| \leq L|z|^p + b(x)|u|^\gamma + a(x) \tag{6.2}$$

with[1] $1 < p \le \gamma < p^*$, and let $u \in W^{1,p}_{\mathrm{loc}}(\Omega)$. We say that u is a *local minimum* of the functional $\mathcal{F}(u)$ if for every $\varphi \in W^{1,p}(\Omega)$, with $K =: \operatorname{supp}\varphi \subset\subset \Omega$ we have

$$\mathcal{F}(u, K) \le \mathcal{F}(u + \varphi, K). \tag{6.3}$$

Remark 6.1 An equivalent form of (6.2), sometimes more useful in computations, is

$$|F(x, u, z)| \le L[|z| + \vartheta(x, u)]^p, \tag{6.4}$$

where

$$\vartheta(x, u)^p = b(x)|u|^\gamma + a(x). \tag{6.5}$$

□

Remark 6.2 In the following we shall always assume that $p < n$. The case $p \ge n$ is simpler in many respects, since by the SOBOLEV immersion theorem the function u belongs to every L^q (if $p = n$) or it is Hölder-continuous (when $p > n$). We leave to the reader the task of making the changes in the statements and in the proofs, necessary to extend the results to these cases. □

As we see from (6.3), the integrals are always computed on domains strictly contained in Ω, and generally speaking a local minimum is not requested to satisfy $\mathcal{F}(u, \Omega) < +\infty$. For instance, a harmonic function in the unit ball B is a local minimum for the DIRICHLET integral

$$\mathcal{D}(u) = \int_B |Du|^2\, dx,$$

but it might not belong to $W^{1,2}(B)$. This is easily seen by considering the function of two variables (given in polar coordinates)

$$u(x, y) = \frac{y}{(x+1)^2 + y^2},$$

which is harmonic in the unit disc B,[2] but whose DIRICHLET integral

$$\int_B \frac{dx\, dy}{[(x+1)^2 + y^2]^2}$$

is infinite.

[1] The assumption $\gamma \ge p$ is not restrictive, since for $\gamma < p$ we have $|u|^\gamma \le |u|^p + 1$.
[2] We remark that u is the imaginary part of the holomorphic function $\frac{1}{1+z}$.

On the other hand, if a local minimum u belongs to $W^{1,p}(\Omega)$, then u minimizes $\mathcal{F}$ among all the functions v in $W^{1,p}(\Omega)$ taking on the boundary the same value as u, that is such that $v - u \in W_0^{1,p}(\Omega)$.

Since regularity problems are essentially local, we can always assume that u belongs to $W^{1,p}(\Omega)$ without loss of generality.

6.2 Quasi-Minima and Differential Quations

A useful generalization of the notion of local minimum is the following.

Definition 6.1 *A function $u \in W_{\text{loc}}^{1,p}(\Omega, \mathbf{R}^N)$ is a* quasi-minimum *of the functional $\mathcal{F}$, with constant $Q \geq 1$ (briefly: a Q-*minimum*), if for every $v \in W_{\text{loc}}^{1,p}(\Omega, \mathbf{R}^N)$, with $K =: \operatorname{supp}(u - v) \subset\subset \Omega$, we have*

$$\mathcal{F}(u, K) \leq Q\mathcal{F}(v, K). \tag{6.6}$$

If moreover $u \in W^{1,p}$, the preceding relation is verified for every v such that $v - u \in W_0^{1,p}(\Omega, \mathbf{R}^N)$.

It is clear from what we have said that a local minimum is a quasi-minimum; actually the local minima are nothing but 1-minima.

Remark 6.3 More generally, one can suppose that instead of (6.6) we have (Hong [1]):

$$\mathcal{F}(u, K) \leq Q\mathcal{F}(v, K) + Q\int_K (|Dv| + \vartheta(x, v))^p \, dx,$$

or, what is the same by (6.4):

$$\mathcal{F}(u, K) \leq Q\int_K (|Dv| + \vartheta(x, v))^p \, dx. \tag{6.7}$$

We could also take the constant Q dependent on the compact set K, without detriment for most of the results we shall prove, since as we have remarked they are essentially of local character. Of course, in this case the various constants entering in the statements would depend on the compact set in question. □

The introduction of quasi-minima is justified by the following results.

Let us begin by recalling the notion of *weak solution* of a partial differential equation.

We shall consider equations in divergence form:

$$\frac{\partial}{\partial x_i} A^i_\alpha(x, u(x), Du(x)) - B_\alpha(x, u(x), Du(x)) = 0\,. \tag{6.8}$$

Definition 6.2 *A function* $u \in W^{1,p}_{\mathrm{loc}}(\Omega, \mathbf{R}^N)$ *is a* weak solution (*or a* solution in the sense of distributions) *of the Eq.* (6.8) *if for every* $\varphi \in W^{1,p}_0(\Omega, \mathbf{R}^N)$ *we have*

$$\int_\Omega \{A^i_\alpha(x, u(x), Du(x))D_i\varphi^\alpha + B_\alpha(x, u(x), Du(x))\varphi^\alpha\}\, dx = 0\,. \tag{6.9}$$

Formally, Eq. (6.9) is obtained from (6.8) multiplying by φ and integrating the first term by parts. On the other hand, in (6.9) we do not assume that the function u has second derivatives, so that it makes sense even for functions $u \in W^{1,p}_{\mathrm{loc}}(\Omega)$. It is clear that the two forms of the equation become equivalent if it is possible to integrate by parts in (6.9), that is when the function u has second derivatives (and of course the coefficients A^i_α are differentiable).

To show the relation between weak solutions and quasi-minima, let us begin by examining the simple case of linear equations:

$$\int a^{ij}_{\alpha\beta}(x) D_j u^\beta D_i\varphi^\alpha\, dx = 0 \quad \forall \varphi \in W^{1,2}_0(\Omega, \mathbf{R}^N)\,, \tag{6.10}$$

in which the coefficients $a_{ij}(x)$ are bounded functions:

$$\|A(x)\| =: \sup_{|\xi|=1} |A(x)\xi| \le M \quad (A(x) = \{a^{ij}_{\alpha\beta}(x)\}) \tag{6.11}$$

and satisfy the conditions of strong ellipticity:

$$a^{ij}_{\alpha\beta}(x)\xi^\alpha_i\xi^\beta_j \ge \nu|\xi|^2\,, \quad \nu > 0\,. \tag{6.12}$$

Now let $v \in W^{1,2}_{\mathrm{loc}}(\Omega, \mathbf{R}^N)$ be a function coinciding with u outside a compact set $K \subset\subset \Omega$. Writing $\varphi = v - u$ in (6.10), we get:

$$\int_K a^{ij}_{\alpha\beta} D_j u^\beta D_i u^\alpha\, dx = \int_K a^{ij}_{\alpha\beta} D_j u^\beta D_i v^\alpha\, dx$$

and using (6.11) and (6.12):

$$\begin{aligned} \nu \int_K |Du|^2\, dx &\le M \int_K |Du||Dv|\, dx \\ &\le M \left(\int_K |Du|^2\, dx\right)^{\frac{1}{2}} \left(\int_K |Dv|^2\, dx\right)^{\frac{1}{2}} \end{aligned}$$

from which it follows at once that u is a Q-minimum of the DIRICHLET integral, with $Q = M^2\nu^{-2}$.

More generally, we can consider weak solutions of (6.9), with the coefficients $A^i_\alpha(x,u,z)$ satisfying the uniform ellipticity condition:

$$A^i_\alpha(x,u,z)z^\alpha_i \geq |z|^p - b_1(x)|u|^\gamma - a_1(x) \tag{6.13}$$

with $1 < p \leq \gamma < p^* = \frac{np}{n-p}$, and the estimates

$$|A(x,u,z)| \leq L|z|^{p-1} + b_2(x)|u|^\sigma + a_2(x) \tag{6.14}$$

with $\sigma = \gamma\frac{p-1}{p}$.

Concerning the term $B(x,u,z)$, we can distinguish two cases. The first, simpler in many accounts, is that of *controlled growth* conditions:

$$|B(x,u,z)| \leq H|z|^\tau + b_3(x)|u|^\delta + a_3(x) \tag{6.15}$$

with $\tau = p\frac{\gamma-1}{\gamma}$ and $\delta = \gamma\frac{p^*-1}{p^*}$.

We shall assume that the functions b_i and a_i are positive, with a_1, $a_2^{\frac{p}{p-1}}$, $a_3^{\frac{p^*}{p^*-1}} \in L^1$, and b_1, $b_2^{\frac{p}{p-1}}$, $b_3^{\frac{p^*}{p^*-1}} \in L^{\frac{p^*}{p^*-\gamma}}$. We have the following:

Theorem 6.1 *Let $u \in W^{1,p}_{\mathrm{loc}}(\Omega, \mathbf{R}^N)$ be a weak solution of the Eq.* (6.8), *with coefficients A and B satisfying conditions* (6.13), (6.14) *and* (6.15). *Then, u is a quasi-minimum of the functional*

$$\mathcal{J}(u,\Omega) = \int_\Omega (|Du|^p + b(x)|u|^\gamma + a(x))\,dx \tag{6.16}$$

with

$$b(x) = b_1 + b_2^{\frac{p}{p-1}} + b_3^{\frac{p^*}{p^*-1}} \in L^{\frac{p^*}{p^*-\gamma}}$$

and

$$a(x) = a_1(x) + a_2(x)^{\frac{p}{p-1}} + a_3(x)^{\frac{p^*}{p^*-1}} + b^{\frac{p^*}{p^*-\gamma}} \in L^1 .$$

Proof. Let $v \in W^{1,p}_{\mathrm{loc}}$, with $K = \mathrm{supp}\,(u-v) \subset\subset \Omega$. Setting $\varphi = u - v$ in (6.9) we get

$$\int_K A^i_\alpha(x,u,Du)D_iu^\alpha\,dx = \int_K A^i_\alpha(x,u,Du)D_iv^\alpha\,dx + \int_K B_\alpha(x,u,Du)(v^\alpha - u^\alpha)\,dx$$

and using relations (6.13)–(6.15):

$$\int_K |Du|^p\,dx \le \int_K b_1|u|^\gamma\,dx + \int_K a_1\,dx$$
$$+ L\int_K |Du|^{p-1}|Dv|\,dx + \int_K b_2|u|^\sigma|Dv|\,dx$$
$$+ \int_K a_2|Dv|\,dx + H\int_K |Du|^\tau|u-v|\,dx$$
$$+ \int_K b_3|u|^\delta|u-v|\,dx + \int_K a_3|u-v|\,dx\,. \qquad (6.17)$$

We have

$$\int_K |Du|^{p-1}|Dv|\,dx \le \epsilon\int_K |Du|^p\,dx + c(\epsilon)\int_K |Dv|^p\,dx\,;$$
$$\int_K b_2|u|^\sigma|Dv|\,dx \le c\left\{\int_K |Dv|^p\,dx + \int_K b_2^{\frac{p}{p-1}}|u|^\gamma\,dx\right\};$$
$$\int_K a_2|Dv|\,dx \le c\left\{\int_K |Dv|^p\,dx + \int_K a_2^{\frac{p}{p-1}}\,dx\right\};$$
$$\int_K |Du|^\tau|u-v|\,dx \le \epsilon\int_K |Du|^p\,dx + c(\epsilon)\int_K |u-v|^\gamma\,dx\,;$$
$$\int_K b_3|u|^\delta|u-v|\,dx \le \epsilon\int_K |u-v|^{p^*}\,dx + c(\epsilon)\int_K b_3^{\frac{p^*}{p^*-1}}|u|^\gamma\,dx\,;$$
$$\int_K a_3|u-v|\,dx \le \epsilon\int_K |u-v|^{p^*}\,dx + c(\epsilon)\int_K a_3^{\frac{p^*}{p^*-1}}\,dx\,.$$

On the other hand:

$$b|u|^\gamma \le c(\gamma)(b|v|^\gamma + b|u-v|^\gamma) \le \epsilon|u-v|^{p^*} + c(\gamma,\epsilon)b^{\frac{p^*}{p^*-\gamma}} + c(\gamma)b|v|^\gamma\,,$$

and by the SOBOLEV theorem:

$$\int_K |u-v|^{p^*}\,dx \le c\left\{\int_K (|Du|^p + |Dv|^p)\,dx\right\}^{\frac{p^*}{p}-1}\int_K (|Du|^p + |Dv|^p)\,dx\,.$$

We remark now that it is possible to assume that

$$\int_K (|Du|^p + b|u|^\gamma)\,dx > \int_K |Dv|^p\,dx$$

since otherwise one would have trivially $\mathcal{J}(u,K) \leq \mathcal{J}(v,K)$. It follows that

$$\int_K |u-v|^{p^*}\,dx \leq c\left\{\int_K (2|Du|^p + b|u|^\gamma)\,dx\right\}^{\frac{p^*}{p}-1} \int_K (|Du|^p + |Dv|^p)\,dx$$

$$\leq c(\|Du\|_p, \|u\|_{p^*}, \|b\|_{\frac{p^*}{p^*-\gamma}}) \int_K (|Du|^p + |Dv|^p)\,dx\,.$$

Introducing all these inequalities in (6.17), after having added to both members the quantity

$$\int_K (b|u|^\gamma + a(x))\,dx\,,$$

and taking ϵ small enough, we easily get the conclusion of the theorem. Note that the constant Q depends on u, as it is permitted, but not on v. □

Let us come now to the second case, in which the term $B(x,u,z)$ satisfies *natural growth* assumptions:[3]

$$|B(x,u,z)| \leq H|z|^p + a_3(x) \tag{6.18}$$

with $0 \leq a_3 \in L^1(\Omega)$.

In this case we have a theorem analogous to the above only for *bounded* solutions of the Eq. (6.9). That explains why we have omitted the term $b_3|u|^\delta$ in (6.18); we can also assume $b_1 = b_2 = 0$ in (6.13) and (6.14), and allow the dependence of the constants L, H and of the functions a_i on $M = \sup u$.

We shall consider separately the case of one equation ($N=1$) and of a system of equations ($N>1$). We will begin from the first one.

Theorem 6.2 *Let $u(x)$ be a bounded solution of* (6.9) ($N=1$), *with conditions* (6.13), (6.14) (*with* $b_1 = b_2 = 0$) *and* (6.18). *Then, u is a quasi-minimum of the functional*

$$\mathcal{H}(u,\Omega) = \int_\Omega (|Du|^p + a(x))\,dx\,, \tag{6.19}$$

[3]The reason for this terminology lies in the fact that when (6.8) is the EULER equation of a functional, we have $A = F_z$ and $B = F_u$. If the function $F(z)$ grows as $|z|^p$, it is natural to expect that A grows as $|z|^{p-1}$, whereas the growth of B remains the same as that of F, that is $|z|^p$. Whence the distinction between natural and non-natural conditions (or conditions of controlled growth).

where

$$a(x) = a_1(x) + a_2(x)^{\frac{p}{p-1}} + a_3(x).$$

Proof. Let $v \in W^{1,p}(\Omega)$ be such that $K =: \operatorname{supp}(u - v) \subset\subset \Omega$, and assume that $|v(x)| \leq M =: \sup_\Omega |u|$. Setting in (6.9)

$$\varphi = (u - v)^+ e^{\lambda(u-v)}$$

($A^+ = \max(A, 0)$), and denoting by S the support of φ, we obtain

$$\int_S A^i D_i u[1 + \lambda(u - v)]e^{\lambda(u-v)}\, dx$$
$$= \int_S A^i D_i v[1 + \lambda(u - v)]e^{\lambda(u-v)}\, dx + \int_S B(u - v)e^{\lambda(u-v)}\, dx,$$

where the coefficients A^i and B are obviously calculated at $(x, u(x), Du(x))$. Using (6.13), (6.14) and (6.18), and remembering that $u - v \geq 0$ on S, we deduce

$$\int_S |Du|^p[1 + \lambda(u - v)]e^{\lambda(u-v)}\, dx$$
$$\leq \int_S a_1[1 + \lambda(u - v)]e^{\lambda(u-v)}\, dx$$
$$+ \int_S (L|Du|^{p-1} + a_2)[1 + \lambda(u - v)]e^{\lambda(u-v)}|Dv|\, dx$$
$$+ \int_S (H|Du|^p + a_3)(u - v)e^{\lambda(u-v)}\, dx.$$

Since $|u|$ and $|v|$ are both bounded by M, we have

$$\int_S |Du|^p[1 + \lambda(u - v)]e^{\lambda(u-v)}\, dx$$
$$\leq c\int_S a_1\, d + c\int_S (L|Du|^{p-1} + a_2)|Dv|\, dx$$
$$+ c\int_S a_3\, dx + \int_S H|Du|^p(u - v)e^{\lambda(u-v)}\, dx.$$

Choosing now $\lambda = H$, the last integral on the right-hand side can be subtracted from the left-hand side. Moreover, by the usual estimate

$AB \leq \epsilon A^{\frac{p}{p-1}} + c(\epsilon)B^p$, we get, summing to both members the integral $\int_S a\,dx$:

$$\int_S (|Du|^p + a)\,dx \leq c \int_S (|Dv|^p + a)\,dx\,. \tag{6.20}$$

Similarly, choosing

$$\varphi = (v-u)^+ e^{\lambda(v-u)}$$

we obtain the inequality

$$\int_T (|Du|^p + a)\,dx \leq c \int_T (|Dv|^p + a)\,dx\,,$$

where $T = \operatorname{supp}\varphi$. The conclusion follows summing the above inequality with (6.20).

Finally, if v does not verity the relation $|v| \leq M$, we set $\bar{v} = \min\{M, \max\{v, -M\}\}$, and we conclude immediately

$$\mathcal{H}(u,K) \leq \mathcal{H}(\bar{v},K) \leq \mathcal{H}(v,K)$$

since $|D\bar{v}| \leq |Dv|$. □

Example 6.1 (FREHSE [3]) We remark that in the preceding theorem the assumption that u is bounded is essential. Actually, as we shall see later, in the scalar case ($N = 1$) every quasi-minimum of the functional $\mathcal{H}$ (or more generally of any regular functional $\mathcal{F}$) is Hölder-continuous, and hence in particular it is bounded. On the other hand, the function

$$u(x) = 12 \log \log |x|^{-1}$$

is a solution of the EULER equation (in short, *an extremal*) of the functional

$$\mathcal{F}(u) = \int_D \left\{1 + \frac{1}{1 + e^u (\log|x|)^{-12}}\right\} |Du|^2\,dx$$

in the disc $D \subset \mathbf{R}^2$ of radius e^{-1}.

Remark 6.4 If $N = 1$, we can consider a *subsolution* of (6.9), that is a function $u(x)$ satisfying the inequality

$$\int_\Omega \{A^i(x,u(x),Du(x))D_i\varphi + B(x,u(x),Du(x))\varphi\}\,dx \leq 0 \tag{6.21}$$

for every $\varphi \geq 0$. If u is a bounded subsolution, we can repeat the preceding proof,[4] and conclude that u is a sub-quasi-minimum of the functional $\mathcal{H}$, or in other words that

$$\mathcal{H}(u, K) \leq Q\mathcal{H}(v, K)$$

for every $v \leq u$, with $K =: \operatorname{supp}(u - v) \subset\subset \Omega$.

Similarly, a bounded supersolution of (6.9) is a super-quasi-minimum of $\mathcal{H}$. □

When, from a single equation, we pass to systems of equations, the boundedness of the solution is no longer sufficient. Actually, already for $N = n = 2$ there exists systems satisfying the conditions of the preceding theorem, and possessing bounded discontinuous solutions, as one can see in the following example.

Example 6.2 (FREHSE [2]) The function

$$u(x) = \begin{pmatrix} \sin\left(\log\log\dfrac{1}{|x|}\right) \\ \cos\left(\log\log\dfrac{1}{|x|}\right) \end{pmatrix}$$

is a weak solution in $\mathbf{R}^2$ of the system

$$\Delta u^\alpha = B^\alpha(u, Du), \quad (\alpha = 1, 2)$$

with

$$B(u, z) = -\begin{pmatrix} u^1 + u^2 \\ u^2 - u^1 \end{pmatrix} |z|^2,$$

verifying natural growth conditions with $p = 2$. We shall see later in this chapter that every quasi-minimum belongs to $W^{1,p+\epsilon}_{\text{loc}}$ for some $\epsilon > 0$. Therefore, if the function u above were a quasi-minimum, it would belong to $W^{1,2+\epsilon}_{\text{loc}}$, and by SOBOLEV theorem it would be Hölder-continuous.

Theorem 6.3 *Let $u \in W^{1,p}_{\text{loc}}(\Omega)$ be a weak solution of the system* (6.9), *with coefficients satisfying natural conditions* (6.13), (6.14) *and* (6.18). *Let $M = \sup|u|$, and assume that*

$$2MH(M) < 1. \tag{6.22}$$

Then, u is a quasi-minimum of the functional (6.19).

[4] Of course, in this case a lower bound for the term B is sufficient. Moreover, in the case of controlled growth we can omit the assumption that u is bounded.

Proof. Let as usual $v \in W^{1,p}_{\mathrm{loc}}(\Omega, \mathbf{R}^N)$, with $K = \operatorname{supp}(u - v) \subset\subset \Omega$. If $|v| \leq M$, we can take $\varphi = u - v$ in (6.9) and make the usual estimates, since the term on the right-hand side containing $|Du|^p$ can be subtracted from the left-hand side by virtue of the assumption $2MH(M) < 1$.

Now let v be an arbitrary function, and let

$$\bar{v} = \begin{cases} v & \text{if } |v| \leq M\,, \\ \dfrac{Mv}{|v|} & \text{if } |v| > M\,. \end{cases}$$

We have $\bar{v} \leq M$ and $|D\bar{v}| \leq 2|Dv|$, and hence

$$\mathcal{H}(u, K) \leq Q\mathcal{H}(\bar{v}, K) \leq 2^p Q\mathcal{H}(v, K)$$

from which the conclusion follows at once. □

Remark 6.5 It is clear that the preceding theorem remains valid if instead of $M = \sup |u|$ we take $M = \sup |u - a|$ for an arbitrary $a \in \mathbf{R}^N$. In particular, we can take $M = \frac{1}{2}\operatorname{osc}(u)$, where $\operatorname{osc}(u)$ is the oscillation of u.

If u is a continuous function, (6.22) is satisfied automatically, provided $K = \operatorname{supp}(u - v)$ is small enough. Consequently, every *continuous* solution of (6.9), with natural growth conditions (6.13), (6.14) and (6.18), is a quasi-minimum of the functional $\mathcal{H}$ in the small. In formulas, it will result in:

$$\mathcal{H}(u, K) \leq Q\mathcal{H}(v, K)$$

whenever $\operatorname{diam}(K)$ is less than a constant ϵ_0 depending only on the modulus of continuity of u, with Q independent of ϵ_0.

This will be largely sufficient to prove all the results of local character, the regularity in the first place. Needless to say, the continuity of u is automatically guaranteed by the SOBOLEV theorem if $p > n$.

The preceding example shows that Theorem 6.19 cannot hold without the assumption $MH(M) < 1$, even if $p = n$. It is not known whether it is possible to replace (6.22) with the weaker assumption $MH(M) < 1$. □

Example 6.3 (DE GIORGI [5]) Contrary to what happens in the scalar case,[5] when $N > 1$ the quasi-minima (and even the minima) of functionals are not necessarily bounded functions. For instance, if $n = N > 2$, the function

$$u^\alpha(x) = x_\alpha |x|^{-\kappa}\,; \quad \kappa = \frac{n}{2}\left\{1 - \frac{1}{\sqrt{(2n-2)^2 + 1}}\right\}$$

[5]See the next chapter, in particular Theorem 7.4.

minimizes the functional

$$\mathcal{F}(u,B)=\int_B A^{ij}_{\alpha\beta}(x)D_iu^\alpha D_ju^\beta\,dx\,,$$

with

$$A^{ij}_{\alpha\beta}(x)=\delta_{\alpha\beta}\delta_{ij}+\left[(n-2)\delta_{\alpha i}+n\frac{x_\alpha x_i}{|x|^2}\right]\left[(n-2)\delta_{\beta j}+n\frac{x_\beta x_j}{|x|^2}\right],$$

among all functions taking the value x on the boundary of the unit ball B.

The proof begins with the remark that the functional $\mathcal{F}$ is convex, and hence u is the (unique) minimum of $\mathcal{F}$ if and only if it is solution of the EULER equation

$$\int_B A^{ij}_{\alpha\beta}(x)D_iu^\alpha D_j\varphi^\beta\,dx=0 \tag{6.23}$$

for every $\varphi\in W^{1,2}_0(B,\mathbf{R}^N)$.

In our case it is easy to verify that

$$D_j[A^{ij}_{\alpha\beta}(x)D_iu^\alpha]=0$$

in $B-\{0\}$. By consequence, (6.23) is satisfied for every with support in $B-\{0\}$.

Assume now that φ has support in B, and let η be a function of class $C^\infty(B)$, with $0\le\eta\le1$, $\eta=0$ in B_R, $\eta=1$ in $B-B_{2R}$ and $|D\eta|\le 2/R$. The function $\eta\varphi^\alpha$ has support in $B-B_R$, and therefore we have

$$\begin{aligned}0&=\int_B A^{ij}_{\alpha\beta}(x)D_iu^\alpha D_j(\eta\varphi^\beta)\,dx\\&=\int_B \eta A^{ij}_{\alpha\beta}(x)D_iu^\alpha D_j\varphi^\beta\,dx+\int_B\varphi^\beta A^{ij}_{\alpha\beta}(x)D_iu^\alpha D_j\eta\,dx\,.\end{aligned}$$

The last integral can be estimated by

$$c\left(\int_B|Du|^2\,dx\right)^{\frac12}R^{\frac{n-2}{2}},$$

and hence it tends to zero with R. Passing to the limit in the preceding relation, we get (6.23) for every $\varphi\in C^\infty_0(B,\mathbf{R}^N)$, and therefore for every $\varphi\in W^{1,2}(B,\mathbf{R}^N)$.

A second example of quasi-minimum arises from *quasi-regular* mappings. We recall that a map $u:\Omega\to\mathbf{R}^n$ is quasi-regular if there exists a constant $A > 0$ such that

$$|Du|^n \leq A \det(Du).$$

If, moreover, u is a homeomorphism, the map is called *quasi-conformal.*

Theorem 6.4 *A quasi-regular map $u \in W^{1,n}(\Omega, \mathbf{R}^n)$ is a quasi-minimum of the functional*

$$\int_\Omega |Du|^n \, dx.$$

Proof. Let φ be a map from Ω to $\mathbf{R}^n$, with support $K \subset\subset \Omega$. We have

$$\int_K \det(Du)\, dx = \int_K \det(Du + D\varphi)\, dx \leq c \int_K |D(u+\varphi)|^n \, dx$$

and the conclusion follows immediately from the definition of quasi-regularity. □

Example 6.4 Another example of a quasi-minimum comes from minima with obstacles. Let $\psi(x)$ be a function in $W^{1,p}(\Omega)$, and assume that $u \in W^{1,p}_{\text{loc}}(\Omega)$ satisfies the inequality $u(x) \geq \psi(x)$ in Ω, and moreover

$$\mathcal{F}(u, K) \leq \mathcal{F}(w, K)$$

for every $w \in W^{1,p}_{\text{loc}}(\Omega)$, with $K = \operatorname{supp}(u - w) \subset\subset \Omega$ and $w \geq \psi$. In other words, u minimizes the functional $\mathcal{F}$ among all the functions whose graph lies above the *obstacle* ψ.

Now let v be a generic function in $W^{1,p}_{\text{loc}}(\Omega)$, (which in general does not lie above the obstacle), with $K = \operatorname{supp}(u - v) \subset\subset \Omega$. Setting $\Sigma = \{x \in \Omega : v(x) \geq \psi(x)\}$ and $w = \max\{v, \psi\}$, we have[6]

$$\begin{aligned}\mathcal{F}(u, K) \leq \mathcal{F}(w, K) &= \mathcal{F}(v, K \cap \Sigma) + \mathcal{F}(\psi, K - \Sigma) \\ &\leq \mathcal{F}(v, K) + \mathcal{F}(\psi, K),\end{aligned}$$

and hence, adding to both members the term $\mathcal{F}(\psi, K)$:

$$\mathcal{G}(u, K) \leq 2\mathcal{G}(v, K),$$

where we have set

$$\mathcal{G}(u, A) = \int_A [F(x, u, Du) + \gamma(x)]\, dx$$

[6]For the sake of simplicity, we assume that $F(x, u, z) \geq 0$. See, anyway, Remark 6.6.

and

$$\gamma(x) = F(x, \psi(x), D\psi(x)) .$$

It is immediately verifiable that $F(x, u, z) + \gamma(x)$ satisfies the same estimates (6.2) as F, with $a + \gamma$ instead of a. In conclusion, a minimum (and also a quasi-minimum) of the functional $\mathcal{F}$, with obstacle ψ, is a *free* quasi-minimum of the functional $\mathcal{G}$.

In the vector case, a similar conclusion can be obtained for the minima of $\mathcal{F}$ confined in a *convex* region Θ. More precisely, let Θ be a convex domain of $\mathbf{R}^N$ with regular boundary, and let $u \in W^{1,p}_{\mathrm{loc}}(\Omega)$ be a function with values in Θ, such that for every w with values in Θ, coinciding with u outside a compact set K, it holds that

$$\mathcal{F}(u, K) \leq \mathcal{F}(w, K) .$$

We can suppose that Θ contains the unit ball B. For $|\xi| \neq 0$, let $R(\xi) > 1$ be such that $\xi|\xi|^{-1}R(\xi) \in \partial\Theta$. The function $R(\xi)$ is regular in $\mathbf{R}^n - \{0\}$, and in particular its derivatives are bounded on ∂B.

Let now v be a generic function of $W^{1,p}_{\mathrm{loc}}(\Omega)$, coinciding with u outside K. Setting

$$w = \begin{cases} v & \text{if } v \in \Theta , \\ \dfrac{v}{|v|} R\left(\dfrac{v}{|v|}\right) & \text{if } v \notin \Theta , \end{cases}$$

we have $w \in \Theta$ and hence, since $w \leq v$ and $|Dw| \leq c|Dv|$:

$$\mathcal{F}(u, K) \leq \mathcal{F}(w, K) \leq Q \int_K (|Dv| + \vartheta(x, v))^p \, dx .$$

As we shall see (see later, Remark 6.9), the above relation is sufficient to prove the results of this chapter.

Remark 6.6 The functional

$$\mathcal{J}(u, \Omega) = \int_\Omega (|Du|^p + b|u|^\gamma + a(x)) \, dx$$

is typical in the theory of quasi-minima, since it is possible to reduce to it all the integrals of the type

$$\mathcal{F}(u, \Omega) = \int_\Omega F(x, u, Du) \, dx ,$$

when the function $F(x,u,z)$ satisfies the inequalities

$$|z|^p - b|u|^\gamma - a(x) \le F(x,u,z) \le L|z|^p + b|u|^\gamma + a(x) \qquad (6.24)$$

with $L \ge 1$, $0 \le a \in L^1(\Omega)$, $\gamma < p^*$ and $b \in L^{\frac{p^*}{p^*-\gamma}}$.

More precisely, if u is a Q-minimum for $\mathcal{F}$, then it is a Q-minimum (with a different constant Q) for the functional $\mathcal{J}$. *Vice versa*, a Q-minimum for $\mathcal{J}$ is also a Q-minimum for $\mathcal{F} + \int (b|u|^\gamma + a)\,dx$.

The proof follows the same lines as that of Theorem 6.1. Taking any function v such that $u - v$ has support $K \subset\subset \Omega$, we have

$$\begin{aligned}\int_K (|Du|^p - b|u|^\gamma - a)\,dx &\le \int_K F(x,u,Du)\,dx \\ &\le Q\int_K F(x,v,Dv)\,dx \\ &\le Q\int_K (L|Dv|^p + b|v|^\gamma + a)\,dx\end{aligned}$$

and hence

$$\mathcal{J}(u,K) \le M\int_K (|Dv|^p + b|v|^\gamma + b|u|^\gamma + a)\,dx$$

and the conclusion follows as in Theorem 6.1.

We remark however that the lower estimate in (6.24) is natural in the scalar case ($N = 1$), much less if $N > 1$. As we shall see, it will be possible to substitute it with a less restrictive assumption. □

6.3 Cubical Quasi-Minima

A definition more general than that of quasi-minima is the following, which involves only integrals on cubes of $\mathbf{R}^n$.

Definition 6.3 *Let $Q > 0$. A function $u \in W^{1,p}_{\mathrm{loc}}(\Omega, \mathbf{R}^N)$ is called a* cubical Q-minimum *for the functional $\mathcal{F}$ if for every cube $Q_R \subset \Omega$ and for every $\varphi \in W^{1,p}_0(Q_R, \mathbf{R}^N)$ we have*

$$\mathcal{F}(u, Q_R) \le Q\mathcal{F}(u + \varphi, Q_R)\,.$$

In a similar way, considering balls of $\mathbf{R}^n$ instead of cubes, we could define spherical quasi-minima; in general one can define a quasi-minimum with respect to a one-parameter family of relatively compact homothetic domains.

It is clear that a Q-minimum is also a cubical, or spherical Q-minimum. In dimension $n > 2$, the two notions do not coincide, as can be seen from the following example.

Example 6.5 (GIAQUINTA and GIUSTI [4]) Let $n > 2$ and let $u(x)$ be a homogeneous function of degree β, $0 > \beta > 1 - \frac{n}{2}$, regular in $\mathbf{R}^n - 0$, without stationary points, and non-constant on the boundary of any cube of $\mathbf{R}^n$.[7]

We shall show that u is a cubical Q-minimum of the DIRICHLET integral

$$\int_\Omega |Du|^2\, dx\,.$$

It will suffice to prove that in any cube $Q_R = Q(x_0, R)$ we have

$$R\int_{Q_R} |Du|^2\, dx \le c\int_{\partial Q_R} |u - u_{\partial Q_R}|^2 dH_{n-1} \tag{6.25}$$

since by (3.42), if $v = u$ su ∂Q_R we have

$$\begin{aligned}\int_{\partial Q_R} |u - u_{\partial Q_R}|^2 dH_{n-1} &= \int_{\partial Q_R} |v - v_{\partial Q_R}|^2 dH_{n-1}\\ &\le cR\int_{Q_R} |Dv|^2\, dx\,.\end{aligned}$$

We can reduce to the cube $Q = Q(0,1)$ setting $x_0 = Ry_0$ and $x = R(y_0 + y)$. Taking into account the homogeneity of u and Du, (6.25) becomes

$$\begin{aligned}\int_Q |Du(y_0 + y)|^2\, dy \le c\Big\{ &⨍_{\partial Q} |u(y_0 + y)|^2 dH_{n-1}(y)\\ &- \Big(⨍_{\partial Q} u(y_0 + y) dH_{n-1}(y)\Big)^2\Big\}.\end{aligned} \tag{6.26}$$

Let $F(y_0)$ be the quantity on the left-hand side of (6.26), and let $G(y_0)$ be that within parentheses on the right. Both F and G are continuous positive functions, since u is not constant on the boundary of any cube of $\mathbf{R}^n$. The ratio F/G is therefore bounded on compact sets and we must only investigate its behavior when $y_0 \to \infty$. We have for every $y \in Q$:

$$u(y_0 + y) = u(y_0) + \langle Du(y_0), y\rangle + \frac{1}{2}\langle D^2u(y_0)y, y\rangle + O(|y_0|^{\beta-3})\,,$$

$$Du(y_0 + y) = Du(y_0) + O(|y_0|^{\beta-2})\,.$$

[7] An example of such a function is $u(x) = |x|^\beta$.

From the last equation we get

$$F(y_0) = |Q||Du(y_0)|^2 + O(|y_0|^{2\beta-3}),$$

and from the first:

$$|u(y_0 + y)|^2 = |u(y_0)|^2 + \langle Du(y_0), y\rangle^2 + 2u(y_0)\langle Du(y_0), y\rangle$$
$$+ u(y_0)\langle D^2u(y_0)y, y\rangle + O(|y_0|^{2\beta-3}),$$

and

$$\fint_{\partial Q} u(y_0 + y)dH_{n-1}(y) = u(y_0) + \frac{1}{2}\fint_{\partial Q}\langle D^2u(y_0)y, y\rangle dH_{n-1}$$
$$+ O(|y_0|^{\beta-3}).$$

It follows:

$$G(y_0) = \fint_{\partial Q}\langle Du(y_0), y\rangle^2 dH_{n-1}(y)O(|y_0|^{2\beta-3})$$
$$\geq c|Du(y_0)|^2 + O(|y_0|^{2\beta-3}).$$

We remark now that since Du is homogeneous of degree $\beta - 1$ and is never zero, we have $|Du(y_0)| \geq c|y_0|^{\beta-1}$, so that in conclusion the ratio F/G is bounded, and the function $u(x)$ is a cubical Q-minimum for the DIRICHLET integral.

On the other hand u is not a Q-minimum, since the function $v = \min\{u, 1\}$ is different from u in the unit ball B, and $\int_B |Dv|^2\,dx = 0$, whereas $\int_B |Du|^2\,dx \neq 0$.

As we have already remarked, the lower inequality (6.24) is rather restrictive, in particular when we are concerned with functions u with values in $\mathbf{R}^N$. For example, if $n = N = 2$, it is not satisfied by the function

$$F(z) = |z|^2 + 2\det(z). \tag{6.27}$$

In fact, condition (6.24) is appropriate in the case of functions $F(x, u, z)$ convex in z, much less so when F is only quasi-convex, as in the above example. In this case, it will be preferable to introduce a condition less simple but more general.

The following definition is a reinforcement of quasi-convexity.

Definition 6.4 *We say that the functional*

$$\tilde{\mathcal{F}}(u, \Omega) = \int_\Omega \tilde{F}(x, u, Du)\,dx$$

is strictly quasi-convex *in* $(x_0, u_0, z_0) \in \Omega \times \mathbf{R}^N \times \mathbf{R}^{nN}$ *if there exists a constant* $\nu > 0$ *such that for every* $\varphi \in W_0^{1,p}(\Omega, \mathbf{R}^N)$ *we have*

$$\int_\Omega [\tilde{F}(x_0, u_0, z_0 + D\varphi) - \tilde{F}(x_0, u_0, z_0)]\, dx \geq \nu \int_\Omega |D\varphi|^p\, dx\,. \tag{6.28}$$

Correspondingly, the function $\tilde{F}(x, u, z)$ *is said to be* strictly quasi-convex in (x_0, u_0, z_0).

Finally, we say that $\tilde{F}$ *is* strictly quasi-convex *if it is so everywhere, with the constant* ν *independent of the point* (x_0, u_0, z_0).

Let us now consider a function $F(x, u, z)$ satisfying the estimate

$$|F(x, u, z)| \leq c(|z| + \vartheta(x, u))^p \tag{6.29}$$

with

$$\vartheta(x, u)^p = b(x)|u|^\gamma + a(x)\,.$$

Instead of (6.24) we shall assume that there exists a function $\tilde{F}(z)$, depending only on z and strictly quasi-convex in 0, such that for every $(x, u, z) \in \Omega \times \mathbf{R}^N \times \mathbf{R}^{nN}$ we have

$$F(x, u, z) \geq \tilde{F}(z) - \vartheta(x, u)^p\,. \tag{6.30}$$

Adding possibly a constant to the function ϑ, we can assume that $\tilde{F}(0) = 0$.

We remark that since $|z|^p$ is strictly quasi-convex in 0, (6.30) is really more general than (6.24); actually it is satisfied by the function (6.27).

Concerning the exponents p and γ, and the functions $b(x)$ and $a(x)$, we shall make the usual assumptions:

$$\text{(i)}\quad 1 < p \leq n\,, \quad p \leq \gamma < p^* = \frac{pn}{n-p}\,, \tag{6.31}$$

$$\text{(ii)}\quad b(x) \in L^{\frac{p^*}{p^*-\gamma}}\,, \quad a(x) \in L^1\,. \tag{6.32}$$

The following theorem holds:

Theorem 6.5 (CACCIOPPOLI's inequality) *Assume that the function* $u \in W^{1,p}(\Omega, \mathbf{R}^N)$ *is a cubical quasi-minimum for the functional*

$$\mathcal{F}(u, \Omega) = \int_\Omega F(x, u(x), Du(x))\, dx \tag{6.33}$$

with $F(x, u, z)$ satisfying (6.29) and (6.30) above. There exists $R_0 > 0$, depending only on u, such that for $R < R_0$ and $Q_R \subset\subset \Omega$ we have:

$$\int_{Q_{R/2}} (|Du|^p + |u|^{p^*})\,dx \le c\left\{\frac{1}{R^p}\int_{Q_R} |u - u_R|^p\,dx \right.$$

$$\left. + |Q_R|\left(\fint_{Q_R} |u|\,dx\right)^{p^*} + \int_{Q_R} g\,dx\right\} \tag{6.34}$$

and moreover

$$\fint_{Q_{R/2}} (|Du|^p + |u|^{p^*})\,dx$$

$$\le c\left\{\left(\fint_{Q_R} (|Du|^p + |u|^{p^*})^m\,dx\right)^{\frac{1}{m}} + \fint_{Q_R} g(x)\,dx\right\}, \tag{6.35}$$

where $m = \frac{n}{p+n} < 1$ and $g = a + b^{\frac{p^*}{p^*-\gamma}}$.

Remark 6.7 The reader can easily convince himself that the same result holds for spherical quasi-minima, and in general for quasi-minima relative to a general family of neighborhoods, provided each of them contains a cube Q_1 and is contained in a cube Q_2 with the ratio of sides bounded.

Moreover, it continues to hold if instead of quasi-minima we deal with functions satisfying (6.7) with $K = Q_R$. □

To the proof of Theorem 6.5 we shall premise the following:

Lemma 6.1 *Let $Z(t)$ be a bonded non-negative function in the interval $[\varrho, R]$. Assume that for $\varrho \le t < s \le R$ we have*

$$Z(t) \le [A(s-t)^{-\alpha} + B(s-t)^{-\beta} + C] + \vartheta Z(s) \tag{6.36}$$

with A, B, $C \ge 0$, $\alpha > \beta > 0$ and $0 \le \vartheta < 1$. Then,

$$Z(\varrho) \le c(\alpha, \vartheta)[A(R-\varrho)^{-\alpha} + B(R-\varrho)^{-\beta} + C]. \tag{6.37}$$

Proof. Consider the sequence t_i such that $t_0 = \varrho$ and

$$t_{i+1} - t_i = (1-\lambda)\lambda^i(R-\varrho)$$

with $0 < \lambda < 1$.

From (6.36) by induction we get

$$Z(\varrho) \le \vartheta^k Z(t_k) + \left[\frac{A}{(1-\lambda)^\alpha (R-\varrho)^\alpha} + \frac{B}{(1-\lambda)^\beta (R-\varrho)^\beta} + C\right] \times \sum_{i=0}^{k-1} \vartheta^i \lambda^{-i\alpha}.$$

Now choose λ in such a way that $\lambda^{-\alpha}\vartheta < 1$. The series on the right-hand side converges, and therefore passing to the limit for $k \to \infty$, we get the conclusion with $c(\alpha, \vartheta) = (1-\lambda)^{-\alpha}(1-\vartheta\lambda^{-\alpha})^{-1}$.

We can now prove the theorem. Let Q_R be a cube strictly contained in Ω, and let $R/2 < t < s \le R$. Let $\eta(x)$ be a function in $C_0^\infty(Q_s)$, with $0 \le \eta \le 1$, $\eta \equiv 1$ in Q_t and $|D\eta| \le \frac{2}{s-t}$. Denoting by u_s the average of u in Q_s:

$$u_s = \fint_{Q_s} u \, dx ,$$

we set $\varphi = \eta(u - u_s)$. We have

$$\begin{aligned} \nu \int_{Q_s} |D\varphi|^p \, dx &\le \int \tilde{F}(D\varphi) \, dx \le \int_{Q_s} F(x, u, D\varphi) \, dx + \int_{Q_s} \vartheta(x, u)^p \, dx \\ &= \int_{Q_s} F(x, u, Du) \, dx + \int_{Q_s} [F(x, u, D\varphi) - F(x, u, Du)] \, dx \\ &\quad + \int_{Q_s} \vartheta(x, u)^p \, dx . \end{aligned} \tag{6.38}$$

Now let $v = u - \varphi = u_s + (1-\eta)(u - u_s)$. From the quasi-minimum property of u and from (6.29) we have

$$\int_{Q_s} F(x, u, Du) \, dx \le c \int_{Q_s} (|Dv| + \vartheta(x, v))^p \, dx . \tag{6.39}$$

We remark now that $D\varphi = Du$ in Q_t, and therefore the second integral on the right-hand side of (6.38) can be estimated by

$$\begin{aligned} &\int_{Q_s - Q_t} (|F(x, u, D\varphi)| + |F(x, u, Du)|) \, dx \\ &\quad \le \int_{Q_s - Q_t} (|Du|^p + |Dv|^p + \vartheta(x, u)^p) \, dx . \end{aligned}$$

Introducing these relations in (6.38), we get

$$\int_{Q_t} |Du|^p\,dx \le c\int_{Q_s-Q_t} |Du|^p\,dx + c\int_{Q_s} |Dv|^p\,dx$$
$$+ \int_{Q_s} (\vartheta(x,u)^p + \vartheta(x,v)^p)\,dx$$
$$\le c\int_{Q_s-Q_t} |Du|^p\,dx + c\int_{Q_s} |Dv|^p\,dx$$
$$+ c\int_{Q_s} [a(x) + b(x)(|v|^\gamma + |u|^\gamma)]\,dx\,. \tag{6.40}$$

We now add to both sides $\int_{Q_s} |u|^{p^*}\,dx$, and we use the inequalities

$$|Dv|^p = |(1-\eta)Du + (u-u_s)D\eta|^p$$
$$\le c[(1-\eta)^p|Du|^p + (s-t)^{-p}|u-u_s|^p]$$

and

$$b|u|^\gamma \le c\left(|u|^{p^*} + b^{\frac{p^*}{p^*-\gamma}}\right),$$

as well as $|v| \le |u-u_s| + |u_s|$ and $|u| \le |u-u_s| + |u_s|$. We obtain thus

$$\int_{Q_t} (|Du|^p + |u|^{p^*})\,dx \le c\biggl\{\int_{Q_s-Q_t} |Du|^p\,dx + \frac{1}{(s-t)^p}\int_{Q_s} |u-u_s|^p\,dx$$
$$+ \int_{Q_s} |u-u_s|^{p^*}\,dx + |Q_s||u_s|^{p^*} + \int_{Q_s} g\,dx\biggr\} \tag{6.41}$$

with $g = a + b^{p^*/(p^*-\gamma)}$.

On the other hand:

$$\int_{Q_s} |u-u_s|^{p^*}\,dx \le c\left(\int_{Q_s} |Du|^p\,dx\right)^{\frac{p^*}{p}} = \chi(s)\int_{Q_s} |Du|^p\,dx\,,$$

where

$$\chi(s) = \left(\int_{Q_s} |Du|^p\,dx\right)^{p/(n-p)}$$

is infinitesimal with s. It follows that, taking R (and therefore s) small enough, this term can be made smaller than $\epsilon\int_{Q_s} |Du|^p\,dx$, which in turn, choosing suitably $\epsilon > 0$, can be subtracted from the left-hand side, leaving on the right the quantity

$$\int_{Q_s - Q_t} |Du|^p \, dx \, .$$

We arrive thus at the inequality

$$\int_{Q_t} (|Du|^p + |u|^{p^*}) \, dx \le c \Big\{ \int_{Q_s - Q_t} |Du|^p \, dx + \frac{1}{(s-t)^p} \int_{Q_s} |u - u_s|^p \, dx + |Q_s||u_s|^{p^*} + \int_{Q_s} g \, dx \Big\}.$$

Moreover, since we have

$$\int_{Q_s} |u - u_s|^p \, dx \le c \int_{Q_R} |u - u_R|^p \, dx \, ,$$

$$|u_s| \le 2^n \fint_{Q_R} |u| \, dx \, ,$$

(remember that $s \ge R/2$), we get

$$\int_{Q_t} (|Du|^p + |u|^{p^*}) \, dx \le c \Big\{ \int_{Q_s - Q_t} (|Du|^p + |u|^{p^*}) \, dx + \frac{1}{(s-t)^p} \int_{Q_R} |u - u_R|^p \, dx + |Q_R| \Big(\fint_{Q_R} |u| \, dx \Big)^{p^*} + \int_{Q_R} g \, dx \Big\}.$$

We now use the "hole filling" method by WIDMAN [1]. We add to both sides the quantity

$$c \int_{Q_t} (|Du|^p + |u|^{p^*}) \, dx$$

and divide by $c+1$; we obtain

$$\int_{Q_t} (|Du|^p + |u|^{p^*}) \, dx \le \vartheta \int_{Q_s} (|Du|^p + |u|^{p^*}) \, dx + \frac{1}{(s-t)^p} \int_{Q_R} |u - u_R|^p \, dx + |Q_R| \Big(\fint_{Q_R} |u| \, dx \Big)^{p^*} + \int_{Q_R} g \, dx$$

with $\vartheta =: \frac{c}{c+1} < 1$.

Applying Lemma 6.1, with

$$Z(t) = \int_{Q_t} (|Du|^p + |u|^{p^*})\, dx$$

and

$$A = \int_{Q_R} |u - u_R|^p\, dx\,, \quad B = 0\,, \quad C = |Q_R| \left(\fint_{Q_R} |u|\, dx \right)^{p^*} + \int_{Q_R} g\, dx\,,$$

we obtain immediately the inequality (6.34).

In order to get (6.35) we must estimate the right-hand side of (6.34). Setting $p_* = \frac{np}{n+p}$, we have $(p_*)^* = p$, and hence, by the SOBOLEV–POINCARÉ inequality (3.32):

$$\int_{Q_R} |u - u_R|^p\, dx \le c \left(\int_{Q_R} |Du|^{p_*}\, dx \right)^{\frac{p}{p_*}} = c \left(\int_{Q_R} |Du|^{pm}\, dx \right)^{\frac{1}{m}}$$

with $m = \frac{n}{p+n}$.

On the other side we have $p^* m > 1$, and therefore

$$\fint_{Q_R} |u|\, dx \le \left(\fint_{Q_R} |u|^{p^* m}\, dx \right)^{\frac{1}{p^* m}}.$$

Introducing these two inequalities[8] into (6.34) we get easily the required estimate (6.35). □

Remark 6.8 In what follows we shall need the above theorem also in a slightly different form. We begin from (6.40), and we estimate $|Dv|^p$ as above, and

$$|v|^\gamma \le c(|u|^\gamma + |u - u_s|^\gamma)\,.$$

We have now

$$\begin{aligned} \int_{Q_s} b(x)(|u - u_s|^\gamma)]\, dx &\le \left(\int_{Q_s} |u - u_s|^{p^*}\, dx \right)^{\frac{p}{p^*}} \\ &\quad \times \left(\int_{Q_s} (b(x)|u - u_s|^{\gamma - p})^{\frac{n}{p}}\, dx \right)^{\frac{p}{n}} \\ &\le c\xi(s) \int_{Q_s} |Du|^p\, dx \end{aligned}$$

[8] If $p_* < 1$, the preceding inequalities continue to hold with $m = \frac{1}{p}$.

with

$$\xi(s) = \left(\int_{Q_s} (b(x)|u - u_s|^{\gamma-p})^{\frac{n}{p}}\,dx\right)^{\frac{p}{n}}$$

$$\le c\left(\int_{Q_s} |u|^{p^*}\,dx\right)^{\frac{\gamma-p}{p^*}} \left(\int_{Q_s} b^{\frac{p^*}{p^*-\gamma}}\,dx\right)^{1-\frac{\gamma}{p^*}}$$

$$\le c\|u\|_{p^*}^{\gamma-p}\|b\|_\sigma s^{n\epsilon}\,.$$

If we choose R (and therefore s) small enough we have $c\xi(R) < \frac{1}{2}$ and hence we can subtract the corresponding term from the left-hand side of (6.40), getting

$$\int_{Q_t} |Du|^p\,dx \le c\left(\int_{Q_s - Q_t} |Du|^p\,dx + \frac{1}{(s-t)^p}\int_{Q_s} |u - u_s|^p\,dx\right)$$
$$+ c\int_{Q_s} (a(x) + b(x)|u|^\gamma)\,dx\,. \tag{6.42}$$

We can now argue as in Theorem 6.5, and we can conclude that

$$\int_{Q_{R/2}} |Du|^p\,dx \le c\left\{\frac{1}{R^p}\int_{Q_R} |u - u_R|^p\,dx + \int_{Q_R} (a + b|u|^\gamma)\,dx\right\}. \tag{6.43}$$

□

Remark 6.9 We note that the above theorem remains valid if we only assume that u verifies

$$\mathcal{F}(u, Q_R) \le Q\int_{Q_R} (|Dv| + \vartheta(x, v))^p\,dx + \epsilon\int_{Q_R} |Du|^p\,dx$$

for every v with $u - v \in W_0^{1,p}(Q_R)$, provided ϵ is small enough. □

Remark 6.10 If the function $F(x, u, z)$ verifies (6.29) and (6.30) with $\gamma < p$ and $b \in L^{\frac{p}{p-\gamma}}$ (in particular this happens if $b = 0$), inequality (6.34) takes the simpler form

$$\int_{Q_{R/2}} |Du|^p\,dx \le c\left\{\frac{1}{R^p}\int_{Q_R} |u - u_R|^p\,dx + \int_{Q_R} a\,dx\right.$$
$$\left. + |Q_R|\left(\fint_{Q_R} |u|\,dx\right)^p\right\} \tag{6.44}$$

and holds for every $R < R_0$, with R_0 independent of u.

Actually in this case we can avoid summing to both sides the quantity $|u|^{p^*}$, and the term $b|u-u_s|^\gamma$ can be estimated by $c(b^{\frac{p}{p-\gamma}}+|u-u_s|^p)$. The integral of the last quantity can be estimated by means of POINCARÉ's inequality (3.33), and can be subtracted from the left-hand side if R is small enough, independently of u.

If moreover $b=0$, (6.44) holds without the last term on the right. In particular, this happens when u is a bounded function, since in this case $\vartheta(x,u)$ can be considered as a function of x only. This always happens if $p>n$, since by SOBOLEV theorem (Theorem 3.11) the function u is Hölder-continuous. □

6.4 L^p Estimates for the Gradient

Setting $f(x)=|Du|^p+|u|^{p^*}$ and writing $2R$ instead of R, inequality (6.35) becomes:

$$\fint_{Q_R} f(x)\,dx \le c\left\{\left(\fint_{Q_{2R}} f^m\,dx\right)^{\frac{1}{m}}+\fint_{Q_{2R}} g(x)\,dx\right\}. \tag{6.45}$$

The purpose of this section is to show how (6.45) implies higher summability of the function f, and hence of the derivatives of u, under the assumption that the function g belongs to some L^r, with $r>1$.

For that, let us begin by considering the case of functions f and g defined in the cube $Q =: Q_{x_0,1}$. Let $d(x)=\operatorname{dist}(x,\partial Q)$, and for $k=0,1,2,\ldots$ define

$$C_k=\left\{x\in Q : \frac{3}{4}2^{-k-1}\le d(x)\le \frac{3}{4}2^{-k}\right\}.$$

Each shell C_k can be divided into a finite family $\mathcal{G}_k$ of equal cubes, each of side $\delta_k=\frac{3}{4}2^{-k-1}$. If to the union of these cubes we add the cube $Q_{1/4}$ concentric to Q, we obtain the whole Q.

Assume now that (6.45) is satisfied for every cube $Q_{2R}\subset\subset Q$. If P is a cube, we denote by $\tilde P$ the cube concentric to P and with double side, so that (6.45) becomes

$$\fint_P f(x)\,dx \le c\left\{\left(\fint_{\tilde P} f^m\,dx\right)^{\frac{1}{m}}+\fint_{\tilde P} g\,dx\right\} \tag{6.46}$$

for every P such that $\tilde P\subset\subset Q$.

If in addition $P \subset C_k$, we have $\frac{3}{16}2^{-k} \le d(x) \le \frac{15}{16}2^{-k}$ for every $x \in \tilde{P}$, and therefore in particular $\tilde{P} \subset\subset Q$, and

$$\fint_P F(x)\,dx \le B\left\{\left(\fint_{\tilde{P}} F^m\,dx\right)^{\frac{1}{m}} + \fint_{\tilde{P}} G\,dx\right\}, \tag{6.47}$$

where we have set

$$F(x) = d(x)^n f(x)\,; \quad G(x) = d(x)^n g(x)\,.$$

It is easy to check that (6.47) holds also for $P \subset Q_{1/4}$.

Lemma 6.2 *For every t with*

$$t > t_0 =: \fint_Q f(x)\,dx\,,$$

setting

$$\Phi_t = \{x \in Q : F(x) > t\}\,; \quad \Gamma_t = \{x \in Q : G(x) > t\}$$

we have

$$\int_{\Phi_t} F\,dx \le c\left\{t^{1-m}\int_{\Phi_t} F^m\,dx + \int_{\Gamma_t} G\,dx\right\}. \tag{6.48}$$

Proof. Let $s = \lambda t$, where λ is a constant that we shall fix during the proof. If $P \in \mathcal{G}_k$ we have

$$\begin{aligned} s > \lambda \fint_Q f(x)\,dx &\ge \lambda\frac{|P|}{|Q|}\fint_P f(x)\,dx \\ &\ge \lambda 4^{-n}\fint_P F(x)\,dx \ge \fint_P F(x)\,dx\,, \end{aligned}$$

whenever $\lambda \ge 4^n$. The above relation remains valid if $P = Q_{1/4}$.

To each of the cubes P we can apply CALDERON–ZYGMUND theorem (Theorem 2.10). In this way we obtain a countable family $\{Q_j\}$ of disjoint subcubes of Q, such that

$$s < \fint_{Q_j} F(x)\,dx \le 2^n s$$

and $F(x) \le s$ in $Q - \cup Q_j$.

From (6.47) we infer that either

$$\fint_{Q_j} F(x)\,dx \le 2B\left(\fint_{\tilde{Q}_j} F^m\,dx\right)^{\frac{1}{m}} \tag{6.49}$$

or

$$\fint_{Q_j} F(x)\,dx \le 2B \fint_{\tilde{Q}_j} G\,dx\,. \tag{6.50}$$

In the first case, we have

$$s \le 2B \left\{ \fint_{\tilde{Q}_j} F^m\,dx \right\}^{\frac{1}{m}}$$

and therefore:

$$s^m |\tilde{Q}_j| \le (2B)^m \int_{\tilde{Q}_j} F^m\,dx\,.$$

Moreover, from the inequality

$$\int_{\tilde{Q}_j} F^m\,dx \le \int_{\tilde{Q}_j \cap \Phi_t} F^m\,dx + t^m |\tilde{Q}_j|$$

we deduce, provided $(2B)^m \lambda^{-m} \le \frac{1}{2}$:

$$|\tilde{Q}_j| \le 2(2B)^m s^{-m} \int_{\tilde{Q}_j \cap \Phi_t} F^m\,dx\,. \tag{6.51}$$

If instead (6.50) holds, we have

$$s|\tilde{Q}_j| \le 2B \int_{\tilde{Q}_j} G\,dx$$

and consequently

$$|\tilde{Q}_j| \le 4Bs^{-1} \int_{\tilde{Q}_j \cap \Gamma_t} G\,dx\,. \tag{6.52}$$

In conclusion, we have in any case:

$$|\tilde{Q}_j| \le \frac{c}{s} \left(t^{1-m} \int_{\tilde{Q}_j \cap \Phi_t} F^m\,dx + \int_{\tilde{Q}_j \cap \Phi_t} G\,dx \right). \tag{6.53}$$

Let us now evaluate the integral of F over Φ_s:

$$\begin{aligned} \int_{\Phi_s} F\,dx &\le \sum_{j=1}^{\infty} \int_{Q_j} F\,dx \le 2^n s \sum_{j=1}^{\infty} |Q_j| \\ &\le 2^n s \left| \bigcup_{j=1}^{\infty} \tilde{Q}_j \right|. \end{aligned} \tag{6.54}$$

We want to estimate the last quantity by means of (6.53). For that, we apply Lemma 2.4 to the family $\{Q_j\}$, and we obtain a countable subfamily of pairwise disjoint cubes $\{\Pi_i\}$ such that, denoting by $\hat{P}$ the cube concentric with P and of quintuple side,

$$\cup \tilde{Q}_j \subset \cup \hat{\Pi}_i \,.$$

We have therefore

$$|\cup \tilde{Q}_j| \leq 5^n \sum_{i=1}^{\infty} |\Pi_i| \,,$$

from which, recalling that the cubes Π_i are disjoint, and using (6.53) and (6.54), we get:

$$\int_{\Phi_s} F\, dx \leq c\left\{t^{1-m}\int_{\Phi_t} F^m\, dx + \int_{\Gamma_t} G(x)\, dx\right\}.$$

On the other hand, we also have

$$\int_{\Phi_t - \Phi_s} F\, dx \leq s^{1-m}\int_{\Phi_t} F^m\, dx \leq ct^{1-m}\int_{\Phi_t} F^m\, dx$$

from which (6.48) follows. □

We need now the following:

Lemma 6.3 *Let $h \geq m \geq 0$ and let $F \in L^h(Q)$. Setting*

$$\varphi(t) = \int_{\Phi(t)} F^m\, dx \,,$$

we have

$$\int_{\Phi_\tau} F^h\, dx = -\int_\tau^\infty t^{h-m}\, d\varphi(t) \,.$$

Proof. We can assume that F is bounded, and that $\varphi(t)$ is continuous at the point τ, since the general result follows by approximation. We have

$$\int_{\Phi_\tau} F^h\, dx = \tau^{h-m}\int_{\Phi_\tau} F^m\, dx + (h-m)\int_{\Phi_\tau} F^m\, dx$$
$$\times \int_\tau^{F(x)} t^{h-m-1}\, dt \,.$$

On the other hand, if χ_t is the characteristic function of Φ_t, we have

$$\int_{\Phi_\tau} F^m\,dx \int_\tau^{F(x)} t^{h-m-1}\,dt = \int_{\Phi_\tau} F^m\,dx \int_\tau^{\infty} t^{h-m-1}\chi_t(x)\,dt$$

$$= \int_\tau^{\infty} t^{h-m-1}\,dt \int_{\Phi_t} F^m\,dx$$

$$= \int_\tau^{\infty} t^{h-m-1}\varphi(t)\,dt$$

and the conclusion follows by integration by parts. □

With the preceding notation, (6.48) can be written in the form

$$-\int_t^\infty \tau^{1-m}\,d\varphi(\tau) \le A[t^{1-m}\varphi(t) + \omega(t)]\,, \tag{6.55}$$

where

$$\omega(t) = \int_{\Gamma_t} G\,dx\,.$$

Proposition 6.1 (GEHRING [1]) *Assume that $\varphi(t)$ is a decreasing function in $[a,+\infty)$, infinitesimal for $t \to +\infty$, and verifying* (6.55) *with $m < 1$ for every $t \ge a$. There exists a real number $r > 1$ such that*

$$-\int_a^\infty u^{r-m}\,d\varphi(u) \le -2a^{r-1}\int_a^\infty u^{1-m}\,d\varphi(u)$$

$$- 2A\int_a^\infty u^{r-1}\,d\omega(u)\,. \tag{6.56}$$

Proof. Let us begin by assuming that $\varphi(s) \equiv 0$ and $\omega(s) \equiv 0$ for $s \ge k-1$.

For $q > 0$ we set

$$I_q(s) = -\int_s^k u^q\,d\varphi(u)\,; \quad I_q = I_q(a)\,;$$

$$\Omega_q = -\int_a^k u^q\,d\omega(u)\,.$$

We have

$$I_{r-m} = -\int_a^k u^{r-1}u^{1-m}\,d\varphi(u) = -\int_a^k u^{r-1}dI_{1-m}(u)$$

$$= a^{r-1}I_{1-m} + (r-1)\int_a^k u^{r-2}I_{1-m}(u)\,du\,.$$

The last integral can be estimated by means of (6.55):

$$I_{r-m} \le a^{r-1} I_{1-m} + A(r-1)\left(\int_a^k u^{r-m-1}\varphi(u)\,du + \int_a^k u^{r-2}\omega(u)\,du\right).$$

On the other hand, integrating by parts, we get

$$\int_a^k u^{r-m-1}\varphi(u)\,du = \frac{I_{r-m}}{r-m} - \frac{a^{r-m}}{r-m}\varphi(a) \le \frac{I_{r-m}}{r-m}$$

and similarly

$$\int_a^k u^{r-2}\omega(u)\,du \le \frac{\Omega_{r-1}}{r-1};$$

it follows that

$$I_{r-m} \le a^{r-1} I_{1-m} + A\frac{r-1}{r-m} I_{r-m} + A\Omega_{r-1}.$$

If we assume now that $A(r-1) \le \frac{r-m}{2}$, we conclude that

$$I_{r-m} \le 2a^{r-1} I_{1-m} + 2A\Omega_{r-1}$$

and (6.56) is proved when $\varphi(t) = \omega(t) = 0$ from some point on.

In the general case, we remark in the first place that

$$-\int_k^T s^{1-m}\,d\varphi(s) \ge -k^{1-m}\int_k^T d\varphi(s) = k^{1-m}[\varphi(k) - \varphi(T)]$$

so that, letting $T \to +\infty$:

$$-\int_k^\infty s^{1-m}\,d\varphi(s) \ge -k^{1-m}\varphi(k). \tag{6.57}$$

Now setting

$$\varphi_k(t) = \begin{cases} \varphi(t) & \text{if } t \le k, \\ 0 & \text{if } t > k \end{cases}$$

(and the analogue for ω_k), and taking into account (6.55), we get for $t \le k$:

$$\begin{aligned} -\int_t^\infty s^{1-m}\,d\varphi_k(s) &= -\int_t^k s^{1-m}\,d\varphi(s) + k^{1-m}\varphi(k) \\ &\le -\int_t^\infty s^{1-m}\,d\varphi(s) \le A(t^{1-m}\varphi_k(t) + \omega_k(t)). \end{aligned}$$

The preceding relation obviously remains valid for $t > k$. For what has been just proved, we therefore have:

$$-\int_a^\infty s^{r-m}\,d\varphi_k(s) \le -2a^{r-1} - \int_a^\infty s^{1-m}\,d\varphi_k(s) - 2A\int_a^\infty s^{r-1}\,d\omega_k(s)$$

$$\le -2a^{r-1} - \int_a^\infty s^{1-m}\,d\varphi(s) - 2A\int_a^\infty s^{r-1}\,d\omega(s)$$

and the conclusion follows letting $k \to +\infty$. □

A simple application of the preceding proposition to Lemma 6.2 leads directly to the inequality

$$\int_{\Phi_a} F^r\,dx \le 2a^{r-1}\int_{\Phi_a} F\,dx + 2A\int_{\Gamma_a} G^r\,dx\,,$$

with $a = \fint_Q f\,dx$.

On the other hand

$$\int_{Q-\Phi_a} F^r\,dx \le a^{r-1}\int_{Q-\Phi_a} F\,dx\,,$$

and therefore in conclusion

$$\int_Q F^r\,dx \le 2a^{r-1}\int_Q F\,dx + 2A\int_Q G^r\,dx\,. \tag{6.58}$$

Coming back to the functions $f(x)$ and $g(x)$, we find:

$$\int_{Q_{1/2}} f^r\,dx \le c\left(a^{r-1}\int_Q f\,dx + \int_Q g^r\,dx\right)$$

or else

$$\fint_{Q_{1/2}} f^r\,dx \le c\left\{\left(\fint_Q f\,dx\right)^r + \fint_Q g^r\,dx\right\}. \tag{6.59}$$

If instead of the cube Q of side 1 we deal with a cube Q_R of side $2R$, we obtain by means of a simple homothety:

Theorem 6.6 (GIAQUINTA and G. MODICA [1]) *Let $f \in L^1(Q_R)$, and assume that for every cube $Q \subset \tilde{Q} \subset\subset Q_R$ we have*

$$\fint_Q f(x)\,dx \le B\left\{\left(\fint_{\tilde{Q}} f^m\,dx\right)^{\frac{1}{m}} + \fint_{\tilde{Q}} g\,dx\right\} \tag{6.60}$$

with $0 < m < 1$. Assume that the function g belongs to $L^s(Q_R)$ for some $s > 1$.

Then there exists an $r > 1$ such that $f \in L^r(Q_{R/2})$, and moreover:

$$\fint_{Q_{R/2}} f^r\,dx \le c\left\{\left(\fint_{Q_R} f\,dx\right)^r + \fint_{Q_R} g^r\,dx\right\}. \tag{6.61}$$

Corollary 6.1 *The conclusion of the preceding theorem holds if we replace assumption* (6.60) *with*

$$\fint_Q f(x)\,dx \le \epsilon \fint_{\lambda Q} f\,dx + B\left\{\left(\fint_{\lambda Q} f^m\,dx\right)^{\frac{1}{m}} + \fint_{\lambda Q} g\,dx\right\} \tag{6.62}$$

with $\lambda \ge 1$, provided ϵ is less than a number ϵ_0 depending only on n and λ.

Proof. Let $t < s \le R$. The cube Q_t can be covered by cubes Q_i of side $r = \frac{s-t}{\lambda}$, in such a way that at most $N(n, \lambda)$ cubes λQ_i intersect. The total number of the cubes Q_i does not exceed $c(s-t)^{-n}(\lambda t)^n$.

From (6.62) for Q_i it follows

$$\int_{Q_i} f\,dx \le \epsilon \int_{\lambda Q_i} f\,dx + c(\lambda)(s-t)^{n-\frac{n}{m}}\left(\int_{\lambda Q_i} f^m\,dx\right)^{\frac{1}{m}} + c\int_{\lambda Q_i} g\,dx\,.$$

Summing over i we get:

$$\int_{Q_t} f\,dx \le \epsilon N \int_{Q_s} f\,dx + c(\lambda)t^n(s-t)^{-\frac{n}{m}}\left(\int_{Q_s} f^m\,dx\right)^{\frac{1}{m}} + cN\int_{Q_s} g\,dx$$

and if $R < t < s < 2R$,

$$\int_{Q_t} f\,dx \le \epsilon N \int_{Q_s} f\,dx + c(\lambda)R^n(s-t)^{-\frac{n}{m}}\left(\int_{Q_{2R}} f^m\,dx\right)^{\frac{1}{m}} + cN\int_{Q_{2R}} g\,dx\,.$$

If $\epsilon N < 1$ we can apply the Lemma 6.1, getting

$$\fint_{Q_R} f\,dx \le c(n,\lambda)\left\{\left(\fint_{Q_{2R}} f^m\,dx\right)^{\frac{1}{m}} + \fint_{Q_{2R}} g\,dx\right\}$$

so that we are reduced to Theorem 6.6. □

Applying the preceding theorem to quasi-minima we obtain:

Theorem 6.7 *Let $u : \Omega \to \mathbf{R}^N$ be a cubical quasi-minimum for the functional*

$$\mathcal{F}(u, \Omega) = \int_\Omega F(x, u(x), Du(x))\,dx\,,$$

and assume that the hypotheses of Theorem 6.5 are satisfied, in particular

$$\tilde{F}(z) - \vartheta(x,u)^p \le F(x,u,z) \le c(|z| + \vartheta(x,u))^p$$

with $\tilde{F}(z)$ *strictly quasi-convex in* 0, $\vartheta(x,u)^p = b(x)|u|^\gamma + a(x)$, *and with the functions* $b \in L^\sigma$, $\sigma > \frac{p^*}{p^*-\gamma}$ *and* $a \in L^s$, $s > 1$.

Then the function $|Du|^p + |u|^{p^*}$ *belongs to* $L^r_{\mathrm{loc}}(\Omega)$ *for some* $r > 1$, *and moreover for every cube* $Q_R \subset Q_{2R} \subset\subset \Omega$ *we have:*

$$\begin{aligned}&\fint_{Q_R} (|Du|^p + |u|^{p^*})^r \, dx \\ &\qquad \le c\left\{ \left(\fint_{Q_{2R}} (|Du|^p + |u|^{p^*}) \, dx \right)^r + \fint_{Q_{2R}} g^r \, dx \right\}, \end{aligned} \tag{6.63}$$

where $g = a + b^{\frac{p^*}{p^*-\gamma}}$.

Remark 6.11 If we use the estimate (6.43) instead of (6.35), and we take into account that the function $g(x) = a(x) + b(x)|u|^\gamma$ belongs to L^r for some $r > 1$, we get the inequality

$$\begin{aligned}\fint_{Q_R} |Du|^{rp} \, dx \le c\left(\fint_{Q_{2R}} |Du|^p \, dx \right)^r \\ + c\fint_{Q_{2R}} (a(x) + b(x)|u|^\gamma)^r \, dx \,. \end{aligned} \tag{6.64}$$

□

Remark 6.12 The estimate (6.63) can be further ameliorated. We have actually

$$\begin{aligned}&\fint_{Q_R} (|Du|^p + |u|^{p^*})^r \, dx \\ &\qquad \le c(q)\left\{ \left(\fint_{Q_{2R}} (|Du|^p + |u|^{p^*})^q \, dx \right)^{\frac{r}{q}} + \fint_{Q_{2R}} g^r \, dx \right\} \end{aligned} \tag{6.65}$$

for every $q > 0$ and every cube $Q_R \subset Q_{2R} \subset\subset \Omega$.

Setting $\psi(x) = |Du|^p + |u|^{p^*}$, let $y \in Q_{\alpha\varrho}$, and write (6.63) for the cube of radius $R = \frac{1-\alpha}{2}\varrho$ and center in y. We have

$$\begin{aligned}\int_{Q(y,\frac{1-\alpha}{2}\varrho)} \psi^r \, dx \le c[(1-\alpha)\varrho]^{n(1-r)} \left(\int_{Q(y,(1-\alpha)\varrho)} \psi \, dx \right)^r \\ + c\int_{Q(y,(1-\alpha)\varrho)} g \, dx \,. \end{aligned}$$

The cube $Q_{\alpha\varrho}$ can be covered by cubes of that sort, in such a way that only a finite number N (independent of α) of cubes of double side intersect. We then have:

$$\int_{Q_{\alpha\varrho}} \psi^r \, dx \leq c[(1-\alpha)\varrho]^{n(1-r)} \left(\int_{Q_\varrho} \psi \, dx \right)^r + c \int_{Q_\varrho} g \, dx \, .$$

On the other side

$$\int_{Q_\varrho} \psi \, dx \leq \left(\int_{Q_\varrho} \psi^r \, dx \right)^{\frac{1-q}{r-q}} \left(\int_{Q_\varrho} \psi^q \, dx \right)^{\frac{r-1}{r-q}}$$

so that, setting $s = \varrho$, $t = \alpha\varrho$ and

$$U_s = \int_{Q_s} \psi^r \, dx$$

we have

$$\begin{aligned} U_t &\leq c(s-t)^{n(1-r)} U_s^{\frac{r(1-q)}{r-q}} \left(\int_{Q_s} \psi^q \, dx \right)^{r\frac{r-1}{r-q}} + c \int_{Q_s} g \, dx \\ &\leq \frac{1}{2} U_s + c(s-t)^{\frac{n(r-q)}{q}} \left(\int_{Q_s} \psi^q \, dx \right)^{\frac{r}{q}} + c \int_{Q_s} g \, dx \\ &\leq \frac{1}{2} U_s + c(s-t)^{\frac{n(r-q)}{q}} \left(\int_{Q_{2R}} \psi^q \, dx \right)^{\frac{r}{q}} + c \int_{Q_{2R}} g \, dx \, . \end{aligned}$$

We can now apply Lemma 6.1 between R and $2R$, thus obtaining the required estimate. □

6.5 Boundary Estimates

A similar result holds for cubical quasi-minima taking prescribed values at the boundary. More precisely, assume that $U(x)$ is a function in $W^{1,t}(\mathbf{R}^n, \mathbf{R}^N)$, with $t > p$, and let $u \in W^{1,p}(\Omega, \mathbf{R}^N)$ be a function such that $u - U \in W_0^{1,p}(\Omega, \mathbf{R}^N)$, and that for every cube $Q_R \subset \mathbf{R}^n$ we have

$$\mathcal{F}(u, \Omega_R) \leq c \int_{\Omega_R} (|Dv| + \vartheta(x, v))^p \, dx$$

for every function v such that $v - u \in W_0^{1,p}(\Omega_R)$ $(\Omega_R = \Omega \cap Q_R)$.

Assume moreover that the function $F(x, u, z)$ satisfies the conditions of the preceding theorem, and that Ω has no internal cusps, that is that there exists a positive constant α_0 such that for every cube Q_R with center on $\partial\Omega$ we have

$$|Q_R - \Omega| \geq \alpha_0 |Q_R| \,. \tag{6.66}$$

In particular the above condition is satisfied if $\partial\Omega$ is Lipschitz-continuous.

Now let as usual $\varrho \leq t < s \leq R$, and let $\eta(x) \in C_0^\infty(Q_s)$, with $0 \leq \eta \leq 1$, $\eta = 1$ on Q_t and $|D\eta| \leq 2/(s-t)$. Setting $u = U$ in $\mathbf{R}^n - \Omega$, $\varphi = \eta(u - U)$ and $v = u - \varphi$, we have as in the proof of Theorem 6.5:

$$\begin{aligned}
\int_{Q_s} |D\varphi|^p \, dx &\leq \int_{Q_s} \tilde{F}(D\varphi)\, dx \leq \int_{Q_s} F(x,u,D\varphi)\, dx + \int_{Q_s} \vartheta(x,u)^p \, dx \\
&= \int_{Q_s} F(x,u,Du)\, dx + \int_{Q_s} [F(x,u,D\varphi) - F(x,u,Du)]\, dx \\
&\quad + \int_{Q_s} \vartheta(x,u)^p \, dx \\
&\leq c\int_{Q_s} (|Dv| + \vartheta(x,v))^p \, dx + c\int_{Q_s} (|Du| + |D\varphi| \\
&\quad + \vartheta(x,u))^{p-1} |Dv| \, dx + \int_{Q_s} \vartheta(x,u)^p \, dx \,,
\end{aligned}$$

where we have taken into account the conclusion of Lemma 5.2.

We now use the estimate $A^{p-1}B \leq \epsilon A^p + c(\epsilon) B^p$. We have

$$\begin{aligned}
&|D\varphi|^{p-1} |Dv| \leq \epsilon |D\varphi|^p + c(\epsilon)|Dv|^p \,, \\
&|Du|^{p-1}|Dv| \leq \epsilon |D(u-U)|^p + c(\epsilon)|Dv|^p + c|DU|^p \,.
\end{aligned}$$

Taking ϵ small enough, the terms with $|D\varphi|^p$ and $|D(u-U)|^p$ can be subtracted from the left-hand side, as in Theorem 6.5. We obtain in this way the inequality

$$\begin{aligned}
\int_{Q_t} |D(u-U)|^p \, dx &\leq c\int_{Q_s - Q_t} |D(u-U)|^p \, dx + c\int_{Q_s} (|Dv|^p + |DU|^p \\
&\quad + |u-U|^{p^*} + |U|^{p^*} + g)\, dx \,,
\end{aligned} \tag{6.67}$$

with $g = a + b^{\frac{p^*}{p^* - \gamma}}$.

On the other hand we have $v = U + (1-\eta)(u-U)$, and therefore

$$|Dv| \leq |DU| + (1-\eta)|D(u-U)| + 2\frac{|u-U|}{s-t}$$

and introducing it in the preceding inequality, we get

$$\int_{Q_t} |D(u-U)|^p\,dx \le c\int_{Q_s-Q_t} |D(u-U)|^p\,dx + \frac{c}{(s-t)^p}\int_{Q_s} |u-U|^p\,dx$$
$$+ c\int_{Q_s} (|DU|^p + |u-U|^{p^*} + |U|^{p^*} + g)\,dx\,.$$

Once again the term $|u-U|^{p^*}$ can be estimated as in Theorem 6.5, applying Theorem 3.16 to the function $u-U$, which is zero in $Q_s - \Omega$, a set of measure greater than $\alpha_0|Q_s|$. It follows that if R, and hence s, is small enough, we have

$$\int_{Q_t} |D(u-U)|^p\,dx \le c\int_{Q_s-Q_t} |D(u-U)|^p\,dx + \frac{c}{(s-t)^p}\int_{Q_s} |u-U|^p\,dx$$
$$+ c\int_{Q_s} (|DU|^p + |U|^{p^*} + g)\,dx\,.$$

Applying at this point the "hole filling" method, we eliminate the first term on the right-hand side, and therefore in conclusion we obtain the CACCIOPPOLI inequality

$$\int_{Q_\varrho} |D(u-U)|^p\,dx \le \frac{c}{(R-\varrho)^p}\int_{Q_R} |u-U|^p\,dx + c\int_{Q_R} g_1\,dx\,, \tag{6.68}$$

in which we have set $g_1 = a + b^{\frac{p^*}{p^*-\gamma}} + |DU|^p + |U|^{p^*}$.

Finally, we estimate the first term on the right by means of (3.29), and we write $2R$ in the place of R, arriving to the inequality[9]

$$⨍_{Q_R} |D(u-U)|^p\,dx \le c\left(⨍_{Q_{2R}} |D(u-U)|^{pm}\,dx\right)^{\frac{1}{m}}$$
$$+ c⨍_{Q_{2R}} g_1\,dx \tag{6.69}$$

with $m = \frac{n}{n+p}$ if $p_* \ge 1$, and $m = \frac{1}{p}$ if $p_* < 1$.

We can now repeat the above argument, and conclude that $D(u-U)$ (and hence Du) belongs to $W^{1,pr}(\Omega_{R/2})$ for some $r > 1$, with the relative estimate. Covering $\partial\Omega$ with a finite number of cubes, and then what remains of Ω with others cubes strictly contained in Ω, we obtain eventually the following:

---

[9]One can also add to both members the term $|u|^{p^*}$, obtaining the analog of (6.35).

Theorem 6.8 *Let Ω be an open set with Lipschitz-continuous boundary, and let $u \in W^{1,p}(\Omega, \mathbf{R}^N)$ be a cubical quasi-minimum for the functional $\mathcal{F}$, verifying the assumptions of Theorem* 6.5, *among all functions taking on $\partial\Omega$ the value $U(x) \in W^{1,t}(\Omega, \mathbf{R}^N)$, $t > p$. Then, Du belongs to $L^{pr}(\Omega)$ for some $r > 1$.*

Remark 6.13 The theorems just proved, and those that we shall prove in the following chapters, continue to hold if $p \geq n$, with suitable changes in the assumptions and in the proofs.

In particular, if $p = n$, Theorems 6.7 and 6.8 imply that the function $u(x)$ is Hölder-continuous in Ω (respectively in $\bar{\Omega}$). □

6.6 Notes and Comments

After a brief mention in [2], the notion of quasi-minimum was introduced for the first time by GIAQUINTA and GIUSTI in [4], where the relation between quasi-minima and elliptic equations in divergence form was also studied. The result of this and of the next chapter show that, at least for what concerns the first stages of the regularity program, quasi-minima represent the natural level of generality, unifying the treatment of different problems, in the first place those relative to the minima of functionals and to the solutions of elliptic partial differential equations, but also to quasi-conformal mappings and to minima with obstacles, each of which would demand otherwise a separate discussion.

The introduction of cubical (or spherical) quasi-minima can be seen at first sight as a gratuitous generality, since we do not know of any significant problems leading to cubical quasi-minima that are not at the same time quasi-minima without specification. On the other hand, it might serve to clarify the scope of different methods introduced for the study of regularity problems.

A substantial part of these methods is based on integral estimates over cubes, or more generally on inequalities concerning integrals over cubes. It is natural that such estimates hold for cubical quasi-minima, so that the relative results (in particular, for what concerns us, those of this chapter and of Chapter 9) will hold for cubical quasi-minima. On the other hand, the method of DE GIORGI, upon which are founded the Hölder-continuity results of Chapter 7, requires estimates on the level set of the solution, *a priori* on completely general sets, and therefore it cannot be extended to cubical quasi-minima.

The existence of cubical quasi-minima that are not quasi-minima, shows that the two methods are substantially different, and that we cannot hope to get results of Hölder regularity using only estimates on given families of sets, such as spheres, cubes, etc.

The use of cubes instead of spheres (or equivalently of the metric $\delta(x,y) = \max_i |x_i - y_i|$ in the place of the ordinary distance), is motivated by the simplicity of the proofs, in particular when we use covering theorems, such as that by CALDERON–ZYGMUND (see Chapter 2), or that, similar in many respects, by KRYLOV and SAFONOV, that we shall prove in the next chapter.

The L^p regularity of the derivatives of solutions to elliptic partial differential equations in divergence form was studied by BOJARSKI [1] and MEYERS [1].

The latter proved that weak solutions of strongly elliptic linear equations

$$\int_\Omega a_{ij}(x) D_j u D_i \varphi \, dx = 0$$

with bounded measurable coefficients, belong to $L^p(\Omega)$ for some $p > 2$.

The method we have used here is based on a generalization, due to GIAQUINTA and G. MODICA [1], of a theorem stated by GEHRING [1] in the course of his research on quasi-conformal mappings.

It is founded on a sort of *reverse* HÖLDER inequality with increasing supports. The spaces of functions verifying these inequalities *on the same cube*; that is those satisfying the estimate

$$\left(\fint_Q |u|^s \, dx \right)^{\frac{1}{s}} \leq c \left(\fint_Q |u|^r \, dx \right)^{\frac{1}{r}}$$

with $r < s$, have been widely studied, in particular for what concerns the dependence of the higher exponent of summability on the constant c of the preceding estimate. Among other things, BOJARSKI [2] has proved that that exponent goes to infinity when $c \to 1$ (see also WIK [2] and D'APUZZO and SBORDONE [1]). However, these results demand estimates on the same cube, and do not apply here.

Our Theorem 6.7 that generalizes MEYERS result is essentially the only general result valid for quasi-minima of functionals dependent on a vector-valued function ($N > 1$). It was proved in a particular case (minima of functionals with $F = F(x,z)$ and convex in z) by ATTOUCH and SBORDONE [1], and in its general form in GIAQUINTA and GIUSTI [4]. It was extended later by LEONETTI [1] to quasi-minima of functionals depending on higher-order derivatives. Example 6.1 shows that a similar result cannot hold for extremals of functionals, even if $F(x,u,z)$ is convex in z and $N = 1$.

Finally, if all the minima of the functional $\mathcal{F}$ are regular functions, and if Q is close enough to 1, the Q-minima of $\mathcal{F}$ are Hölder-continuous functions. For instance, if u is a cubical Q-minimum of the DIRICHLET functional, and if $Q \leq \frac{n-1}{n-2+2\alpha}$, $(0 < \alpha < \frac{1}{2})$, then $u \in C^{0,\alpha}(\Omega)$ (ZIEMER [2]).

Chapter 7

Hölder Continuity

The aim of this chapter is to prove that in the scalar case ($N = 1$) the quasi-minima and the ω-minima (see later, Section 7.7) of regular functionals of the calculus of variations are Hölder-continuous functions. The main result is a version of the fundamental theorem of DE GIORGI [1] and NASH [1] concerning the regularity of solutions of linear elliptic equations with discontinuous coefficients, a result that was later generalized among others by LADYŽENSKAYA and URAL'CEVA [2] to bounded solutions to non-linear elliptic equations. We shall prove, following GIAQUINTA and GIUSTI [2], that the same technique applies to quasi-minima of functionals. Since, as we have shown in the preceding chapter, weak solutions of elliptic equations in divergence form are quasi-minima of suitable functionals, this chapter contains in a unified form the regularity theory for elliptic partial differential equations and for minima of regular functionals of the calculus of variations.

7.1 Caccioppoli's Inequality

Let us consider the functional

$$\mathcal{F}(u, \Omega) = \int_{\Omega} F(x, u, Du)dx \tag{7.1}$$

in which, as usual, $F(x, u, z)$ is a Caratheodory function satisfying the inequalities

$$|z|^p - b(x)|u|^\gamma - a(x) \le F(x, u, z) \le L|z|^p + b(x)|u|^\gamma + a(x) \tag{7.2}$$

where[1] $1 < p \leq \gamma < p^* = \frac{pn}{n-p}$, and $a(x)$ and $b(x)$ are two non-negative functions, belonging respectively to $L^s(\Omega)$ and $L^\sigma(\Omega)$, with $s > \frac{n}{p}$ and $\sigma > \frac{p^*}{p^*-\gamma}$. We shall assume that $\frac{1}{s} = \frac{p}{n} - \epsilon$ and $\frac{1}{\sigma} = 1 - \frac{\gamma}{p^*} - \epsilon$ for some $\epsilon > 0$.

Our first result concerns *sub-quasi-minima* of the functional $\mathcal{F}$. We recall that $u \in W^{1,p}_{\text{loc}}(\Omega)$ is a sub-quasi-minimum if for every non-positive function $\varphi \in W^{1,p}(\Omega)$, with support $K \subset \Omega$, we have $\mathcal{F}(u, K) \leq Q\mathcal{F}(u + \varphi, K)$.

Similarly, u is a super-quasi-minimum if the preceding relation holds for every $\varphi \geq 0$. A quasi-minimum is at the same time a super- and a sub-quasi-minimum.

If $u(x)$ belongs to $W^{1,p}_{\text{loc}}(\Omega)$, k is a real number, and Q_R is a cube strictly contained in Ω, we set

$$A(k, R) = \{x \in Q_R : u(x) > k\}, \tag{7.3}$$

$$B(k, R) = \{x \in Q_R : u(x) < k\}. \tag{7.4}$$

We have $|A(k, R)| = |Q_R| - |B(k, R)|$ for almost every k, so that when necessary we can assume without loss of generality that all the values k under consideration will satisfy this relation.

The next theorem is a variation of Caccioppoli's inequality.

Theorem 7.1 *Let $u \in W^{1,p}(\Omega)$ be a sub-quasi-minimum of the functional* (7.1), *and let conditions* (7.2) *hold. Then there exists $R_0 > 0$ (depending on $\|u\|_{p^*}$ and $\|b\|_\sigma$) such that for every $x_0 \in \Omega$, every ϱ, R, with $0 < \varrho < R < \min(R_0, \text{dist}(x_0, \partial\Omega))$ and every $k \geq 0$ we have:*

$$\int_{A(k,\varrho)} |Du|^p dx \leq \frac{c}{(R-\varrho)^p} \int_{A(k,R)} (u-k)^p dx + c(\|a\|_s + k^p R^{-n\epsilon})|A(k, R)|^{1-\frac{p}{n}+\epsilon}. \tag{7.5}$$

Proof. Let η be a function in $C_0^\infty(Q_R)$, with $0 \leq \eta \leq 1$, $\eta = 1$ on Q_ϱ and $|D\eta| \leq \frac{2}{R-\varrho}$. Setting $w = (u-k)^+ = \max(u-k, 0)$, the function $v = u - \eta w$ is not greater than u, and differs from u at most in $A(k, R)$. It follows that

[1] If $p \geq n$ we can take γ arbitrarily. On the other hand in this case, taking into account the results of the preceding chapter and of the Sobolev immersion theorem, every quasi-minimum of the functional $\mathcal{F}$ is automatically a Hölder-continuous function, so that it will be sufficient to discuss only the case $p < n$.

$\mathcal{F}(u, A(k,R)) \le Q\mathcal{F}(v, A(k,R))$, and hence

$$\int_{A(k,R)} |Du|^p dx \le Q\left(\int_{A(k,R)} |Dv|^p dx + \int_{A(k,R)} [b(|u|^\gamma + |v|^\gamma) + a] dx\right). \tag{7.6}$$

Let us evaluate the right-hand side. For $x \in A(k,R)$ we have $u = u(1-\eta)+\eta(w+k)$, $v = u(1-\eta)+\eta k$, and hence $Dv = (1-\eta)Du-(u-k)D\eta$, and

$$|u|^\gamma + |v|^\gamma \le c(\gamma)\{(\eta w)^\gamma + |u|^\gamma(1-\eta)^\gamma + \eta^\gamma k^\gamma\}$$

$$|Dv|^p \le c(p)\left\{(1-\eta)^p|Du|^p + \frac{1}{(R-\varrho)^p}(u-k)^p\right\}.$$

Adding to both members of (7.6) the term $\int_{A(k,R)} b|u|^\gamma dx$, and using the above inequalities, we get

$$\begin{aligned}\int_{A(k,R)} (|Du|^p + b|u|^\gamma) dx \\ \le c\Bigg\{\int_{A(k,R)} (1-\eta)^p(|Du|^p + b|u|^\gamma)dx \\ + \frac{1}{(R-\varrho)^p}\int_{A(k,R)} w^p dx + \int_{A(k,R)} (b(\eta w)^\gamma + bk^\gamma + a)dx\Bigg\}.\end{aligned} \tag{7.7}$$

On the other hand

$$\begin{aligned}&\int_{A(k,R)} b(\eta w)^\gamma dx \\ &\le \left(\int_{A(k,R)} (\eta w)^{p^*} dx\right)^{\frac{p}{p^*}} \left(\int_{Q_R} [b(\eta w)^{\gamma-p}]^{\frac{n}{p}} dx\right)^{\frac{p}{n}} \\ &\le c\xi(R)\int_{A(k,R)} |D(\eta w)|^p dx \\ &\le c\xi(R)\left(\int_{A(k,R)} |Du|^p dx + \frac{1}{(R-\varrho)^p}\int_{A(k,R)} w^p dx\right),\end{aligned}$$

where we have set

$$\xi(R) = \left(\int_{Q_R} [b(\eta w)^{\gamma-p}]^{n/p} dx\right)^{\frac{p}{n}}.$$

We have

$$\xi(R) \leq \left(\int_{Q_R} |u|^{p^*} dx\right)^{\frac{\gamma-p}{p^*}} \left(\int_{Q_R} b^{\frac{p^*}{p^*-\gamma}}\right)^{1-\frac{\gamma}{p^*}}$$

and therefore

$$\xi(R) \leq \|u\|_{p^*}^{\gamma-p} \|b\|_\sigma |Q_R|^\epsilon .$$

If we choose R small enough (depending only on $\|u\|_{p^*}$ and $\|b\|_\sigma$), the quantity under examination can be partially subtracted from the left-hand side of (7.7), leaving on the right-hand side the term

$$\frac{c}{(R-\varrho)^p} \int_{A(k,R)} w^p dx .$$

We obtain therefore, recalling that $\eta = 1$ in Q_ϱ:

$$\int_{A(k,\varrho)} (|Du|^p + b|u|^\gamma) dx \leq c\left\{ \int_{A(k,R)-A(k,\varrho)} (|Du|^p + b|u|^\gamma) dx \right.$$
$$\left. + \frac{1}{(R-\varrho)^p} \int_{A(k,R)} w^p dx + \int_{A(k,R)} (bk^\gamma + a) dx \right\}$$

At this point we can argue as in the preceding chapter, summing to both sides the quantity on the left multiplied by c, and making use of Lemma 6.1; we arrive in this way to the inequality

$$\int_{A(k,\varrho)} (|Du|^p + b|u|^\gamma) dx \leq \frac{c}{(R-\varrho)^p} \int_{A(k,R)} w^p dx + c \int_{A(k,R)} (bk^\gamma + a) dx .$$

We remark now that

$$k^{p^*} |A(k,R)| \leq \int_{Q_R} |u|^{p^*} dx ,$$

and hence

$$\begin{aligned} k^\gamma \int_{A(k,R)} b dx &\leq k^\gamma \|b\|_\sigma |A(k,R)|^{1-\frac{1}{\sigma}} \\ &= \|b\|_\sigma (k^{p^*} |A(k,R)|)^{\frac{\gamma-p}{p^*}} k^p |A(k,R)|^{1-\frac{p}{n}+\epsilon} \\ &\leq \|b\|_\sigma \|u\|_{p^*}^{\gamma-p} |Q_R|^\epsilon k^p R^{-n\epsilon} |A(k,R)|^{1-\frac{p}{n}+\epsilon} \\ &\leq k^p R^{-n\epsilon} |A(k,R)|^{1-\frac{p}{n}+\epsilon} \end{aligned}$$

since, having chosen $R < R_0$ we can assume that $\|b\|_\sigma \|u\|_{p^*}^{\gamma-p} |Q_R|^\epsilon \leq 1$.

We have moreover

$$\int_{A(k,R)} a dx \leq \|a\|_s |A(k,R)|^{1-\frac{1}{s}} = \|a\|_s |A(k,R)|^{1-\frac{p}{n}+\epsilon}$$

and the results follows at once. □

Remark 7.1 If $u(x)$ is a super-quasi-minimum for the functional $\mathcal{F}$, $-u(x)$ will be a sub-quasi-minimum for the functional

$$\bar{\mathcal{F}}(v,\Omega) = \int_\Omega \bar{F}(x,v,Dv)dx$$

with $\bar{F}(x,u,z) = F(x,-u,-z)$. Since $\bar{F}$ satisfies conditions (7.2), we conclude that Caccioppoli's inequality (7.5) holds for the function $-u$, with k replaced by $-k$; we have therefore for every $k < 0$:

$$\int_{B(k,\varrho)} |Du|^p dx \leq \frac{c}{(R-\varrho)^p} \int_{B(k,R)} (k-u)^p dx$$
$$+ c(\|a\|_s + |k|^p R^{-n\epsilon}) |B(k,R)|^{1-\frac{p}{n}+\epsilon}, \qquad (7.8)$$

a relation valid for $0 < \varrho < R < \min(R_0, \operatorname{dist}(x_0, \partial\Omega))$. □

Remark 7.2 Caccioppoli's inequalities (7.5) and (7.8) hold if u belongs to $W^{1,p}_{\text{loc}}$. Of course, in this case one must assume that $Q_R \subset \Sigma \subset\subset \Omega$, and the radius R_0 will depend on Σ. On the other hand, as we have often repeated, when dealing with local results the assumptions $u \in W^{1,p}$ and $u \in W^{1,p}_{\text{loc}}$ are equivalent, since it is always possible to restrict ourselves to an arbitrary fixed open set $\Sigma \subset\subset \Omega$. □

Remark 7.3 The same inequalities remain valid if u is a sub-quasi-minimum (or a super-quasi-minimum) with Dirichlet conditions on $\partial\Omega$; more precisely if for every function $v \leq u$, with $v - u \in W^{1,p}_0(\Omega)$, and $K = \operatorname{supp}(v-u)$, we have

$$\mathcal{F}(u,K) \leq Q\mathcal{F}(v,K).$$

If the trace of u on $\partial\Omega$ is a bounded function, we can repeat the proof of the preceding theorem, provided we take $k > \sup_{\partial\Omega\cap Q_R} u$; in fact in this case the function $\eta(u-k)^+$ belongs to $W^{1,p}_0(Q_r)$.

In a similar way, (7.8) will be valid for every $k < \inf_{\partial\Omega\cap Q_R} u$. □

7.2 De Giorgi Classes

The results of the preceding section suggest the definition of new classes of functions.

Definition 7.1 *Let $u \in W^{1,p}_{\mathrm{loc}}(\Omega)$. We say that u belongs to the* De Giorgi class $DG_p^+ = DG_p^+(\Omega, H, \chi, \epsilon, R_0, \kappa_0)$ *if for every couple of concentric cubes $Q_\varrho \subset Q_R \subset\subset \Omega$, with $R < R_0$, and for every $k \geq \kappa_0 \geq 0$ we have*

$$\int_{A(k,\varrho)} |Du|^p dx \leq \frac{H}{(R-\varrho)^p} \int_{A(k,R)} (u-k)^p dx + H(\chi^p + k^p R^{-n\epsilon})|A(k,R)|^{1-\frac{p}{n}+\epsilon} . \tag{7.9}$$

We can define similarly DG_p^- to be the class of functions u such that $-u \in DG_p^+$. More explicitly, they are the functions in $W^{1,p}_{\mathrm{loc}}(\Omega)$ such that for every $\varrho < R \leq R_0$ and $k \leq -\kappa_0$ one has

$$\int_{B(k,\varrho)} |Du|^p dx \leq \frac{H}{(R-\varrho)^p} \int_{B(k,R)} (k-u)^p dx + H(\chi^p + |k|^p R^{-n\epsilon})|B(k,R)|^{1-\frac{p}{n}+\epsilon} . \tag{7.10}$$

It is clear that if a function u satisfies (7.9) or (7.10) with some ϵ, it will verify them with any positive $\epsilon' < \epsilon$. Consequently, we shall always assume $\epsilon \leq \frac{p}{n}$.

Finally, we shall indicate by DG_p the class of the functions belonging both to DG_p^+ and DG_p^-:

$$DG_p =: DG_p^+ \cap DG_p^- .$$

A rather surprising characteristic of De Giorgi classes is that (7.9) and (7.10) contain practically all the information deriving from the minimum properties of the function u, at least for what concerns its Hölder continuity.

Before beginning the study of the properties of the functions in $DG_p^\pm$, we shall make some remarks that will simplify considerably the following proofs.

Remark 7.4 If we set $v = u + \chi R^\beta$ ($\beta = \frac{n\epsilon}{p}$) and $h = k + \chi R^\beta$ in (7.9), and $v = u - \chi R^\beta$ and $h = k - \chi R^\beta$ in (7.10), we get respectively:

$$\int_{A(h,\varrho)} |Dv|^p dx \leq \frac{H}{(R-\varrho)^p} \int_{A(h,R)} (v-h)^p dx + H(\chi^p + k^p R^{-n\epsilon})|A(h,R)|^{1-\frac{p}{n}+\epsilon} ,$$

$$\int_{B(h,\varrho)} |Dv|^p dx \le \frac{H}{(R-\varrho)^p} \int_{B(h,R)} (h-v)^p dx$$
$$+ H(\chi^p + k^p R^{-n\epsilon})|B(h,R)|^{1-\frac{p}{n}+\epsilon}$$

and hence

$$\int_{A(h,\varrho)} |Dv|^p dx \le \frac{H}{(R-\varrho)^p} \int_{A(h,R)} (v-h)^p dx$$
$$+ H h^p R^{-n\epsilon} |A(h,R)|^{1-\frac{p}{n}+\epsilon}, \qquad (7.11)$$

$$\int_{B(h,\varrho)} |Dv|^p dx \le \frac{H}{(R-\varrho)^p} \int_{B(h,R)} (h-v)^p dx$$
$$+ H|h|^p R^{-n\epsilon} |B(h,R)|^{1-\frac{p}{n}+\epsilon}. \qquad (7.12)$$

Of course, the first relation will be valid for $h \ge h_0 = \kappa_0 + \chi R^\beta$, and the second for $h \le -h_0 = -\kappa_0 - \chi R^\beta$. □

Remark 7.5 By means of a homothety we can reduce to the case $R = 1$. More precisely, let $s < r < R$, and let us write (7.11) and (7.12) (with y in the place of x) for radii s and r. Making the change of variables $y = Rx$ and setting $s = \sigma R$, $t = \tau R$ and $w(x) = v(y)$, the function w satisfies the relations

$$\int_{A(h,\sigma)} |Dw|^p dx \le \frac{H}{(\tau-\sigma)^p} \int_{A(h,\tau)} (w-h)^p dx$$
$$+ H h^p \tau^{-n\epsilon} |A(h,\tau)|^{1-\frac{p}{n}+\epsilon},$$

$$\int_{B(h,\sigma)} |Dw|^p dx \le \frac{H}{(\tau-\sigma)^p} \int_{B(h,\tau)} (h-w)^p dx$$
$$+ H|h|^p \tau^{-n\epsilon} |B(h,\tau)|^{1-\frac{p}{n}+\epsilon}.$$

In particular, if $\tau \ge \frac{1}{2}$:

$$\int_{A(h,\sigma)} |Dw|^p dx \le \frac{H_1}{(\tau-\sigma)^p} \int_{A(h,\tau)} (w-h)^p dx$$
$$+ H_1 h^p |A(h,\tau)|^{1-\frac{p}{n}+\epsilon}, \qquad (7.13)$$

$$\int_{B(h,\sigma)} |Dw|^p dx \le \frac{H_1}{(\tau-\sigma)^p} \int_{B(h,\tau)} (h-w)^p dx$$
$$+ H_1 |h|^p |B(h,\tau)|^{1-\frac{p}{n}+\epsilon}. \qquad (7.14)$$

In conclusion, we can assume that (7.13) is satisfied for every $h \geq h_0 = \kappa_0 + \chi R^\beta$, and (7.14) for $h \leq -h_0 = -\kappa_0 - \chi R^\beta$. We can come back to the general case with a suitable homothety, writing $u + \chi R^\beta$ or $u - \chi R^\beta$ instead of u. □

The following lemma will be quite useful later.

Lemma 7.1 *Let $\alpha > 0$ and let $\{x_i\}$ be a sequence of real positive numbers, such that*

$$x_{i+1} \leq C B^i x_i^{1+\alpha}$$

with $C > 0$ and $B > 1$.

If $x_0 \leq C^{-\frac{1}{\alpha}} B^{-\frac{1}{\alpha^2}}$, we have

$$x_i \leq B^{-\frac{i}{\alpha}} x_0 \tag{7.15}$$

and hence in particular

$$\lim_{i \to \infty} x_i = 0 \, .$$

Proof. We proceed by induction. The inequality (7.15) is obviously true for $i = 0$. Assume now that it holds for i. We have

$$x_{i+1} \leq C B^{i(1-\frac{1+\alpha}{\alpha})} x_0^{1+\alpha} = (C B^{\frac{1}{\alpha}} x_0^{\alpha}) B^{-\frac{i+1}{\alpha}} x_0$$

and (7.15) follows immediately for $i+1$. □

We are now able to prove the following:

Theorem 7.2 *Let $u(x)$ be a function of DG_p^+. Then, u is locally bounded from above in Ω, and for every $x_0 \in \Omega$ and $R \leq \min(R_0, \operatorname{dist}(x_0, \partial\Omega))$ we have:*

$$\sup_{Q_{\frac{R}{2}}} u(x) \leq c \left\{ \left(\fint_{Q_R} u_+^p \, dx \right)^{\frac{1}{p}} + \kappa_0 + \chi R^\beta \right\} . \tag{7.16}$$

Proof. We can suppose $R = 1$, and that (7.13) is satisfied for every $h \geq h_0$. For $\frac{1}{2} \leq \sigma < \tau \leq 1$, let $\eta(x)$ be a function of class $C_0^\infty(Q_{\frac{\sigma+\tau}{2}})$ with $\eta = 1$ on Q_σ and $|D\eta| \leq \frac{4}{\tau - \sigma}$. Setting $\zeta = \eta(w-k)^+$, $k \geq h_0$, we have

$$\int_{A(k,\sigma)} (w-k)^p dx$$

$$\leq \int \zeta^p dx \leq \left(\int \zeta^{p^*} dx \right)^{\frac{p}{p^*}} |A(k,\tau)|^{1-\frac{p}{p^*}}$$

$$\leq c|A(k,\tau)|^{\frac{p}{n}} \int |D\zeta|^p dx$$

$$\leq c\left(\int_{A(k,\frac{\sigma+\tau}{2})} |Dw|^p dx + \frac{1}{(\tau-\sigma)^p}\int_{A(k,\frac{\sigma+\tau}{2})} (w-k)^p dx\right)|A(k,\tau)|^{\frac{p}{n}}.$$

Introducing inequality (7.13) in the preceding one, we obtain for $k \geq h_0$,

$$\int_{A(k,\sigma)} (w-k)^p dx \leq \frac{c|A(k,\tau)|^{\frac{p}{n}}}{(\tau-\sigma)^p}\int_{A(k,\tau)} (w-k)^p dx + ck^p|A(k,\tau)|^{1+\epsilon}. \tag{7.17}$$

We remark now that if $h < k$ we have

$$\int_{A(h,\tau)} (w-h)^p dx \geq (k-h)^p |A(k,\tau)| \tag{7.18}$$

and moreover

$$\int_{A(k,\tau)} (w-k)^p dx \leq \int_{A(k,\tau)} (w-h)^p dx \leq \int_{A(h,\tau)} (w-h)^p dx.$$

Introducing these relations in (7.17) we obtain:

$$\int_{A(k,\sigma)} (w-k)^p dx \leq c\left(\int_{A(h,\tau)} (w-h)^p dx\right)^{1+\epsilon} \times \frac{1}{(k-h)^{p\epsilon}}\left(\frac{1}{(\tau-\sigma)^p} + \frac{k^p}{(k-h)^p}\right) \tag{7.19}$$

where we have used the assumption $\epsilon \leq \frac{p}{n}$ and the fact that $|A(k,\tau)| \leq |Q_1|$.

Let now $d \geq h_0$ be a number that we shall fix later, and consider the sequences

$$k_i = 2d(1-2^{-i-1}),$$

$$\sigma_i = \frac{1}{2}(1+2^{-i}).$$

Writing (7.19) for $\sigma = \sigma_{i+1}$, $\tau = \sigma_i$, $k = k_{i+1}$ and $h = k_i$, and setting

$$\Phi_i = d^{-p}\int_{A(k_i,\sigma_i)} (w-k_i)^p dx,$$

we obtain the relation

$$\Phi_{i+1} \leq C2^{ip(1+\epsilon)}\Phi_i^{1+\epsilon}.$$

We can now apply the preceding lemma with $B = 2^{p(1+\epsilon)}$. If Φ_0 is less than a suitable constant, a condition which is satisfied if

$$d \geq c\left(\int_{Q_1} w_+^p\, dx\right)^{\frac{1}{p}}, \tag{7.20}$$

we have

$$\lim_{i\to\infty} \Phi_i = 0$$

and hence $|A(2d, \frac{1}{2})| = 0$, that we can write in the form

$$\sup_{Q_{1/2}} w \leq 2d\,.$$

The conditions imposed on d can be satisfied setting

$$d = h_0 + c\left(\int_{Q_1} w_+^p\, dx\right)^{\frac{1}{p}},$$

and therefore:

$$\sup_{Q_{1/2}} w \leq c\left(\int_{Q_1} w_+^p\, dx\right)^{\frac{1}{p}} + 2h_0\,.$$

The conclusion follows coming back to the function $u(x) = w(\frac{x}{R}) - \chi R^\beta$.

□

If instead of the cube $Q_{\frac{R}{2}}$ we want an estimate over the cube Q_{tR}, $t < 1$, we can use the following

Corollary 7.1 *With the assumptions of the preceding theorem, we have*

$$\sup_{Q_{tR}} u(x) \leq c\left\{\left(\frac{1}{(1-t)^n}\int_{Q_R} u_+^p\, dx\right)^{\frac{1}{p}} + \kappa_0 + \chi R^\beta\right\}. \tag{7.21}$$

Proof. Let x_1 be a point in Q_{tR} such that

$$\sup_{Q_{tR}} u(x) = \sup_{Q(x_1, \frac{1-t}{2}R)} u(x)\,.$$

By the preceding theorem we have

$$\sup_{Q_{tR}} u(x) \leq c\left\{\left(\fint_{Q_{(1-t)R}} u_+^p\, dx\right)^{\frac{1}{p}} + \kappa_0 + \chi[(1-t)R]^\beta\right\}$$

from which (7.21) follows at once. □

The following result is also a consequence of Theorem 7.2.

Theorem 7.3 *With the assumptions of Theorem 7.2, for every $q > 0$ there exists a constant $c(q)$ such that*

$$\sup_{Q_\varrho} u \leq c(q) \left\{ \left(\frac{1}{(R-\varrho)^n} \int_{Q_R} u_+^q \, dx \right)^{\frac{1}{q}} + \kappa_0 + \chi R^\beta \right\} \tag{7.22}$$

for every $\varrho < R \leq \min(R_0, \operatorname{dist}(x_0, \partial\Omega))$

Proof. Let $U_\tau = \sup_{Q_\tau} u$. From (7.21) we have for $\varrho \leq \sigma < \tau \leq R$:

$$U_\sigma \leq c \left\{ \left(\frac{1}{(\tau-\sigma)^n} \int_{Q_\tau} u_+^p \, dx \right)^{\frac{1}{p}} + \kappa_0 + \chi \tau^\beta \right\}$$

and hence

$$\begin{aligned} U_\sigma &\leq c \left\{ \left(\frac{1}{(\tau-\sigma)^n} \int_{Q_\tau} u_+^q \, dx \right)^{\frac{1}{p}} U_\tau^{1-\frac{q}{p}} + \kappa_0 + \chi \tau^\beta \right\} \\ &\leq \frac{1}{2} U_\tau + c(q) \left(\frac{1}{(\tau-\sigma)^n} \int_{Q_R} u_+^q \, dx \right)^{\frac{1}{q}} + c(\kappa_0 + \chi R^\beta) . \end{aligned}$$

An application of Lemma 6.1 leads immediately to the conclusion. □

Similar results hold for functions $u \in DG_p^-$. It will be sufficient to remark that in this case one has $-u \in DG_p^+$, and to write for instance (7.22) for $-u$. If moreover the function u belongs to $DG_p = DG_p^+ \cap DG_p^-$, we have

$$\sup_{Q_\varrho} |u| \leq c(q) \left\{ \left(\frac{1}{(R-\varrho)^n} \int_{Q_R} |u|^q \, dx \right)^{\frac{1}{q}} + \kappa_0 + \chi R^\beta \right\} \tag{7.23}$$

for every $\varrho < R \leq \min(R_0, \operatorname{dist}(x_0, \partial\Omega))$.

Finally, if Σ is an open set strictly contained in Ω, covering Σ with cubes of side R, with $2R = \min(R_0, \frac{1}{2}\operatorname{dist}(\Sigma, \partial\Omega))$ and writing (7.23) for radii R and $2R$, we immediately get:

Theorem 7.4 *Let $u(x)$ be a function in the De Giorgi class DG_p, and let $\Sigma \subset\subset \Omega$. For every $q > 0$ there exists a constant $c = c(q, \Sigma)$ such that*

$$\sup_{\Sigma} |u| \leq c \left\{ \left(\int_\Omega |u|^q \, dx \right)^{\frac{1}{q}} + \kappa_0 + \chi \right\} . \tag{7.24}$$

We can now state a first regularity result for quasi-minima of regular functionals of the calculus of variations, which follows at once from the above, if we remark that by Caccioppoli's inequality a sub-quasi-minimum belongs to DG_p^+ with $\kappa_0 = 0$.

Theorem 7.5 *Let $u(x) \in W^{1,p}(\Omega)$ be a sub-quasi-minimum of a functional $\mathcal{F}$ satisfying conditions (7.2). Then, u is locally bounded above in Ω. Moreover, for every $q > 0$ there exists a constant $c(q)$, depending also on $\|u\|_{1,p}$ and $\|b\|_\sigma$, such that for every $\varrho < R < \min(R_0, \operatorname{dist}(x_0, \partial\Omega))$ we have*

$$\sup_{Q_\varrho} u \le c(q) \left\{ \left(\frac{1}{(R-\varrho)^n} \int_{Q_R} u_+^q \, dx \right)^{\frac{1}{q}} + \|a\|_s^{1/p} R^\beta \right\}. \tag{7.25}$$

Similarly, every super-quasi-minimum of $\mathcal{F}$ is locally bounded below, with an estimate analogous to (7.25). Finally, every quasi-minimum is locally bounded in Ω, and we have

$$\sup_{Q_\varrho} |u| \le c(q) \left\{ \left(\frac{1}{(R-\varrho)^n} \int_{Q_R} |u|^q dx \right)^{\frac{1}{q}} + \|a\|_s^{1/p} R^\beta \right\}. \tag{7.26}$$

Remark 7.6 If a function $u(x)$ in the class DG^+ $[DG^-]$ belongs to $W^{1,p}(\Omega)$ and if its trace on $\partial\Omega$ is a function bounded from above [below], then u is bounded from above [below] in Ω. Actually, if Q_R is a cube intersecting $\partial\Omega$, and if we set $\Omega_R = Q_R \cap \Omega$ and $\Sigma_R = Q_R \cap \partial\Omega$, the function

$$w = \eta(u-k)^+$$

belongs to $W_0^{1,p}(Q_R)$ whenever $k \ge \kappa_0 \ge \sup_{\Sigma_R} u$. We have then

$$\sup_{\Omega_\varrho} u \le c(q) \left\{ \left(\frac{1}{(R-\varrho)^n} \int_{\Omega_R} u_+^q \, dx \right)^{\frac{1}{q}} + \sup_{\Sigma_R} u + \chi R^\beta \right\}. \tag{7.27}$$

A similar inequality holds for $u \in DG^-$:

$$\inf_{\Omega_\varrho} u \ge c(q) \left\{ - \left(\frac{1}{(R-\varrho)^n} \int_{\Omega_R} u_-^q \, dx \right)^{\frac{1}{q}} + \inf_{\Sigma_R} u - \chi R^\beta \right\}. \tag{7.28}$$

In particular, if $u \in DG_p$ setting $\operatorname{osc}(u, A) = \sup_A u - \inf_A u$, we have

$$\operatorname{osc}(u,\Omega_\varrho) \le c(q)\left\{\left(\frac{1}{(R-\varrho)^n}\int_{\Omega_R}|u|^q dx\right)^{\frac{1}{q}} + \operatorname{osc}(u,\Sigma_R) + \chi R^\beta\right\}, \tag{7.29}$$

an estimate that will be useful later. □

7.3 Quasi-Minima

The results of the preceding section can be used to prove the Hölder-continuity of the quasi-minima of regular functionals of the calculus of variations, and hence of the solutions of partial differential equations of elliptic type.

We remark first that, once the local boundedness of the Q-minima of the functional

$$\mathcal{F}(u,\Omega) = \int_\Omega F(x,u,Du)dx$$

has been proved, one can assume that the function $F(x,u,z)$ satisfies instead of (7.2), the inequalities

$$|z|^p - \alpha(x,M) \le F(x,u,z) \le L(M)|z|^p + \alpha(x,M), \tag{7.30}$$

where $M \ge \sup|u|$ and $L(M)$ and $\alpha(x,M)$ are increasing functions of M. In what follows it will be sufficient to take $M = 2\sup|u|$.

In particular, following the proof of Theorem 7.1 one can see easily that the Q-minima of $\mathcal{F}$ satisfy for every $\varrho < R \le R_0$ and for every $k \in \mathbf{R}$, the estimates

$$\int_{A(k,\varrho)}|Du|^p dx \le \frac{H}{(R-\varrho)^p}\int_{A(k,R)}(u-k)^p dx + H\chi^p|A(k,R)|^{1-\frac{p}{n}+\epsilon} \tag{7.31}$$

$$\int_{B(k,\varrho)}|Du|^p dx \le \frac{H}{(R-\varrho)^p}\int_{B(k,R)}(k-u)^p dx + H\chi^p|B(k,R)|^{1-\frac{p}{n}+\epsilon}, \tag{7.32}$$

that is inequalities (7.9) and (7.10) without the term k^p, but with H and χ dependent on M. As we have already remarked, it is possible to assume that $\epsilon \le \frac{p}{n}$.

It follows that the preceding estimates hold with the same constants for $u - \vartheta$, at least as long as $|\vartheta| + \sup|u| \le M$.

Remark 7.7 If the function $F(x, u, z)$ satisfies inequalities (7.2), we can take L independent of M and $\alpha(x, M) = a(x) + b(x)M^\gamma$ in (7.30), and hence we have (7.31) and (7.32) with H independent of M and

$$\chi^p = \|\alpha(x, M)\|_s = \|a(x) + b(x)M^\gamma\|_s . \tag{7.33}$$

□

We have the following proposition, analogous to Theorem 7.2:

Proposition 7.1 *Let $u(x)$ be a bounded function, verifying (7.31) for every $k \in \mathbf{R}$. Then if $|k_0| + \sup|u| \leq M$ we have*

$$\sup_{Q_{\frac{R}{2}}} u \leq c\left(\frac{1}{R^n}\int_{A(k_0,R)}(u-k_0)^p dx\right)^{\frac{1}{p}}\left(\frac{|A(k_0,R)|}{R^n}\right)^{\frac{\alpha}{p}} + k_0 + c\chi R^\beta , \tag{7.34}$$

where α is the positive solution of the equation $\alpha^2 + \alpha = \epsilon$.

Proof. We can suppose $k_0 = 0$. Repeating with the necessary changes the proof of Theorem 7.2, and using (7.31) instead of (7.13), we get in the place of (7.17) the estimate

$$\int_{A(k,\varrho)}(u-k)^p dx \leq \frac{c|A(k,r)|^{\frac{p}{n}}}{(r-\varrho)^p}\int_{A(k,r)}(u-k)^p dx + c\chi^p|A(k,r)|^{1+\epsilon} \tag{7.35}$$

for every $\varrho < r \leq R$. Moreover, for each $h < k$ and every $\varrho \leq r$:

$$|A(k,\varrho)| \leq (k-h)^{-p}U(h,r) , \tag{7.36}$$

where

$$U(k,t) = \int_{A(k,t)}(u-k)^p dx .$$

Recalling that $n\epsilon = p\beta$, we immediately get from (7.35) and (7.36):

$$U(k,\varrho) \leq c(r-\varrho)^{-p}U(h,r)|A(h,r)|^{\frac{p}{n}} + c\chi^p(k-h)^{-p}U(k,r)|A(k,r)|^\epsilon$$

$$\leq c\left[\left(\frac{r}{r-\varrho}\right)^p + \left(\frac{\chi r^\beta}{k-h}\right)^p\right] r^{-n\epsilon}U(h,r)|A(h,r)|^\epsilon .$$

Raising both members of (7.36) to the power α, and multiplying each member with the corresponding member of the last inequality, we get

$$U(k,\varrho)|A(k,\varrho)|^{\alpha}$$

$$\leq c\left[\left(\frac{r}{r-\varrho}\right)^{p}+\left(\frac{\chi r^{\beta}}{k-h}\right)^{p}\right]\frac{r^{-n\epsilon}}{(k-h)^{p\alpha}}U(h,r)^{1+\alpha}|A(h,r)|^{\epsilon}.$$

Let us now choose α in such a way that $\alpha(1+\alpha)=\epsilon$, and let us define

$$\varphi(k,t)=U(k,t)|A(k,t)|^{\alpha}.$$

For $\varrho<r\leq R$ and $h<k$ we have:

$$\varphi(k,\varrho)\leq c\left[\left(\frac{r}{r-\varrho}\right)^{p}+\left(\frac{\chi r^{\beta}}{k-h}\right)^{p}\right]\frac{r^{-n\epsilon}}{(k-h)^{p\alpha}}\varphi(h,r)^{1+\alpha}. \qquad (7.37)$$

Let now $d\geq\chi R^{\beta}$ be a constant that we shall fix later, and define

$$k_i=d(1-2^{-i}),$$

$$r_i=\frac{R}{2}(1+2^{-i}).$$

From (7.37) with $\varrho=r_{i+1}$, $r=r_i$, $k=k_{i+1}$ and $h=k_i$ we get

$$\varphi_{i+1}\leq cd^{-p\alpha}2^{pi(1+\alpha)}R^{-n\epsilon}\varphi_i^{1+\alpha},$$

where

$$\varphi_i=\varphi(k_i,\sigma_i).$$

We can now apply Lemma 7.1. Choosing

$$d\geq cR^{-\frac{n\epsilon}{\alpha p}}\varphi_0^{\frac{1}{p}}$$

with the constant c large enough, we can conclude that the sequence φ_i tends to zero, and hence

$$\varphi\left(d,\frac{R}{2}\right)=0.$$

The conditions imposed on d will be satisfied taking

$$d=\chi R^{\beta}+cR^{-\frac{n\epsilon}{\alpha p}}\varphi_0^{\frac{1}{p}},$$

and hence, recalling the choice of α, we arrive at

$$\sup_{Q_{\frac{R}{2}}}u(x)\leq d=c\left(R^{-n}\int_{A(0,R)}u^{p}dx\right)^{\frac{1}{p}}\left(\frac{|A(0,R)|}{R^{n}}\right)^{\frac{\alpha}{p}}+\chi R^{\beta}.$$

The conclusion follows at once writing $u-k_0$ instead of u. □

We must now evaluate the measure of the set $A(k,R)$, when k is close to the maximum of u. For that we need the following lemma, which, with suitable changes, will also be useful later.

Lemma 7.2 *Let u be a bounded function, satisfying (7.31) (with $p>1$) for every $k \in \mathbf{R}$, and let $2k_0 = M(2R) + m(2R) =: \sup_{Q_{2R}} u + \inf_{Q_{2R}} u$. Assume that $|A(k_0,R)| \leq \gamma |Q_R|$ for some $\gamma < 1$. If for an integer ν, it holds that*

$$\operatorname{osc}(u, 2R) \geq 2^{\nu+1} \chi R^{\beta}, \tag{7.38}$$

then, setting $k_\nu = M(2R) - 2^{-\nu-1}\operatorname{osc}(u,2R)$, we have

$$|A(k_\nu, R)| \leq c\nu^{-\frac{n(p-1)}{p(n-1)}} |Q_R| . \tag{7.39}$$

Proof. For $k_0 < h < k$ let us define

$$v(x) = \begin{cases} k-h & \text{if} \quad u \geq k, \\ u-h & \text{if} \quad h < u < k, \\ 0 & \text{if} \quad u \leq h. \end{cases}$$

We have $v = 0$ in $Q_R - A(k_0, R)$, and since the measure of this set is greater than $(1-\gamma)|Q_R|$, we can apply the Sobolev inequality, obtaining

$$\left(\int_{Q_R} v^{\frac{n}{n-1}} dx \right)^{1-\frac{1}{n}} \leq c \int_\Delta |Dv| dx = c \int_\Delta |Du| dx$$

in which $\Delta = A(h,R) - A(k,R)$. We therefore have

$$\begin{aligned} (k-h)|A(k,R)|^{1-\frac{1}{n}} &\leq \left(\int_{Q_R} v^{\frac{n}{n-1}} dx \right)^{1-\frac{1}{n}} \\ &\leq c|\Delta|^{1-\frac{1}{p}} \left(\int_{A(h,R)} |Du|^p dx \right)^{\frac{1}{p}} . \end{aligned} \tag{7.40}$$

On the other hand, from (7.31) we deduce

$$\begin{aligned} \int_{A(h,R)} |Du|^p dx &\leq \frac{c}{R^p} \int_{A(h,2R)} (u-h)^p dx + c\chi^p |A(k,2R)|^{1-\frac{p}{n}+\epsilon} \\ &\leq cR^{n-p}(M(2R)-h)^p + c\chi^p R^{n-p+n\epsilon} . \end{aligned}$$

For $h \leq k_\nu$, we have $M(2R) - h \geq M(2R) - k_\nu \geq \chi R^\beta$, and hence

$$(k-h)|A(k,R)|^{1-\frac{1}{n}} \leq c|\Delta|^{1-\frac{1}{p}} R^{\frac{n-p}{p}} (M(2R)-h) .$$

Writing the above inequality for the levels $k = k_i = M(2R) - 2^{-i-1}\text{osc}(u, 2R)$ and $h = k_{i-1}$, and raising to the power $\frac{p}{p-1}$, we get

$$|A(k_\nu, R)|^{\frac{p(n-1)}{n(p-1)}} \leq |A(k_i, R)|^{\frac{p(n-1)}{n(p-1)}} \leq cR^{\frac{n-p}{p-1}}|\Delta_i|$$

with $\Delta_i = A(k_i, R) - A(k_{i-1}, R)$.

We now sum over i from 1 to ν, obtaining

$$\nu|A(k_\nu, R)|^{\frac{p(n-1)}{n(p-1)}} \leq cR^{\frac{n-p}{p-1}}|A(k_0, R)| \leq cR^{\frac{p(n-1)}{(p-1)}},$$

and (7.39) follows at once. □

Finally, we shall need the following algebraic lemma.

Lemma 7.3 *Let $\varphi(t)$ be a positive function, and assume that there exists a constant q and a number τ, $0 < \tau < 1$ such that for every $R < R_0$*

$$\varphi(\tau R) \leq \tau^\delta \varphi(R) + BR^\beta \tag{7.41}$$

with $0 < \beta < \delta$, and

$$\varphi(t) \leq q\varphi(\tau^k R)$$

for every t in the interval $(\tau^{k+1}R, \tau^k R)$.[2]

Then, for every $\varrho < R \leq R_0$ we have

$$\varphi(\varrho) \leq C\left\{\left(\frac{\varrho}{R}\right)^\beta \varphi(R) + B\varrho^\beta\right\}, \tag{7.42}$$

where C is a constant depending only on q, τ, δ and β.

Proof. Starting from (7.41) we prove by induction that

$$\varphi(\tau^{k+1}R) \leq \tau^{(k+1)\delta}\varphi(R) + BR^\beta\tau^{k\beta}\sum_{i=0}^{k}\tau^{i(\delta-\beta)}$$

and hence, since the series on the right-hand side is convergent,

$$\varphi(\tau^{k+1}R) \leq \tau^{(k+1)\delta}\varphi(R) + cBR^\beta\tau^{k\beta}.$$

Choosing now k in such a way that $\tau^{k+1}R < \varrho \leq \tau^k R$ we arrive immediately at the desired result. □

Remark 7.8 If instead $\beta \geq \delta$, we can estimate BR^β by means of $BR^{\delta-\epsilon}R_0^{\beta-\delta+\epsilon}$ and we get (7.42) with $\delta - \epsilon$ instead of β. Of course, in this case the constant C will depend on ϵ as well. □

[2]In particular, this inequality holds with $q = 1$ if φ is non-decreasing.

Theorem 7.6 *Let $u(x)$ be a bounded function, satisfying (7.31) and (7.32) with $p > 1$ for every $k \in \mathbf{R}$. Then, u is (locally) Hölder-continuous in Ω.*

Proof. Let, as above, $2k_0 = M(2R) + m(2R)$. We can assume without loss of generality that $|A(k_0, R)| \leq \frac{1}{2}|Q_R|$, since otherwise we would have $|B(k_0, R)| = |Q_R| - |A(k_0, R)| \leq \frac{1}{2}|Q_R|$, and it will be sufficient to write $-u$ instead of u.

Setting $k_\nu = M(2R) - 2^{-\nu-1}\mathrm{osc}(u, 2R)$, we have $k_\nu > k_0$. We can write (7.34) with k_ν instead of k_0:

$$\sup_{Q_{\frac{R}{2}}}(u - k_\nu) \leq \left(\frac{c}{R^n}\int_{A(k_\nu,R)}(u - k_\nu)^p dx\right)^{\frac{1}{p}}\left(\frac{|A(k_\nu, R)|}{R^n}\right)^{\frac{\alpha}{p}} + c\chi R^\beta$$

$$\leq c\sup_{Q_R}(u - k_\nu)\left(\frac{|A(k_\nu, R)|}{R^n}\right)^{\frac{\alpha+1}{p}} + c\chi R^\beta . \tag{7.43}$$

Let us now choose the integer ν in such a way that

$$c\nu^{-\frac{n(p-1)}{p(n-1)}} \leq \frac{1}{2} .$$

If $\mathrm{osc}(u, 2R) \geq 2^{\nu+1}\chi R^\beta$ we deduce from (7.39)

$$M\left(\frac{R}{2}\right) - k_\nu \leq \frac{1}{2}(M(2R) - k_\nu) + c\chi R^\beta$$

so that, subtracting from both members the quantity $m(\frac{R}{2})$,

$$\mathrm{osc}\left(u, \frac{R}{2}\right) \leq \left(1 - \frac{1}{2^{\nu+2}}\right)\mathrm{osc}(u, 2R) + c\chi R^\beta .$$

In conclusion, either the function $\mathrm{osc}(u, R)$ satisfies the above relation, or else

$$\mathrm{osc}(u, 2R) \leq 2^{\nu+1}\chi R^\beta .$$

In any case, we have

$$\mathrm{osc}\left(u, \frac{R}{2}\right) \leq \left(1 - \frac{1}{2^{\nu+2}}\right)\mathrm{osc}(u, 2R) + c2^\nu\chi R^\beta . \tag{7.44}$$

We can now apply the preceding lemma with $\tau = 1/4$ and $\delta = \log_\tau(1 - 2^{-\nu-2})$. Decreasing if necessary the value of β, we can assume that $\beta < \delta$.

We therefore have

$$\operatorname{osc}(u,\varrho) \le c\left\{\left(\frac{\varrho}{R}\right)^{\beta}\operatorname{osc}(u,R) + \chi\varrho^{\beta}\right\} \tag{7.45}$$

for every $\varrho < R \le \min(R_0, \operatorname{dist}(x_0, \partial\Omega))$. □

Remark 7.9 Note that the exponent δ does not depend on the center of the cube Q_R, and therefore $u \in C^{0,\beta}(\Omega)$. Of course, the norm of u in $\Sigma \subset\subset \Omega$ can diverge when $\Sigma \to \Omega$. □

The following result will be useful later.

Theorem 7.7 *Let $u(x) \in W^{1,p}_{\text{loc}}(\Omega)$ satisfy (7.31) and (7.32) with $p > 1$ for every $k \in \mathbf{R}$, and let $Q_{4R} \subset \Omega$. Then, for every $\varrho < R$ we have*

$$\int_{Q_\varrho} |u - u_\varrho|^p dx \le c\left(\frac{\varrho}{R}\right)^{n+p\beta}\int_{Q_R}|u - u_R|^p dx + c\chi^p\varrho^{n+p\beta}, \tag{7.46}$$

$$\int_{Q_\varrho} |Du|^p dx \le c\left(\frac{\varrho}{R}\right)^{n-p+p\beta}\int_{Q_R}|Du|^p dx + c\chi^p\varrho^{n-p+p\beta}. \tag{7.47}$$

Proof. We shall prove (7.46) first. Let $\vartheta < \sup|u|$. We have remarked that the functions $u - \vartheta$ and $\vartheta - u$ satisfy (7.31) and (7.32), and hence by Theorem 7.2 we have

$$\sup_{Q_r}[u - \vartheta] \le c\left\{\left(\fint_{Q_{2r}}(u-\vartheta)_+^p dx\right)^{\frac{1}{p}} + \chi r^{\beta}\right\}, \tag{7.48}$$

$$\sup_{Q_r}[\vartheta - u] \le c\left\{\left(\fint_{Q_{2r}}(\vartheta-u)_+^p dx\right)^{\frac{1}{p}} + \chi r^{\beta}\right\}. \tag{7.49}$$

Summing both sides, if $\inf_{Q_r} u \le \vartheta \le \sup_{Q_r} u$, whence in particular if $\vartheta = u_r =: \fint_{Q_r} u\,dx$, we get

$$\begin{aligned}\operatorname{osc}(u,r) &\le c\left\{\left(\fint_{Q_{2r}}|u-u_r|^p dx\right)^{\frac{1}{p}} + 2\chi r^{\beta}\right\}\\ &\le c\left\{\left(\fint_{Q_{2r}}|u-u_{2r}|^p dx\right)^{\frac{1}{p}} + 2\chi r^{\beta}\right\},\end{aligned} \tag{7.50}$$

where in the last passage we have taken into account Remark 3.4, and in particular (3.36).

On the other hand

$$\fint_{Q_t} |u - u_t|^p dx \le \operatorname{osc}(u,t)^p$$

and hence, taking (7.45) into account, we get for $\varrho < r = \frac{R}{2}$:

$$\fint_{Q_\varrho} |u - u_\varrho|^p dx \le \operatorname{osc}(u,\varrho)^p \le c\left\{\left(\frac{\varrho}{r}\right)^{p\beta} \operatorname{osc}(u,r)^p + \chi^p \varrho^{p\beta}\right\}$$

$$\le c\left\{\left(\frac{\varrho}{R}\right)^{p\beta} \fint_{Q_R} |u - u_R|^p dx + \chi^p \varrho^{p\beta}\right\},$$

that is (7.46) for $\varrho < \frac{R}{2}$.

We shall prove (7.47) by remarking that writing (7.31) and (7.32) between $\frac{\varrho}{2}$ and ϱ, with $k = u_\varrho$, and summing both sides, we get

$$\varrho^p \int_{Q_{\varrho/2}} |Du|^p dx \le c\left\{\int_{Q_\varrho} |u - u_\varrho|^p dx + \chi^p \varrho^{n+p\beta}\right\}.$$

On the other hand we have

$$\int_{Q_R} |u - u_R|^p dx \le cR^p \int_{Q_R} |Du|^p dx,$$

and introducing these relations into (7.46) we get (7.47), this time for $\varrho < \frac{R}{4}$.

Finally, we remark that both the estimates hold for every $\varrho < R$, possibly with a different constant. For instance if $\varrho \ge \frac{R}{4}$ we have

$$\int_{Q_\varrho} |u - u_\varrho|^p dx \le c\int_{Q_R} |u - u_R|^p dx \le c4^{n+p\beta}\left(\frac{\varrho}{R}\right)^{n+p\beta} \int_{Q_R} |u - u_R|^p dx.$$

A similar argument proves (7.47) for every $\varrho < R$. □

7.4 Boundary Regularity

When the function $u(x)$ is a quasi-minimum with Dirichlet conditions on $\partial\Omega$, and its trace on $\partial\Omega$ is a Hölder-continuous function, it is possible to extend Theorem 7.6, proving the Hölder-continuity of u up to the boundary.

Theorem 7.8 *Let Ω be an open set in $\mathbf{R}^n$ with Lipschitz-continuous boundary, and let $u \in W^{1,p}(\Omega)$ be a quasi-minimum for the functional $\mathcal{F}$. Assume that the trace of u on $\partial\Omega$ be a Hölder-continuous function. Then, u is Hölder-continuous in $\bar{\Omega}$.*

Proof. For $t < 1$ we define:

$$
\begin{aligned}
M(t) &= \sup\nolimits_{\Omega_t} u\,, & M_S(t) &= \sup\nolimits_{\partial\Omega\cap Q_t} u\,;\\
m(t) &= \inf\nolimits_{\Omega_t} u\,, & m_S(t) &= \inf\nolimits_{\partial\Omega\cap Q_t} u\,;\\
\operatorname{osc}(u,t) &= M(t) - m(t)\,, & \operatorname{osc}_S(u,t) &= M_S(t) - m_S(t)\,.
\end{aligned}
$$

Let Q_R be a cube such that Q_{2R} meets $\partial\Omega$. It is not restrictive to assume that

$$M(2R) - M_S(2R) \geq m_S(2R) - m(2R)\,,$$

since otherwise we can reduce to that case by changing u into $-u$.

Let us assume that

$$\operatorname{osc}(u, 2R) \geq 2\operatorname{osc}_S(u, 2R)\,. \tag{7.51}$$

We have then

$$
\begin{aligned}
\operatorname{osc}(u, 2R) &= M(2R) - M_S(2R) + \operatorname{osc}_S(u, 2R) + m_S(2R) - m(2R)\\
&\leq 2[M(2R) - M_S(2R)] + \frac{1}{2}\operatorname{osc}(u, 2R)
\end{aligned}
$$

and therefore

$$M(2R) - \frac{1}{4}\operatorname{osc}(u, 2R) \geq M_S(2R)\,.$$

It follows that for $\nu \geq 1$ it holds that $k_\nu =: M(2R) - 2^{-\nu-1}\operatorname{osc}(u, 2R) \geq M_S(2R)$, and hence, taking Remark 7.3 into account, we have Caccioppoli's inequality (7.5) for every $k \geq k_\nu$. Consequently,

$$
\begin{aligned}
\sup_{Q_{\frac{R}{2}}} u \leq{}& c\left(\frac{1}{R^n}\int_{A(k_\nu,R)} (u - k_\nu)^p dx\right)^{\frac{1}{p}} \left(\frac{|A(k_\nu, R)|}{R^n}\right)^{\frac{\alpha}{p}}\\
&+ k_\nu + c\chi R^\beta\,.
\end{aligned} \tag{7.52}
$$

On the other hand, if $h > M_S(2R)$, the function $v(x)$ defined in Lemma 7.2 is zero on $\partial\Omega \cap Q_{2R}$, so that it can be extended to a function on Q_R setting it to be zero in $Q_R - \Omega$. Since the boundary of Ω is Lipschitz-continuous, we have $v = 0$ in a set whose measure is greater than $\gamma|Q_R|$, so that the conclusion of Lemma 7.2 holds in this case too. Arguing as in Theorem 7.6, we arrive to the relation

$$\operatorname{osc}\left(u, \frac{R}{2}\right) \leq \left(1 - \frac{1}{2^{\nu+2}}\right) \operatorname{osc}(u, 2R) + c2^{\nu}\chi R^{\beta} ,$$

which holds whenever $\operatorname{osc}(u, 2R) \geq 2\operatorname{osc}_S(u, 2R)$.

In any case we have

$$\operatorname{osc}\left(u, \frac{R}{2}\right) \leq \left(1 - \frac{1}{2^{\nu+2}}\right) \operatorname{osc}(u, 2R) + c2^{\nu}\chi R^{\beta} + 2\operatorname{osc}_S(u, 2R) ,$$

and since by assumption[3] $\operatorname{osc}_S(u, 2R) \leq BR^{\beta}$:

$$\operatorname{osc}\left(u, \frac{R}{2}\right) \leq \left(1 - \frac{1}{2^{\nu+2}}\right) \operatorname{osc}(u, 2R) + c(2^{\nu}\chi + B)R^{\beta} . \tag{7.53}$$

Comparing with (7.44), we conclude that the last inequality holds also when Q_{2R} does not intersects $\partial\Omega$. The conclusion then follows arguing as in Theorem 7.6. In particular, assuming that $\beta < \delta =: -\log_4(1 + 2^{-\nu-2})$, we have for every $\varrho < R$:

$$\operatorname{osc}(u, \varrho) \leq c\left(\frac{\varrho}{R}\right)^{\beta} \operatorname{osc}(u, R) + c(\chi + B)\varrho^{\beta} . \tag{7.54}$$

□

From the last inequality we can deduce the analogous of (7.46) and (7.47) for the Dirichlet problem with zero boundary data. In this case we have $B = 0$, and hence:

$$\fint_{\Omega_\varrho} |u|^p dx \leq c\operatorname{osc}(u, \varrho)^p \leq c\left(\frac{\varrho}{R}\right)^{p\beta} \operatorname{osc}\left(u, \frac{R}{2}\right)^p + c\chi^p \varrho^{p\beta} .$$

The right-hand side can be estimated using (7.29) with $\varrho = \frac{R}{2}$. We obtain thus the analogous of (7.46):

$$\int_{\Omega_\varrho} |u|^p dx \leq c\left(\frac{\varrho}{R}\right)^{n+p\beta} \int_{\Omega_R} |u|^p dx + c\chi^p \varrho^{n+p\beta} . \tag{7.55}$$

From this inequality, arguing as in Theorem 7.7, we easily get

$$\int_{\Omega_\varrho} |Du|^p dx \leq c\left(\frac{\varrho}{R}\right)^{n-p+p\beta} \int_{\Omega_R} |Du|^p dx + c\chi^p \varrho^{n-p+p\beta} . \tag{7.56}$$

[3]It is not restrictive to assume that the trace of u is Hölder-continuous with exponent β.

7.5 The Harnack Inequality

When the function $u \in DG_p^-$ is *positive*, we can estimate its infimum with greater accuracy. As usual, we can assume that u satisfies estimates (7.14), that is (7.10) with $\chi = 0$ and[4] $R = 1$.

Lemma 7.4 *Let $u \in DG_p^-$ with $\kappa_0 = 0$, and assume that u is positive in the cube $Q = Q_1$. There exists a positive constant γ_0 such that if $|B(\vartheta, 1)| \leq \gamma_0 |Q|$ for some $\vartheta > 0$, then*

$$\inf_{Q_{1/2}} u \geq \frac{\vartheta}{2}\,.$$

Proof. We can argue as in Lemma 7.2. For $h < k < \vartheta$ we set

$$v(x) = \begin{cases} 0 & \text{if} \quad u \geq k\,, \\ k - u & \text{if} \quad h < u < k\,, \\ k - h & \text{if} \quad u \leq h\,. \end{cases}$$

Let $\frac{1}{2} \leq \varrho \leq 1$; we have $v = 0$ in $Q_\varrho - B(k, \varrho)$, and since $|B(k, \varrho)| \leq |B(\vartheta, 1)| \leq \gamma_0 |Q|$ and $|Q_\varrho| \geq 2^{-n}|Q|$, we get $|Q_\varrho - B(k, \varrho)| \geq (2^{-n} - \gamma_0)|Q_\varrho|$. It follows that if $\gamma_0 \leq 2^{-n-1}$ we can apply the Sobolev inequality, obtaining

$$\left(\int_{Q_\varrho} v^{\frac{n}{n-1}} dx \right)^{1-\frac{1}{n}} \leq c \int_\Delta |Dv| dx\,,$$

where $\Delta = B(k, \varrho) - B(h, \varrho)$. We therefore have

$$\begin{aligned} (k - h)|B(h, \varrho)|^{1-\frac{1}{n}} &\leq \left(\int_{Q_\varrho} v^{\frac{n}{n-1}} dx \right)^{1-\frac{1}{n}} \\ &\leq c|\Delta|^{1-\frac{1}{p}} \left(\int_{B(k,\varrho)} |Du|^p dx \right)^{\frac{1}{p}} . \end{aligned} \tag{7.57}$$

On the other hand from (7.14) we get

[4] Note that since u is bounded we might use (7.32), namely (7.10) without the term k^p, instead of (7.14). In this case, however, we could not assume $\chi = 0$, since, with the exception of the homogeneous case, that we shall discuss later, it is not possible at the same time to eliminate the term k^p and to assume $\chi = 0$.

$$\int_{B(k,\varrho)} |Du|^p dx \le \frac{c}{(R-\varrho)^p} \int_{B(k,R)} (k-u)^p dx$$
$$+ ck^p |B(k,R)|^{1-\frac{p}{n}+\epsilon} \le \frac{c}{(R-\varrho)^p} k^p |B(k,R)|^{1-\frac{p}{n}+\epsilon} \tag{7.58}$$

since $u > 0$ and $\epsilon \le \frac{p}{n}$. From (7.57) we deduce:

$$(k-h)|B(h,\varrho)|^{1-\frac{1}{n}} \le \frac{ck}{R-\varrho} |B(k,R)|^{1-\frac{1}{n}+\frac{\epsilon}{p}}. \tag{7.59}$$

Consider now the sequence of radii $r_i = \frac{1}{2}(1+2^{-i})$ and the corresponding levels $k_i = \frac{\vartheta}{2}(1+2^{-i})$. Setting $B_i = |B(k_i, r_i)|$, we get from (7.59):

$$2^{-i-1} B_{i+1}^{1-\frac{1}{n}} \le c 2^{i+1} B_i^{1-\frac{1}{n}+\frac{\epsilon}{p}}$$

and in conclusion

$$B_{i+1} \le C 4^{\frac{ni}{n-1}} B_i^{1+\alpha},$$

where we have set $\alpha = \frac{\epsilon n}{p(n+1)}$. Applying the Lemma 7.1 we get $\lim_{i\to\infty} B_i = 0$, that is $u \ge \frac{\vartheta}{2}$ in $Q_{1/2}$, provided

$$B_0 = |B(\vartheta, 1)| \le C^{-\frac{1}{\alpha}} 4^{-\frac{n}{(n-1)\alpha^2}} =: \gamma_1 |Q|.$$

The conclusion of the lemma then follows by setting $\gamma_0 = \min(2^{-n-1}, \gamma_1)$. □

The next lemma is an improvement of the preceding one.

Lemma 7.5 *Let $u \in DG_p^-$ with $\kappa_0 = 0$, and assume that u be positive in the cube Q_2. For every γ, $0 < \gamma < 1$, there exists a constant $\lambda(\gamma) > 0$ such that if $|B(\vartheta, 1)| \le \gamma |Q_1|$ for some $\vartheta > 0$, then*

$$\inf_{Q_{1/2}} u \ge \lambda(\gamma)\vartheta.$$

Proof. Setting $\varrho = 1$ and $R = 2$ in the inequality (7.58) above, we get

$$\int_{B(k,1)} |Du|^p dx \le ck^p,$$

which introduced in (7.57) gives:

$$(k-h)^{\frac{p}{p-1}} |B(h,1)|^{\frac{p(n-1)}{n(p-1)}} \le ck^{\frac{p}{p-1}} (|B(k,1)| - |B(h,1)|). \tag{7.60}$$

Consider now the sequence of levels $k_i = \vartheta 2^{-i}$, with $i \leq \nu$. Setting $b_i = |B(k_i, 1)|$ we have, since $b_{i-1} \geq b_i \geq b_\nu$,

$$(\vartheta 2^{-i-1})^{\frac{p}{p-1}} b_\nu^{\frac{p(n-1)}{n(p-1)}} \leq c(\vartheta 2^{-i})^{\frac{p}{p-1}} (b_i - b_{i+1}) .$$

Simplifying and summing over i between 0 and ν, we get

$$(\nu + 1) b_\nu^{\frac{p(n-1)}{n(p-1)}} \leq c|Q| = c|Q|^{\frac{p(n-1)}{n(p-1)}} ,$$

where as usual $Q = Q_1$, and hence

$$b_\nu \leq \left(\frac{c}{\nu + 1} \right)^{\frac{n(p-1)}{p(n-1)}} |Q| .$$

Taking ν large enough we have then

$$b_\nu = |B(\vartheta 2^{-\nu}, 1)| \leq \gamma_0 |Q|$$

so that by the preceding lemma $u \geq \vartheta 2^{-\nu-1}$ in $Q_{1/2}$. □

The above lemma can be further generalized. Actually, if $|B(\vartheta, 1)| \leq \gamma |Q|$, and if $T > 1/2$, we have

$$|A(\vartheta, 2T)| \geq |A(\vartheta, 1)| \geq (1 - \gamma)|Q| \geq \frac{1-\gamma}{(2T)^n} |Q_{2T}|$$

and hence

$$|B(\vartheta, 2T)| \leq \left(1 - \frac{1-\gamma}{(2T)^n} \right) |Q_{2T}| .$$

We can therefore apply the preceding lemma, and conclude the following:

Lemma 7.6 *For every $\gamma \in (0, 1)$ and for every $T > 1/2$ there exists a positive constant $\mu(\gamma, T)$ such that if $u > 0$ in Q_{2T} and if $|B(\vartheta, 1)| \leq \gamma |Q|$ for some $\vartheta > 0$, then*

$$\inf_{Q_T} u \geq \mu(\gamma, T) \vartheta .$$

We have $\mu(\gamma, T) = \lambda(1 - \frac{1-\gamma}{(2T)^n})$. In the following we shall use the above lemma with $T = 1$.

Remark 7.10 The three lemmas just proved obviously hold for every cube Q_R with $R < R_0$. In particular, if u is positive in Q_{2TR} and if

$|B(\vartheta, R)| \leq \gamma |Q_R|$, then

$$\inf_{Q_{TR}} u(x) \geq \mu(\gamma, T)\vartheta . \qquad \square$$

Finally, we shall need the following covering theorem, whose analogy with the theorem of Calderon and Zygmund (Theorem 2.10) is evident.

Proposition 7.2 (KRYLOV and SAFONOV [1]) *Let $E \subset Q_R \subset \mathbf{R}^n$ be a measurable set, and let $0 < \delta < 1$. Moreover let*

$$E_\delta = \bigcup_{\substack{x \in Q_R \\ \varrho > 0}} \{Q(x, 3\varrho) \cap Q_R : |Q(x, 3\varrho) \cap E| \geq \delta |Q_\varrho|\} .$$

Then, either $|E| \geq \delta |Q_R|$, in which case $E_\delta = Q_R$, or else

$$|E_\delta| \geq \frac{1}{\delta} |E| .$$

Proof. For every $x \in Q_R$ we have $Q(x, 3R) \supset Q_R$ and hence, if $|E| \geq \delta |Q_R|$ it follows $|Q(x, 3R) \cap E| = |E| \geq \delta |Q_R|$, and hence $E_\delta = Q_R$.

Let us assume now that $|E| < \delta |Q_R|$.

Let us divide Q_R into 2^n equal cubes. If for any one of these subcubes Q we have

$$|Q \cap E| \geq \delta |Q| , \tag{7.61}$$

we say that Q_R is *final*,[5] and we do not divide Q any further. We repeat the preceding procedure for all nonfinal cubes Q, and we call F_δ the union of all the final cubes.

If Q is final, and if $Q(x, \varrho)$ is the subcube of Q for which (7.61) holds, we have $|Q(x, 3\varrho) \cap E| \geq \delta |Q_\varrho|$, and hence $Q(x, 3\varrho) \cap Q_R \subset E_\delta$. Since $Q \subset Q(x, 3\varrho) \cap Q_R$, we have also $Q \subset E_\delta$, and hence $F_\delta \subset E_\delta$.

On the other hand, almost every $x \in E$ belongs to F_δ. Actually, if $x \notin F_\delta$ there exists a sequence of cubes $Q_i \ni x$ with sides $\varrho_i \to 0$ and such that $|Q_i \cap E| \leq \delta |Q_i|$. This means that E has upper density less than 1 at the point $x \in E$, and this can happen only on a set of zero measure.

We have therefore, denoting by $\mathcal{M}$ the family of all final cubes,

$$|E| = |E \cap F_\delta| = \sum_{Q \in \mathcal{M}} |Q \cap E| \leq \delta \sum_{Q \in \mathcal{M}} |Q| = \delta |F_\delta| \leq \delta |E_\delta|$$

and the proposition is proved. $\square$

[5] Of course, due to the assumption $|E| < \delta |Q_R|$, the cube Q_R cannot be final; it is only the starting point of the successive divisions.

We can now prove the main theorem of this section.

Theorem 7.9 (DI BENEDETTO and TRUDINGER [1]) *Let $u(x)$ be a positive function, belonging to the De Giorgi class $DG_p^-(\Omega)$, with $\kappa_0 = 0$. There exists $R_0 > 0$, an exponent $q > 0$ and a constant c such that if $x_0 \in \Omega$ and $R < \min(R_0, \frac{1}{6\sqrt{n}}\mathrm{dist}(x_0, \partial\Omega))$, we have*

$$\inf_{Q(x_0,\frac{R}{2})} u(x) \geq c\left\{\left(\fint_{Q(x_0,R)} u^q dx\right)^{\frac{1}{q}} - \chi R^\alpha\right\} \tag{7.62}$$

Proof. As usual, we can assume $\chi = 0$ and $R = 1$. For a fixed δ (for instance $\delta = \frac{1}{2}$), we apply the preceding proposition to the set

$$A_t^{i-1} = A(t\mu^{i-1}, 1) = \{x \in Q : u(x) > t\mu^{i-1}\},$$

where $\gamma = 1 - 3^{-n}\delta$ and $\mu = \mu(\gamma, 2)$ is the constant of Lemma 7.6.

Assume now that for some $z \in Q$ we have $Q(z, 3\varrho) \cap Q \subset (A_t^{i-1})_\delta$, and hence

$$|A(t\mu^{i-1}, 1) \cap Q(z, 3\varrho)| \geq \delta|Q_\varrho| = \frac{\delta}{3^n}|Q_{3\varrho}|.$$

By Lemma 7.6 with $T = 2$, taking into account the remark immediately following it, we deduce $u \geq \mu\mu^{i-1}t = \mu^i t$ in $Q(z, 3\varrho)$, and hence

$$(A_t^{i-1})_\delta \subset A_t^i.$$

By the preceding proposition, we must have either $A_t^i = Q$ or $|A_t^i| \geq \delta^{-1}|A_t^{i-1}|$. In any case, we can conclude that if for some integer s we have

$$|A_t^0| = |A(t,1)| > \delta^s|Q| \tag{7.63}$$

then

$$|A_t^{s-1}| > \delta^{-1}|A_t^{s-2}| > \cdots > \delta^{1-s}|A_t^0| > \delta|Q|$$

and therefore $A_t^s = Q$, so that in conclusion

$$u(x) \geq \mu^s t \quad \text{in} \quad Q.$$

We now choose s in such a way that (7.63) is satisfied; for instance let s be the smallest integer such that

$$s \geq \frac{1}{\log\delta}\log\frac{|A(t,1)|}{|Q|}.$$

With this choice of s we get

$$\inf_Q u(x) \geq \mu^s t = ct \left(\frac{|A(t,1)|}{|Q|} \right)^{\frac{\log \mu}{\log \delta}},$$

or equivalently, setting $\xi = \inf_Q u(x)$ and $a = \frac{\log \delta}{\log \mu}$:

$$|A(t,1)| \leq c|Q|\xi^a t^{-a}. \tag{7.64}$$

On the other hand

$$\int_Q u^q dx = q \int_0^\infty t^{q-1}|A(t,1)|dt = q \int_\xi^\infty t^{q-1}|A(t,1)|dt + |Q|\xi^q.$$

Introducing the estimate (7.64) we get

$$\fint_Q u^q dx \leq cq\xi^a \int_\xi^\infty t^{q-a-1}dt + \xi^q.$$

If we choose $q < a$ the integral converges, and hence

$$\fint_Q u^q dx \leq c\xi^q$$

so that

$$\inf_{Q_{1/2}} u(x) \geq \inf_Q u(x) \geq c \left(\fint_Q u^q dx \right)^{\frac{1}{q}}.$$

The conclusion follows coming back to a generic R and writing $u + \chi R^\alpha$ instead of u. □

Joining the above result with Theorem 7.3 we get the following HARNACK inequality:

Theorem 7.10 *Let $u(x)$ be a positive function belonging to De Giorgi class $DG_p(\Omega)$ with $\kappa_0 = 0$, and let ϱ be a number less than $\frac{R_0}{2}$, and such that the cube of side 6ϱ is contained in Ω.*

Then

$$\sup_{Q_\varrho} u \leq c(\inf_{Q_\varrho} u + \chi\varrho^\alpha). \tag{7.65}$$

Remark 7.11 The last inequality can be proved directly, starting from the regularity results of the preceding sections (in particular from (7.45)) and from Lemma 7.6. The proof that follows is an adjustment of that given by Di Benedetto [1] in the parabolic case. □

Proof. We shall begin by showing that

$$u(x_0) \le c \inf_{Q(x_0,R)} u(x)\,, \tag{7.66}$$

or equivalently

$$v(x) =: \frac{u(x)}{u(x_0)} \ge c > 0$$

in $Q(x_0, R)$.

As usual, we can assume $\chi = 0$ and $R = 1$.

It is easily seen that the function v satisfies the inequalities (7.13) and (7.14) for every $h > 0$, with the some constants as u. In particular Theorem 7.6 will hold for v, with the estimate

$$\mathrm{osc}_{Q(x,\varrho)} v \le c\,\mathrm{osc}_{Q(x,R)} v \left(\frac{\varrho}{R}\right)^{\beta} \le c\|v\|_{\infty,Q(x,R)} \left(\frac{\varrho}{R}\right)^{\beta} \tag{7.67}$$

for every $x \in \Omega$ and every $\varrho < R < \frac{1}{2}\mathrm{dist}(x, \partial\Omega)$.

Now let $K_\tau = (1-\tau)^{-\delta}$, where $\delta > 0$ will be chosen later, and let τ_0 be the largest value of τ for which $\|v\|_{\infty,Q(x_0,\tau)} \ge K_\tau$. Since the left-hand side of the preceding relation is bounded, and the right-hand side diverges as $\tau \to 1$, we have $0 \le \tau_0 < 1$.

Let $\bar{x} \in \overline{Q(x_0, \tau_0)}$ be such that $v(\bar{x}) = \|v\|_{\infty,Q(x_0,\tau_0)} \ge (1-\tau_0)^{-\delta}$. We have

$$\|v\|_{\infty,Q(\bar{x},\frac{1-\tau_0}{2})} \le \|v\|_{\infty,Q(x_0,\frac{1+\tau_0}{2})} < K_{\frac{1+\tau_0}{2}} = 2^{\delta}(1-\tau_0)^{-\delta}\,.$$

On the other hand, using (7.45) with $\chi = 0$, $R = \frac{1-\tau_0}{2}$ and $\varrho = \epsilon R$ ($\epsilon < 1$), we get

$$\mathrm{osc}_{Q(\bar{x},\frac{1-\tau_0}{2}\epsilon)} v \le c\|v\|_{\infty,Q(\bar{x},\frac{1-\tau_0}{2})}\epsilon^{\beta} \le c2^{\delta}(1-\tau_0)^{-\delta}\epsilon^{\beta}$$

and hence

$$v(x) \ge v(\bar{x}) - \mathrm{osc}_{Q(\bar{x},\frac{1-\tau_0}{2}\epsilon)} v \ge (1-\tau_0)^{-\delta}(1 - c2^{\delta}\epsilon^{\beta})$$

for every $x \in Q(\bar{x}, \frac{1-\tau_0}{2}\epsilon)$. Choosing $\epsilon = c^{-1}2^{-\delta-1}$, we obtain

$$v(x) \ge \frac{1}{2}(1-\tau_0)^{-\delta} \quad \text{in} \quad Q\left(\bar{x}, \frac{1-\tau_0}{2}\epsilon\right)\,. \tag{7.68}$$

We can now apply the Lemma 7.6 (or better the remark following it) with $R = \frac{1-\tau_0}{2}\epsilon$, $T = 2$ and $\vartheta = \frac{1}{2}(1-\tau_0)^{-\delta}$. We have $\gamma = 0$, and therefore

$$v(x) \geq \frac{\mu}{2}(1-\tau_0)^{-\delta} \quad \text{in} \quad Q(\bar{x}, (1-\tau_0)\epsilon),$$

with $\mu = \mu(0, 2)$.

Iterating the above argument, we get for every integer ν:

$$v(x) \geq \frac{\mu^\nu}{2}(1-\tau_0)^{-\delta} \quad \text{in} \quad Q(\bar{x}, 2^{\nu-1}(1-\tau_0)\epsilon). \tag{7.69}$$

Let ν be such that $2 \leq 2^{\nu-1}(1-\tau_0)\epsilon < 4$. We have

$$\mu^\nu \geq \left(\frac{8}{\epsilon(1-\tau_0)}\right)^{\log_2 \mu}$$

and hence

$$v(x) \geq \frac{1}{2}\left(\frac{8}{\epsilon}\right)^{\log_2 \mu} (1-\tau_0)^{-\delta-\log_2 \mu} \quad \text{in} \quad Q(\bar{x}, 2) \supset Q(x_0, 1).$$

We now choose $\delta = -\log_2 \mu$; we have $\epsilon = \frac{\mu}{2c}$ and therefore

$$v(x) \geq \frac{1}{2}\left(\frac{16c}{\mu}\right)^{\log_2 \mu} \quad \text{in} \quad Q(x_0, 1),$$

which gives the estimate (7.66).

Let now $Q_\varrho = Q(x_1, \varrho)$ be a cube contained in Ω, with ϱ small enough, and let $x_0 \in \bar{Q}_\varrho$ be such that $u(x_0) = \sup_{Q_\varrho} u(x)$. Taking $R = 3\varrho$, we have from (7.66):

$$\sup_{Q_\varrho} u(x) \leq c \inf_{Q(x_0,R)} u(x) \leq c \inf_{Q_\varrho} u(x),$$

and hence (7.65) with $\chi = 0$. The general case follows as usual by writing $u + \chi R^\beta$ instead of u. $\square$

7.6 The Homogeneous Case

Of particular interest is the case when the function u belongs to a homogeneous De Giorgi class DGO_p, that is when it satisfies the relations

$$\int_{A(k,\varrho)} |Du|^p dx \leq \frac{H}{(R-\varrho)^p} \int_{A(k,R)} (u-k)^p dx, \tag{7.70}$$

$$\int_{B(k,\varrho)} |Du|^p dx \leq \frac{H}{(R-\varrho)^p} \int_{B(k,R)} (k-u)^p dx. \tag{7.71}$$

In particular, this happens when u is a quasi-minimum of the functional

$$\mathcal{F}(u,\Omega) = \int_\Omega F(x,u(x),Du(x))dx$$

with

$$|z|^p \le F(x,u,z) \le L|z|^p .$$

The most significative example of this situation is that of quadratic functionals

$$F(x,u,z) = a_{ij}(x,u)z_i z_j ,$$

with the coefficients a_{ij} bounded and satisfying the ellipticity condition

$$a_{ij}(x,u)\xi_i\xi_j \ge \nu|\xi|^2 , \quad \nu > 0 .$$

For positive functions $u \in DGO_p$ we have the estimate (7.65) with $\chi = 0$, and therefore:

Theorem 7.11 (HARNACK's inequality) *Let Ω be a bounded connected open set in $\mathbf{R}^n$, and let $\Sigma \subset\subset \Omega$. Let $u(x)$ be a positive function in $DGO_p(\Omega)$, $(p > 1)$. There exists a constant $C(\Sigma,\Omega)$ such that*

$$\sup_\Sigma u \le C \inf_\Sigma u \tag{7.72}$$

Proof. Let $Q_1, Q_2, \ldots, Q_N$ be a finite family of cubes, such that any two consecutive cubes Q_i and Q_{i+1} have non-empty intersection and that

$$\sup_\Sigma u = \sup_{Q_1} u ; \quad \inf_\Sigma u = \inf_{Q_N} u .$$

We can assume that each of these cubes has side $\varrho < R_0$, and that the cubes of side 6ϱ are contained in Ω.

For each cube Q_i we can write the inequality (7.65) with $\chi = 0$:

$$\sup_{Q_i} u \le c \inf_{Q_i} u .$$

On the other hand, since any two consecutive cubes intersect, we have

$$\inf_{Q_i} u \le \sup_{Q_{i+1}} u ,$$

and the conclusion follows at once. □

If moreover Ω is a cube of side R, and Σ is a concentric cube of side τR, the constant C depends only on τ but is independent of R, as is easily seen by homothety.

Two consequences of Harnack's inequality are particularly worthy of note.

Theorem 7.12 (Strong maximum principle) *Let Ω be a connected set, and let $u(x)$ be a function in the class $DGO_p^-(\Omega)$. If u has an interior minimum point, then u is constant in Ω.*

Proof. We note that (7.70) and (7.71) do not change if we write $u + \lambda$ instead of u. As a consequence, adding possibly a constant to u, we can assume that $\min_\Omega u = 0$. Let E be the set of the points of Ω in which u assumes its minimum value 0. By assumption, E is non-empty.

For $\epsilon > 0$, the function $u + \epsilon$ is positive, and we can apply Theorem 7.9. If Q is a cube with center in a point of E and side small enough, we have

$$\epsilon = \inf_Q (u+\epsilon) \geq c \left(\fint_Q (u+\epsilon)^p dx \right)^{\frac{1}{p}} \geq c \left(\fint_Q u^p dx \right)^{\frac{1}{p}}$$

and therefore, since $\epsilon > 0$ is arbitrary, we must have $u = 0$ in Q.

It follows that E is an open set; since by the continuity of u it is also closed, we conclude that $E = \Omega$. □

In a similar way one can prove that if $u \in DGO_p^+(\Omega)$ has an interior maximum point, then it is constant. Finally, if $u \in DGO_p(\Omega)$, it must assume both its maximum and its minimum only on the boundary of Ω, unless it is constant.

Theorem 7.13 (LIOUVILLE) *Let $u \in DGO_p(\mathbf{R}^n)$, and assume that u is bounded below. Then, u is constant.*

Proof. Let $\lambda = \inf u > -\infty$. Writing $u - \lambda$ instead of u, we can suppose $\lambda = 0$, and therefore by the preceding theorem $u > 0$ in $\mathbf{R}^n$. From Harnack's inequality we have for every $R > 0$:

$$\sup_{Q_R} u \leq c \inf_{Q_R} u$$

If we let R go to infinity, the right-hand side tends to zero, and hence $u \equiv 0$. □

7.7 ω-Minima

As we have seen in the preceding chapter (Example 6.5), the regularity Hölder continuity, do not hold for cubical quasi-minima. Generally speaking, estimates on cubes only are not sufficient to prove regularity.

The question can be posed whether estimates on cubes are sufficient in some special cases, of course more general than 1-minima (i.e. quasi-minima with $Q = 1$), for which the difference between estimates on cubes and on general sets disappears. The answer is positive for a particular sort of cubical quasi-minima, the ω-minima according to the following:

Definition 7.2 *Let $\omega : \mathbf{R}_+ \to \mathbf{R}_+$ be a continuous, bounded, increasing and concave function, with $\omega(0) = 0$. We say that a function $u \in W^{1,p}_{\text{loc}}(\Omega)$ is an* ω-minimum *for the functional $\mathcal{F}$ if for every cube $Q_R \subset\subset \Omega$, and every $w \in W^{1,p}(Q_R)$ with $w = u$ on ∂Q_R, we have*

$$\mathcal{F}(u, Q_R) \leq [1 + \omega(R)]\mathcal{F}(w, Q_R). \tag{7.73}$$

It is clear that an ω-minimum is also a cubical quasi-minimum, and that the difference between the two lies only in the behavior for very small sides. It is not known whether an ω-minimum is also a quasi-minimum. Nevertheless, it can be proved that ω-minima are Hölder continuous functions. The present section will be dedicated to the proof of this result.

As above, we shall assume that the integrand $F(x, u, z)$ is a Caratheodory function satisfying the inequalities

$$|z|^p - b(x)|u|^\gamma - a(x) \leq F(x, u, z) \leq L|z|^p + b(x)|u|^\gamma + a(x) \tag{7.74}$$

with $p < n$, $1 < p \leq \gamma < p^* = \frac{pn}{n-p}$, $a(x)$ and $b(x)$ being two non-negative functions, belonging respectively to $L^s(\Omega)$ and $L^\sigma(\Omega)$, with $s > \frac{n}{p}$ and $\sigma > \frac{p^*}{p^*-\gamma}$.

We have seen in the preceding chapter (Remark 6.11) that the above assumptions imply the existence of a $r > 1$ such that every spherical quasi-minimum u belongs to $W^{1,pr}_{\text{loc}}(\Omega)$, with the estimate

$$\begin{aligned} \fint_{Q_R} |Du|^{rp} dx &\leq c \left(\fint_{Q_{2R}} |Du|^p dx \right)^r \\ &\quad + c \fint_{Q_{2R}} (a(x) + b(x)|u|^\gamma)^r dx. \end{aligned} \tag{7.75}$$

It is evident that we can suppose r as close to 1 as we wish, so that we can assume without loss of generality that $r\gamma < p^*$.

As above, we shall assume that $\frac{1}{s} = \frac{p}{n} - \epsilon$ and $\frac{1}{\sigma} = 1 - \frac{\gamma}{p^*} - \epsilon$ for some $\epsilon > 0$. It is easily seen that we can take ϵ as small as we like, without prejudice to the results of the above sections, the constants being independent of ϵ, as far as it remains bounded away from zero.

Since, as we have remarked, it is not known whether ω-minima are quasi-minima, we cannot apply the regularity result of the preceding sections directly to ω-minima. The proof will be achieved by comparing ω-minima of the functional $\mathcal{F}$ on cubes of side R with quasi-minima of the functional

$$\mathcal{G}(v, Q_R) = \int_{Q_R} (|Dv|^p + a(x) + \Lambda + b(x)|v|^\gamma)dx \tag{7.76}$$

with a suitable constant Λ depending on R.

We shall begin with an estimate concerning quasi-minima of the above functional. For simplicity, we set $\alpha(x) = a(x) + \Lambda$.

Theorem 7.14 *Let v be a Q-minimum of the functional $\mathcal{G}$ in a cube of side R. Then, there exists a positive constant R_0, depending only on a, b and on the norm of v in $W^{1,p}$ such that for every $\varrho < R \leq R_0$ we have*

$$\int_{Q_\varrho} (|Dv|^p + \varrho^{-\mu}|v|^\gamma)dx \leq c\left(\frac{\varrho}{R}\right)^{n-p+n\epsilon} \left\{ \int_{Q_R} (|Dv|^p + R^{-\mu}|v|^\gamma)dx \right.$$
$$\left. + \|\alpha\|_s R^{n-p+n\epsilon} + (\|\alpha\|_s R^{n-p+n\epsilon})^{\frac{\gamma}{p}} R^{n\epsilon} \right\}, \tag{7.77}$$

where $\mu = \frac{n}{\sigma}$, and c is a constant depending on a, b and on the norm of v in $W^{1,p}$

Proof. It will be simpler to prove the estimate (7.77) with $2R$ instead of R. Suppose first that $\varrho < R$. We use the inequality (7.47)

$$\int_{Q_\varrho} |Dv|^p dx \leq c\left(\frac{\varrho}{R}\right)^{n-p+p\beta} \left\{ \int_{Q_R} |Dv|^p dx + \chi^p R^{n-p+p\beta} \right\},$$

where, according to the Remark 7.7,

$$\chi^p = \|\alpha + bM^\gamma\|_s \leq \|\alpha\|_s + \|b\|_s M^\gamma .$$

The quantity $M = 2\sup_{Q_R}|v|$ can be estimated by means of (7.26) with $q = \gamma$:

$$\sup_{Q_R} |v|^\gamma \leq c\left\{ \fint_{Q_{2R}} |v|^\gamma dx + \|a\|_s^{\frac{\gamma}{p}} R^{\beta\gamma} \right\},$$

where $\beta = \frac{n\epsilon}{p}$. Moreover, we have

$$\|b\|_s \le c\|b\|_\sigma R^{n(\frac{1}{s}-\frac{1}{\sigma})},$$

and therefore

$$\chi^p \le \|\alpha\|_s + c\|b\|_\sigma \left(R^{n(\frac{1}{s}-\frac{1}{\sigma}-1)} \int_{Q_{2R}} |v|^\gamma dx + \|\alpha\|_s^{\frac{\gamma}{p}} R^{n(\frac{1}{s}-\frac{1}{\sigma}+\frac{\beta\gamma}{n})} \right).$$

In conclusion

$$\int_{Q_\varrho} |Dv|^p dx \le c\left(\frac{\varrho}{R}\right)^{n-p+p\beta} \left\{ \int_{Q_{2R}} |Dv|^p dx + R^{-\mu} \int_{Q_{2R}} |v|^\gamma dx \right.$$
$$\left. + \|\alpha\|_s R^{n-p+n\epsilon} + \|\alpha\|_s^{\frac{\gamma}{p}} R^{\frac{n\gamma}{p^*}+n\epsilon+\beta\gamma} \right\}. \tag{7.78}$$

On the other hand

$$\varrho^{-\mu} \int_{Q_\varrho} |v|^\gamma dx \le c\varrho^{n(1-\frac{1}{\sigma})} M^\gamma = c\varrho^{n(\frac{\gamma}{p^*}+\epsilon)} M^\gamma$$
$$\le c\left(\frac{\varrho}{R}\right)^{n(\frac{\gamma}{p^*}+\epsilon)} \left\{ R^{-\mu} \int_{Q_{2R}} |v|^\gamma dx + \|\alpha\|_s^{\frac{\gamma}{p}} R^{\frac{n\gamma}{p^*}+n\epsilon+\beta\gamma} \right\}$$
$$\le c\left(\frac{\varrho}{R}\right)^{n-p+n\epsilon} \left\{ R^{-\mu} \int_{Q_{2R}} |v|^\gamma dx + \|\alpha\|_s^{\frac{\gamma}{p}} R^{\frac{n\gamma}{p^*}+n\epsilon+\beta\gamma} \right\},$$

where in the last passage we have used the inequality $\frac{n\gamma}{p^*} \ge \frac{np}{p^*} = n-p$.

Adding the last inequality to (7.78) we get (7.77) for $\varrho < R$. On the other hand if $R \le \varrho < 2R$, (7.77) holds trivially, possibly with a different constant c, and hence the theorem is proved. □

Let now u be an ω-minimum of the functional $\mathcal{F}$. We shall get an estimate similar to (7.77) by comparing u with a suitable Q-minimum v of the functional $\mathcal{G}$. For the construction of such v we shall use Ekeland's variational principle of Section 5.4. We define a metric space X as the set of all functions $w \in W^{1,p}(Q_R)$ such that $u - w \in W_0^{1,p}(Q_R)$, and

$$\int_{Q_R} (|Dw|^p + b|w|^\gamma) dx \le \int_{Q_R} (|Du|^p + b|u|^\gamma) dx.$$

We equip X with the metric

$$d(w,v) = C_R \int_{Q_R} |Dw - Dv| dx, \tag{7.79}$$

where C_R is a constant that we shall choose later. A simple application of Fatou's lemma shows that X is complete.

For $\delta > 0$, let v_δ be a function in X such that

$$\mathcal{F}(v_\delta, Q_R) \leq \inf_X \mathcal{F} + \delta\,.$$

We have

$$\mathcal{F}(u, Q_R) \leq (1 + \omega(R))\mathcal{F}(v_\delta, Q_R) \leq \inf_X \mathcal{F} + \delta + \omega[\mathcal{F}(u, Q_R) + \delta]$$

from which, letting $\delta \to 0$, we get

$$\mathcal{F}(u) \leq \inf_X \mathcal{F} + \omega\mathcal{F}(u) \leq \inf_X \mathcal{F} + c\omega R^n \fint_{Q_R} (|Du|^p + b|u|^\gamma + a)dx$$

in which we have written $\mathcal{F}(u)$ instead of $\mathcal{F}(u, Q_R)$.

By Ekeland's principle, there exists a function $v \in X$, in particular such that

$$\int_{Q_R} (|Dv|^p + b|v|^\gamma)dx \leq \int_{Q_R} (|Du|^p + b|u|^\gamma)dx\,,$$

satisfying

$$\int_{Q_R} |Dv - Du|dx \leq C_R^{-1} \tag{7.80}$$

and such that

$$\mathcal{F}(v) \leq \mathcal{F}(w) + c\omega R^n C_R \fint_{Q_R} (|Du|^p + b|u|^\gamma + a)dx \int_{Q_R} |Dv - Du|dx\,.$$

If we now choose

$$C_R^{-1} = R^n \omega \left(\fint_{Q_R} (|Du|^p + b|u|^\gamma + a)dx \right)^{\frac{1}{p}}$$

the last inequality becomes

$$\mathcal{F}(v) \leq \mathcal{F}(w) + \Lambda^{1-\frac{1}{p}} \int_{Q_R} |Dv - Du|dx \tag{7.81}$$

with

$$\Lambda = \Lambda(R) = c\fint_{Q_R} (|Du|^p + b|u|^\gamma + a)dx\,. \tag{7.82}$$

We remark that we have

$$\int_{Q_R} |Dv|^p dx \le \int_{Q_R} (|Du|^p + b|u|^\gamma) dx \le c_1 \tag{7.83}$$

$$\int_{Q_R} |v|^{p^*} dx \le c \int_{Q_R} |v-u|^{p^*} dx + c \int_{Q_R} |u|^{p^*} dx$$

$$\le c \left(\int_{Q_R} (|Du| + |Dv|)^p dx \right)^{\frac{p^*}{p}} + c \int_{Q_R} |u|^{p^*} dx \le c_2 \tag{7.84}$$

with c_1 and c_2 independent of R.

We shall prove now that v is a Q-minimum of the functional $\mathcal{G}$.

Let $w \in W^{1,p}(Q_R)$, with $\varphi = v - w \in W_0^{1,p}$, and let $K = \operatorname{supp} \varphi$. If $w \in X$, we have from (7.81)

$$\int_K (|Dv|^p - b|v|^\gamma - a) dx \le \mathcal{F}(v, K) \le \mathcal{F}(w, K) + \Lambda^{1-\frac{1}{p}} \int_K |D\varphi| dx$$

and hence

$$\int_K |Dv|^p dx$$

$$\le \int_K (|Dw|^p + b|w|^\gamma + a) dx + \int_K (b|v|^\gamma + a) dx + \Lambda^{1-\frac{1}{p}} \int_K |D\varphi| dx \, .$$

We estimate

$$\Lambda^{1-\frac{1}{p}} |D\varphi| \le \epsilon |D\varphi|^p + c_\epsilon \Lambda \le c\epsilon (|Dv|^p + |Dw|^p) + c_\epsilon \Lambda$$

so that, taking ϵ small enough

$$\int_K (|Dv|^p + b|v|^\gamma + a + \Lambda) dx \le c \int_K (|Dw|^p + b|w|^\gamma + a + b|v|^\gamma + \Lambda) dx \, .$$

On the other hand, if $w \notin X$, we have

$$\int_K (|Dv|^p + b|v|^\gamma) dx \le \int_K (|Du|^p + b|u|^\gamma) dx \le \int_K (|Dw|^p + b|w|^\gamma) dx$$

so that the above inequality holds for every $w \in W^{1,p}(Q_R)$ with $w - v \in W_0^{1,p}(Q_R)$.

Arguing as in Theorem 6.1 (see Remark 6.6), we conclude that

$$\int_K (|Dv|^p + b|v|^\gamma + a + \Lambda) dx \le Q \int_K (|Dw|^p + b|w|^\gamma + a + \Lambda) dx$$

and hence v is a Q-minimum of the functional $\mathcal{G}$. The constant Q will depend on v only through its norm $\|Dv\|_p + \|v\|_{p^*}$, and hence in the last instance it depends only on u and it is independent of R.

Let us estimate now the quantity

$$\fint_{Q_{\frac{R}{2}}} |Du - Dv|^p dx\,.$$

Let $r > 1$ be the exponent in (6.64), and let $\vartheta = \frac{r(p-1)}{rp-1}$. We have

$$\left(\fint_{Q_{\frac{R}{2}}} |Du - Dv|^p dx\right)^{\frac{1}{p}}$$

$$\leq \left(\fint_{Q_{\frac{R}{2}}} |Du - Dv|^{rp} dx\right)^{\frac{\vartheta}{rp}} \left(\fint_{Q_{\frac{R}{2}}} |Du - Dv| dx\right)^{1-\vartheta}.$$

We estimate the two factors separately. For the first, we use (6.64) both for u and v, and we get

$$\left(\fint_{Q_{\frac{R}{2}}} |Du - Dv|^{rp} dx\right)^{\frac{1}{rp}}$$

$$\leq c\left(\fint_{Q_{\frac{R}{2}}} |Du|^{rp} dx\right)^{\frac{1}{rp}} + c\left(\fint_{Q_{\frac{R}{2}}} |Dv|^{rp} dx\right)^{\frac{1}{rp}}$$

$$\leq c\left(\fint_{Q_R} |Du|^p dx\right)^{\frac{1}{p}} + c\left(\fint_{Q_R} (b|u|^\gamma + a)^r dx\right)^{\frac{1}{rp}}$$

$$+ c\left(\fint_{Q_R} |Dv|^p dx\right)^{\frac{1}{p}} + c\left(\fint_{Q_R} (b|v|^\gamma + a + \Lambda)^r dx\right)^{\frac{1}{rp}}$$

and therefore

$$\left(\fint_{Q_{\frac{R}{2}}} |Du - Dv|^{rp} dx\right)^{\frac{1}{rp}}$$

$$\leq c\left(\fint_{Q_R} |Du|^p dx\right)^{\frac{1}{p}} + c\left(\fint_{Q_R} (b|u|^\gamma)^r dx\right)^{\frac{1}{rp}}$$

$$+ c\left(\fint_{Q_R} (b|v|^\gamma)^r dx\right)^{\frac{1}{rp}} + c\left(\fint_{Q_R} a^r dx\right)^{\frac{1}{rp}} + c\Lambda^{\frac{1}{p}}\,.$$

On the other hand

$$\fint_{Q_{\frac{R}{2}}} |Du - Dv| dx \le cR^{-n} C_R = c\omega(R)\Lambda^{\frac{1}{p}}$$

and

$$\Lambda^{\frac{1}{p}} \le c\left(\fint_{Q_R} |Du|^p dx\right)^{\frac{1}{p}} + c\left(\fint_{Q_R} (b|u|^\gamma)^r dx\right)^{\frac{1}{rp}} + c\left(\fint_{Q_R} a^r dx\right)^{\frac{1}{rp}},$$

so that in conclusion

$$\fint_{Q_{\frac{R}{2}}} |Du - Dv|^p dx \le c\omega(R)^\kappa \left\{ \fint_{Q_R} |Du|^p dx + \left(\fint_{Q_R} b|u|^\gamma)^r dx\right)^{\frac{1}{r}} \right.$$

$$\left. + c\left(\fint_{Q_R} (b|v|^\gamma)^r dx\right)^{\frac{1}{r}} + R^{n-p+n\epsilon} \|a\|_s \right\} \qquad (7.85)$$

with some exponent $\kappa > 0$.

Finally, we estimate the integrals containing the function b. We have

$$\left(\fint_{Q_R} (b|u|^\gamma)^r dx\right)^{\frac{1}{r}} \le c\left(\fint_{Q_R} (b|u - u_R|^\gamma)^r dx\right)^{\frac{1}{r}} + c\left(\fint_{Q_R} (b|u_R|^\gamma)^r dx\right)^{\frac{1}{r}}.$$

Now

$$|u_R|^\gamma \left(\fint_{Q_R} b^r dx\right)^{\frac{1}{r}} \le \left(\fint_{Q_R} b^\sigma dx\right)^{\frac{1}{\sigma}} \fint_{Q_R} |u|^\gamma dx$$

and

$$\left(\int_{Q_R} (b|u - u_R|^\gamma)^r dx\right)^{\frac{1}{r}} \le \left(\int_{Q_R} (|u - u_R|^{p^*} dx\right)^{\frac{\gamma}{p^*}} \left(\int_{Q_R} b^{\frac{rp^*}{p^* - r\gamma}} dx\right)^{\frac{1}{r} - \frac{\gamma}{p^*}}$$

$$= \left(\int_{Q_R} (|u - u_R|^{p^*} dx\right)^{\frac{p}{p^*}} \left(\int_{Q_R} (|u - u_R|^{p^*} dx\right)^{\frac{\gamma - p}{p^*}}$$

$$\times \left(\int_{Q_R} b^{\frac{rp^*}{p^* - r\gamma}} dx\right)^{\frac{1}{r} - \frac{\gamma}{p^*}}$$

$$\le c\int_{Q_R} |Du|^p dx.$$

In a similar way

$$\left(\fint_{Q_R} (b|v|^\gamma)^r dx\right)^{\frac{1}{r}} \le c\left(\fint_{Q_R} (b|u - v|^\gamma)^r dx\right)^{\frac{1}{r}} + c\left(\fint_{Q_R} (b|u|^\gamma)^r dx\right)^{\frac{1}{r}}.$$

The last term on the right-hand side has already been treated; the first can be estimated as above by

$$c\int_{Q_R}(|Du|^p + R^{-\mu}|u|^\gamma)dx\,.$$

In conclusion,

$$\begin{aligned}&\fint_{Q_{\frac{R}{2}}}|Du - Dv|^p dx\\ &\quad\le c\omega(R)^\kappa\left\{\fint_{Q_R}(|Du|^p + R^{-\mu}|u|^\gamma)dx + \|a\|_s R^{n-p+n\epsilon}\right\}.\end{aligned}\tag{7.86}$$

With the help of the above estimate, we can transfer inequality (7.77) to the ω-minimum u. We have for $\varrho < \frac{R}{2}$:

$$\begin{aligned}&\int_{Q_\varrho}(|Du|^p + \varrho^{-\mu}|u|^\gamma)dx\\ &\qquad\le c\int_{Q_\varrho}(|Dv|^p + \varrho^{-\mu}|v|^\gamma)dx\\ &\qquad\quad + c\int_{Q_\varrho}(|Du - Dv|^p + \varrho^{-\mu}|u - v|^\gamma)dx\\ &\qquad\le c\left(\frac{\varrho}{R}\right)^{n-p+n\epsilon}\left\{\int_{Q_R}(|Dv|^p + R^{-\mu}|v|^\gamma)dx + \Theta(R)\right\}\\ &\qquad\quad + c\int_{Q_\varrho}(|Du - Dv|^p + \varrho^{-\mu}|u - v|^\gamma)dx\\ &\qquad\le c\left(\frac{\varrho}{R}\right)^{n-p+n\epsilon}\left\{\int_{Q_R}(|Du|^p + R^{-\mu}|u|^\gamma)dx + \Theta(R)\right\}\\ &\qquad\quad + c\int_{Q_{\frac{R}{2}}}(|Du - Dv|^p + \varrho^{-\mu}|u - v|^\gamma)dx\,,\end{aligned}$$

where

$$\Theta(R) = \|\alpha\|_s R^{n-p+n\epsilon} + (\|\alpha\|_s R^{n-p+n\epsilon})^{\frac{\gamma}{p}} R^{n\epsilon}\,.$$

Let us estimate the single terms on the right-hand side, beginning with Θ. We have

$$\|\alpha\|_s \le \|a\|_s + R^{\frac{n}{s}}\Lambda$$

and therefore

$$\|\alpha\|_s R^{n-p+n\epsilon} \le \|a\|_s R^{n-p+n\epsilon} + R^n \Lambda .$$

Recalling now that

$$\Lambda R^n = c \int_{Q_R} (|Du|^p + b|u|^\gamma + a) dx \le c ,$$

we get easily

$$\begin{aligned} \Theta(R) &\le c(\|a\|_s R^{n-p+n\epsilon} + \Lambda R^n) \\ &\le c \left(\int_{Q_R} (|Du|^p + R^{-\mu}|u|^\gamma) dx + \|a\|_s R^{n-p+n\epsilon} \right) . \end{aligned} \tag{7.87}$$

The term $\int_{Q_{\frac{R}{2}}} |Du - Dv|^p dx$ has been already considered in (7.86). The remaining term can be treated as follows:

$$\begin{aligned} \int_{Q_{\frac{R}{2}}} |u - v|^\gamma) dx &\le cR^n \left(\fint_{Q_R} |u - v|^{p^*} dx \right)^{\frac{\gamma}{p^*}} \\ &\le cR^{n(1-\frac{\gamma}{p^*})} \left(\int_{Q_R} |Du - Dv|^p dx \right)^{\frac{\gamma}{p}} \\ &\le cR^{\mu+n\epsilon} \int_{Q_R} (|Du|^p + b|u|^\gamma + a) dx \end{aligned}$$

since

$$\int_{Q_R} (|Du|^p + b|u|^\gamma + a) dx \le c .$$

Putting together all these inequalities, we get the following:

Theorem 7.15 *Let u be an ω-minimum of the functional $\mathcal{F}$. There exists a $R_0 > 0$ such that for every cube $Q_R \subset\subset \Omega$, with $R < R_0$,*

$$\begin{aligned} \int_{Q_\varrho} (|Du|^p + \varrho^{-\mu}|u|^\gamma) dx \le c \left\{ \left(\frac{\varrho}{R}\right)^{n-p+n\epsilon} + \omega(R)^\kappa + \left(\frac{R}{\varrho}\right)^\mu R^{n\epsilon} \right\} \\ \times \int_{Q_R} (|Du|^p + R^{-\mu}|u|^\gamma) dx + c\|a\|_s R^{n-p+n\epsilon} . \end{aligned} \tag{7.88}$$

The inequality (7.88) is the main tool in the proof of the Hölder continuity of u. Writing for the sake of simplicity

$$\varphi(\varrho) = \int_{Q_\varrho} (|Du|^p + \varrho^{-\mu}|u|^\gamma)dx\,,$$

and $4\delta = n\epsilon$, it becomes

$$\varphi(\varrho) \leq c\left\{\left(\frac{\varrho}{R}\right)^{n-p+4\delta} + \omega(R)^\kappa + \left(\frac{R}{\varrho}\right)^\mu R^{4\delta}\right\}\varphi(R) + c\|a\|_s R^{n-p+4\delta}\,.$$

Let τ be such that $2c\tau^{2\delta} = 1$, and let R_0 be such that

$$\omega(R)^\kappa + \tau^{-\mu}R^{4\delta} \leq \frac{1}{2}\tau^{n-p+2\delta}$$

for every $R < R_0$. Choosing $\varrho = \tau R$, the preceding inequality becomes

$$\varphi(\tau R) \leq \tau^{n-p+2\delta}\varphi(R) + c\|a\|_s R^{n-p+\delta}\,.$$

Since for every $t \in (\tau^{k+1}R, \tau^k R)$ we have $\varphi(t) \leq \tau^{-\mu}\varphi(\tau^k R)$, we can apply Lemma 7.3, from which we obtain

$$\varphi(\varrho) \leq c\left\{\left(\frac{\varrho}{R}\right)^{n-p+\delta}\varphi(R) + \|a\|_s \varrho^{n-p+\delta}\right\}. \tag{7.89}$$

Let now $\Sigma \subset\subset \Omega$ be an open set with smooth boundary, and let $R_0 < \operatorname{dist}(\Sigma, \partial\Omega)$. If Q_ϱ is any cube centered on Σ and of side $\varrho < R_0$, we have from (7.89)

$$\int_{Q_\varrho} |u - u_\varrho|^p dx \leq c\varrho^p \int_{Q_\varrho} |Du|^p dx \leq c\varrho^{n+\delta}(\|u\|_{1,p} + \|a\|_s)\,,$$

and hence, by Theorem 2.9, u is Hölder-continuous in Σ. We have therefore proved the following:

Theorem 7.16 *Let*

$$\mathcal{F}(u, A) = \int_A F(x, u(x), Du(x))dx$$

with the function $F(x, u, z)$ satisfying conditions (7.74). Every ω-minimum of $\mathcal{F}$ belongs to $C^{0,\delta}(\Omega)$ for some $\delta > 0$.

7.8 Boundary Regularity

With arguments similar to those of the preceding section we can prove the Hölder continuity up to the boundary for ω-minima with prescribed boundary values.

We assign the boundary data as the trace of a function $U \in W^{1,m}(\Omega)$, with $m > n$. In particular, assuming that $\partial\Omega$ is Lipschitz-continuous, we get from Sobolev imbedding Theorem 3.11 that $U \in C^{0,\alpha}(\bar{\Omega})$, with $\alpha = 1 - \frac{n}{m}$. In this case, we can immediately reduce to zero boundary values setting $w = u - U$ and

$$\tilde{F}(x, w, z) = F(x, w + U(x), z + DU(x)) .$$

It is easily seen that if u is a ω-minimum of the functional $\mathcal{F}$, w is a ω-minimum of the corresponding functional

$$\tilde{\mathcal{F}}(w, \Omega) = \int_\Omega \tilde{F}(x, w, Dw)dx .$$

Moreover, the function $\tilde{F}$ satisfies

$$c|z|^p - b(x)|w|^\gamma - \tilde{a}(x) \le \tilde{F}(x, w, z) \le L|z|^p + b(x)|w|^\gamma + \tilde{a}(x) \tag{7.90}$$

with a suitable positive constant c and with

$$\tilde{a}(x) = a(x) + |DU(x)|^p + b(x)|U(x)|^\gamma \in L^s$$

for some $s > \frac{n}{p}$.

We can therefore forget about the boundary values, and consider only the homogeneous case $u = 0$ on $\partial\Omega$.

We need estimates similar to those of the preceding section, in which the cubes Q_r are replaced by the sets $\Omega_r = Q_r \cap \Omega$, the cubes Q_r being centered on $\partial\Omega$. Since $\partial\Omega$ is Lipschitz-continuous, we have $|Q_r - \Omega| \ge \alpha_0 |Q_r|$ for some $\alpha_0 > 0$, and therefore Theorem 6.8 holds, and we have the estimate

$$\fint_{\Omega_{\frac{R}{2}}} |Du|^{rp} dx \le c \left(\fint_{\Omega_R} |Du|^p dx \right)^r + c \fint_{\Omega_R} (a(x) + b(x)|u|^\gamma)^r dx \tag{7.91}$$

for every cubical Q-minimum.

If now v is as in the preceding section, we have $v = 0$ on $\partial\Omega \cap Q_R$, and we can replace estimates (7.22) and (7.47) with (7.29) and (7.56), that is

$$\sup_{\Omega_\varrho} |v| \le c(q) \left\{ \left(\frac{1}{(R-\varrho)^n} \int_{\Omega_R} |u|^q dx \right)^{\frac{1}{q}} + \|a\|_s R^\beta \right\} , \tag{7.92}$$

$$\int_{\Omega_\varrho} |Du|^p dx \leq c\left(\frac{\varrho}{R}\right)^{n-p+p\beta} \int_{\Omega_R} |Du|^p dx + c\chi^p \varrho^{n-p+p\beta}. \qquad (7.93)$$

We can then proceed as above; actually here the estimates are made simpler from the fact that both the Q-minimum v and the ω-minimum u are zero on $\partial\Omega \cap Q_R$, so that the terms $\int_{\Omega_R} b|u|^\gamma dx$ and similar can be estimated directly in terms of the integral of the gradient. We obtain in conclusion an estimate analogous to (7.88), namely

$$\int_{\Omega_\varrho} (|Du|^p + \varrho^{-\mu}|u|^\gamma)dx$$

$$\leq c\left\{\left(\frac{\varrho}{R}\right)^{n-p+n\epsilon} + \omega(R)^\kappa + \left(\frac{R}{\varrho}\right)^\mu R^{n\epsilon}\right\} \int_{\Omega_R} (|Du|^p + R^{-\mu}|u|^\gamma)dx$$

$$+ c\|a\|_s R^{n-p+n\epsilon}. \qquad (7.94)$$

valid for any concentric cubes centered on $\partial\Omega$, and for every $\varrho < R \leq R_0$.

Let now x_0 be any point of $\bar{\Omega}$, and let as above

$$\varphi(\varrho) = \int_{\Omega_\varrho(x_0)} (|Du|^p + \varrho^{-\mu}|u|^\gamma)dx.$$

We distinguish four cases.

Case 1. $Q(x_0, R_0) \subset \Omega$. In this case all the cubes involved are contained in Ω, and the results of the preceding section hold. In particular we have

$$\varphi(x_0, \varrho) \leq c\left\{\left(\frac{\varrho}{R_0}\right)^{n-p+\delta} \varphi(x_0, R_0) + \|a\|_s \varrho^{n-p+\delta}\right\}. \qquad (7.95)$$

Case 2. $x_0 \in \partial\Omega$. In this case we can argue as in the preceding section, starting from (7.94), and we conclude that the preceding estimate holds in this case too.

Case 3. $Q(x_0, \varrho)$ intersects $\partial\Omega$. If $x_1 \in \partial\Omega \cap Q(x_0, \varrho)$, we have $\Omega(x_0, \varrho) \subset \Omega(x_1, 2\varrho)$, and hence

$$\varphi(x_0, \varrho) \leq c\varphi(x_1, 2\varrho) \leq c\left\{\left(\frac{\varrho}{R_0}\right)^{n-p+\delta} \varphi(x_1, R_0) + \|a\|_s \varrho^{n-p+\delta}\right\}.$$

Case 4. If none of the above situations is verified, let r be the largest side of the cube centered at x_0 and contained in Ω. We have

$$\varphi(x_0, \varrho) \leq c\left\{\left(\frac{\varrho}{r}\right)^{n-p+\delta} \varphi(x_0, r) + \|a\|_s \varrho^{n-p+\delta}\right\}.$$

As above, there is a point $x_1 \in \partial\Omega$ such that $Q(x_0, r) \subset \Omega(x_1, 2r)$, and we can continue as in Case 3, getting

$$\varphi(x_0, \varrho) \leq c\left\{\left(\frac{\varrho}{R_0}\right)^{n-p+\delta} \varphi(x_1, R_0) + \|a\|_s \varrho^{n-p+\delta}\right\}.$$

In any case, we can therefore conclude that

$$\int_{\Omega_\varrho} |Du|^p dx \leq c\varrho^{n-p+\delta}(\|u\|_{1,p} + \|a\|_s),$$

from which the Hölder continuity of u in $\bar{\Omega}$ follows immediately.

7.9 Notes and Comments

The core of this chapter is Theorem 7.6, where we have proved that the functions in De Giorgi's classes DG^p are Hölder-continuous. This result was proved by De Giorgi in his famous paper [1], which opened the way to the regularity of solutions of elliptic equations with bounded measurable coefficients, and for minima of regular functionals in the calculus of variations. De Giorgi's theorem was later generalized by various authors, so as to cover the most general solutions of non-linear equations in divergence form. We note in particular the papers by STAMPACCHIA [1, 2, 4] and the book by LADYŽENSKAYA and URAL'CEVA [1].

Almost at the same time, a different proof of the regularity of solutions to parabolic and elliptic equations was given by NASH [1].

Slightly later, MOSER [2] proved Harnack's inequality, thus extending to solutions of linear equations in divergence form a classical result for harmonic functions. Starting from Harnack's inequality, Moser gave a new proof of the Hölder-continuity of solutions of elliptic equations.

Moser's proof goes as follows. Let u be a solution of the elliptic equation

$$\int_\Omega a_{ij}(x) D_j u D_i \varphi dx = 0$$

for every $\varphi \in W_0^{1,2}(\Omega)$, and assume that every positive solution w in Q_{2R} satisfies Harnack's inequality

$$\sup_{Q_R} w \leq c \inf_{Q_R} w.$$

Setting $M(R) = \sup_{Q_R} u$ and $m(R) = \inf_{Q_R} u$, we can apply Harnack's inequality to the functions $M(2R) - u$ and $u - m(2R)$, obtaining

$$M(2R) - M(R) \leq c[M(2R) - m(R)]\,,$$

$$M(R) - m(2R) \leq c[m(R) - m(2R)]\,.$$

Summing these inequalities, and setting $\text{osc}(u, R) = M(R) - m(R)$, we get

$$\text{osc}(u, 2R) + \text{osc}(u, R) \leq c[\text{osc}(u, 2R) - \text{osc}(u, R)]$$

and whence

$$\text{osc}(u, R) \leq \frac{c-1}{c+1}\text{osc}(u, 2R) =: \gamma\text{osc}(u, 2R)\,,$$

with $\gamma < 1$. By induction, writing R instead of $2R$:

$$\text{osc}(u, 2^{-k}R) \leq \gamma^k \text{osc}(u, R)\,.$$

Let now $\varrho < R$. We can choose k so that $2^{-k-1}R < \varrho \leq 2^{-k}R$, getting

$$\text{osc}(u, \varrho) \leq c\left(\frac{\varrho}{R}\right)^{\beta} \text{osc}(u, R)$$

with $\beta = -\frac{\log\gamma}{\log 2} > 0$, whence the Hölder-continuity of the function u.

The extension of the method of De Giorgi to minima (and quasi-minima) of functionals, independently of their Euler equation, was made by GIAQUINTA and GIUSTI [2], after FREHSE [3] had studied a particular case, under rather restrictive hypotheses.

For what concerns boundary regularity, ZIEMER [1] proved the continuity of quasi-minima at every boundary point satisfying a WIENER condition, thus extending, although not in the maximum of generality, well-known results for elliptic equations in divergence form.

Harnack's inequality was proved by DI BENEDETTO and TRUDINGER [1] for functions in De Giorgi classes, and hence for quasi-minima of integral functionals

$$\mathcal{F}(u, \Omega) = \int_\Omega F(x, u, Du)dx$$

We have also given a second proof of that result, obtained by means of an idea, introduced by DI BENEDETTO [1] in his extension of Harnack's inequality to De Giorgi classes of parabolic type. The same idea leads to the proof of a Harnack's inequality for De Giorgi classes relative to Hörmander vector fields (MARCHI [3]).

The notion of ω-minimum was introduced by ANZELLOTTI [1]. The Hölder continuity of ω-minima was proved by DOLCINI, ESPOSITO and FUSCO [1] in the special case of integrand F satisfying

$$|z|^p \leq F(x, u, z) \leq L(1 + |z|^p)$$

and later by ESPOSITO and MINGIONE [1] in the general case.

Chapter 8

First Derivatives

The results of the preceding chapter are the most general one can obtain for arbitrary scalar quasi-minima of regular functionals

$$\mathcal{F}(u,\Omega) = \int_\Omega F(x,u,Du)\,dx\,. \tag{8.1}$$

Actually, we cannot expect that the regularity of a quasi-minimum (or even of a minimum) of a functional of the calculus of variations, in the sole assumptions of the preceding chapters, goes beyond the Hölder-continuity theorem we have proved in the preceding chapter. The following example is characteristic of the general situation.

Example 8.1 The function $u(x) = x_1|x|^{-\alpha}$, $0 < \alpha < 1$, is a weak solution of the elliptic differential equation

$$\int_B a^{ij}(x) D_j u D_i \varphi\,dx = 0 \quad \forall\,\varphi \in C_0^\infty(B)\,,$$

where B is the unit ball in $\mathbf{R}^n$, $n \geq 2$, and

$$a^{ij}(x) = \delta^{ij} + \frac{\alpha(n-\alpha)}{(1-\alpha)(n-1-\alpha)} \frac{x_i x_j}{|x|^2}\,.$$

The proof of the above assertion can be obtained by first checking that the function u is a solution of the equation

$$D_i[a^{ij}(x) D_j u] = 0$$

in $B - \{0\}$, and then arguing as in Example 6.3.

It follows that $u(x)$ minimizes the functional

$$\int_B \left(|Du|^2 + \sigma \left\langle \frac{x}{|x|}, Du \right\rangle^2 \right) dx$$

with $\sigma =: \frac{\alpha(n-\alpha)}{(1-\alpha)(n-1-\alpha)} > 0$, whose integrand $F(x,z)$ satisfies the assumptions

$$|z|^2 \leq F(x,z) \leq (1+\sigma)|z|^2 .$$

In particular, u is a Q-minimum of the DIRICHLET functional, with $Q = 1 + \sigma$. We note that σ can be as close to zero as we wish.

Consequently, if we want to obtain regularity results for the first derivatives, we must abandon the notion of quasi-minimum, whose duty was performed in the preceding chapters, and we must consider the minima, or more generally the ω-minima of the functional (8.1). Moreover, we must assume that the function $F(x,u,z)$ is regular enough; in particular that it has first and second derivatives with respect to z.

In this chapter we shall consider the scalar case, and we shall prove the Hölder-continuity of the first derivatives of the ω-minima (hence in particular of the minima) of the integral (8.1), and of the solutions of elliptic equations in divergence form.

The core of the chapter is the study of the minima of functionals depending only on the gradient

$$\mathcal{F}(u,\Omega) = \int_\Omega F(Du(x))\, dx \,, \tag{8.2}$$

or more generally of the weak solutions of elliptic equations of the form

$$D_i A^i(Du) = 0 \,. \tag{8.3}$$

Once suitable estimates for these functions have been obtained, we shall consider the general case of ω-minima of the functional (8.1), in which the dependence on x and u will be considered as a perturbation, and we shall prove the Hölder-continuity of the first derivatives. The same results hold for weak solutions of elliptic equations of the type

$$D_i A^i(x,u,Du) = B(x,u,Du) \,.$$

8.1 The Difference Quotients

Before beginning the study of the functional (8.1), we shall prove some results that will be useful later.

Definition 8.1 *Let $f(x)$ be a function defined in an open set $\Omega \subset \mathbf{R}^n$, and let h be a real number. We call the* difference quotient *of f with respect to x_s the function*

$$\Delta_{s,h} f(x) = \frac{f(x+he_s) - f(x)}{h},$$

where e_s denotes the direction of the x_s axis.

When no confusion can arise, we shall omit the index s, and we shall write simply Δ_h instead of $\Delta_{s,h}$.

The function $\Delta_{s,h} f$ is defined in the set

$$\Delta_{s,h}\Omega =: \{x \in \Omega : x + he_s \in \Omega\},$$

and hence in the set

$$\Omega_{|h|} = \{x \in \Omega : \operatorname{dist}(x, \partial\Omega) > |h|\}.$$

The following properties of the difference quotients are immediate:

(i) If $f \in W^{1,p}(\Omega)$, then $\Delta_h f \in W^{1,p}(\Omega_{|h|})$, and

$$D_i(\Delta_h f) = \Delta_h(D_i f). \tag{8.4}$$

(ii) If at least one of the functions f or g has support contained in $\Omega_{|h|}$, then

$$\int_\Omega f \Delta_h g \, dx = -\int_\Omega g \Delta_{-h} f \, dx. \tag{8.5}$$

(iii) We have

$$\Delta_h(fg)(x) = f(x+he_s)\Delta_h g(x) + g(x)\Delta_h f(x). \tag{8.6}$$

Remark 8.1 It follows immediately from (ii) that the derivatives $D_s g$ of a Lipschitz-continuous function g, which exists almost everywhere as limits of the difference quotient $\Delta_{s,h} g$ coincide with its weak derivatives. In fact, if f is a test function, we can pass to the limit in (ii), getting

$$\int f D_s g \, dx = -\int g D_s f \, dx.$$

In other words, we have $\operatorname{Lip}(\Omega) = W^{1,\infty}(\Omega)$. □

Lemma 8.1 *There exists a constant $c(n)$ such that if $v \in W^{1,p}(\Omega)$, $\Sigma \subset\subset \Omega$ and $|h| < h_0 = \frac{1}{10\sqrt{n}} \operatorname{dist}(\Sigma, \partial\Omega)$,*

$$\|\Delta_{s,h} v\|_{p,\Sigma} \leq c \|D_s v\|_{p,\Omega} \,. \tag{8.7}$$

Proof. We can assume $s = n$. Let us show first inequality (8.7) when Σ is the cube Q_R. We have for almost every $x \in \Sigma$:

$$\Delta_h v(x) = \frac{1}{h} \int_{x_n}^{x_n+h} D_n v(\bar{x}, t)\, dt \,,$$

where $\bar{x} = (x_1, \ldots, x_{n-1})$. Now let $f(t) = D_n v(\bar{x}, t)$. From the HÖLDER inequality we get:

$$\left| \frac{1}{h} \int_z^{z+h} f(t)\, dt \right|^p \leq \frac{1}{h} \int_0^h |f(z+t)|^p\, dt$$

and hence

$$\begin{aligned} \int_{-R}^{R} \left| \frac{1}{h} \int_z^{z+h} f(t)\, dt \right|^p dz &\leq \frac{1}{h} \int_{-R}^{R} dz \int_0^h |f(z+t)|^p\, dt \\ &= \frac{1}{h} \int_0^h dt \int_{-R}^{R} |f(z+t)|^p\, dz \\ &\leq \int_{-R-h_0}^{R+h_0} |f(z)|^p\, dz \,. \end{aligned}$$

Denoting by K_R the projection of Q_R on $\mathbf{R}^{n-1}$, we have:

$$\begin{aligned} \int_{Q_R} |\Delta_h v|^p\, dx &= \int_{K_R} d\bar{x} \int_{-R}^{R} dx_n \left| \frac{1}{h} \int_{x_n}^{x_n+h} D_n v(\bar{x}, t)\, dt \right|^p \\ &\leq \int_{K_R} d\bar{x} \int_{-R-h_0}^{R+h_0} |D_n v(\bar{x}, x_n)|^p\, dx_n \\ &\leq \int_{Q_{R+h_0}} |D_n v|^p\, dx \,, \end{aligned}$$

and (8.7) is proved in the case of a cube.

Now let $\Sigma \subset\subset \Omega$. The set $\bar{\Sigma}$ is contained in the union of a finite number of cubes Q_i of side $2R = 2h_0$, without interior points in common. For each of them we can write the preceding inequality:

$$\int_{Q_R} |\Delta_h v|^p \, dx \le \int_{Q_{2R}} |D_n v|^p \, dx \, .$$

Since at most 5^n cubes of double side overlap, we have immediately the result with $c(n) = 5^n$. □

The preceding proposition has a converse.

Lemma 8.2 *Let $v \in L^p(\Omega)$, $1 < p < \infty$, and assume that there exists a constant K such that for every h small enough we have*

$$\|\Delta_{s,h} v\|_{p,\Omega_{|h|}} \le K \, .$$

Then, $D_s v \in L^p(\Omega)$, and

$$\|D_s v\|_{p,\Omega} \le K \, .$$

Moreover, when $h \to 0$, $\Delta_{s,h} v \to D_s v$ in $L^p_{\text{loc}}(\Omega)$.

Proof. Let h_i be a sequence converging to zero, and let

$$g_i = \begin{cases} \Delta_{h_i} v & \text{in } \Omega_{|h_i|} \\ 0 & \text{in } \Omega - \Omega_{|h_i|} \, . \end{cases}$$

The sequence g_i is bounded in $L^p(\Omega)$ and therefore, since that space is reflexive, we can extract a subsequence weakly convergent to a function $g \in L^p(\Omega)$, with $\|g\|_{p,\Omega} \le K$.

Let us show that $g = D_s v$. If $\varphi \in C_0^\infty(\Omega)$, we have

$$\int g\varphi \, dx = \lim_{i \to \infty} \int \varphi \Delta_{h_i} v \, dx = - \lim_{i \to \infty} \int v \Delta_{h_i} \varphi \, dx$$

$$= - \int v D_s \varphi \, dx$$

since $\Delta_h \varphi \to D_s \varphi$ uniformly.

In order to prove the last statement, let $w \in C^{1,p}(\Omega)$. We have

$$\Delta_h v - D_s v = \Delta_h (v - w) + \Delta_h w - D_s w + D_s (w - v)$$

and hence from the preceding lemma:

$$\|\Delta_h v - D_s v\|_{p,\Sigma} \le \|\Delta_h w - D_s w\|_{p,\Sigma} + c \|D_s (w - v)\|_{p,\Omega} \, .$$

The conclusion follows by remarking that $C^{1,p}(\Omega)$ is dense in $W^{1,p}(\Omega)$, and that if $w \in C^{1,p}(\Omega)$, $\Delta_h w \to D_s w$ uniformly on compact sets. □

Finally, we shall use the following lemma:

Lemma 8.3 *Let ξ and π be two vectors in $\mathbf{R}^n$, and let*

$$Z(t) = (1 + |(1-t)\xi + t\pi|^2)^{\frac{1}{2}} .$$

For every $s > -1$ and $r > 0$ there exists two constants $c_1(s,r)$ and $c_2(s,r)$ such that

$$c_1(1 + |\xi|^2 + |\pi|^2)^{\frac{s}{2}} \le \int_0^1 (1-t)^r Z(t)^s \, dt \le c_2(1 + |\xi|^2 + |\pi|^2)^{\frac{s}{2}} .$$

Proof. Since we have trivially $Z(t)^2 \le c(1 + |\xi|^2 + |\pi|^2)$, we need only an estimate from below.

If $|\pi| > |\xi|$, we have for $\frac{2}{3} \le t \le 1$,

$$|(1-t)\xi + t\pi| \ge t|\pi| - (1-t)|\xi| \ge \frac{2}{3}|\pi| - \frac{1}{3}|\xi| \ge \frac{1}{3}|\pi|$$

and therefore

$$Z(t)^2 \ge c(1 + |\xi|^2 + |\pi|^2) .$$

The same inequality holds for $0 \le t \le \frac{1}{3}$ if $|\xi| \ge |\pi|$. The required estimates follow at once. □

8.2 Second Derivatives

In this section we shall consider weak solutions of elliptic equations in divergence form

$$\int A^i(Du) D_i\varphi \, dx = 0 \quad \forall \varphi \in C_0^\infty(\Omega) \tag{8.8}$$

with coefficients depending only on the gradient, and we will show that they have second derivatives in Ω.

For that purpose, it will be necessary to assume that the functions $A^i(z)$ are of class C^1, and that they satisfy the inequalities:

$$|A^i| + V(z)|A^{ij}| \le LV(z)^{p-1} , \tag{8.9}$$

$$A^i(z)z_i \ge \nu V(z)^p - c , \tag{8.10}$$

$$A^{ij}(z)\xi_i\xi_j \ge \nu V(z)^{p-2}|\xi|^2 \tag{8.11}$$

with $\nu > 0$, where we have set[1]

$$V(z) = \sqrt{1 + |z|^2}\,, \tag{8.12}$$

$$A^{ij}(z) = \frac{\partial A^i(z)}{\partial z_j}\,. \tag{8.13}$$

We remark that multiplying the coefficients by ν^{-1}, we can assume that $\nu = 1$.

The main result of this section is given by the following:

Theorem 8.1 *Let $p > 1$ and let $u \in W^{1,p}(\Omega)$ be a weak solution of the equation*

$$\int A^i(Du) D_i \varphi \, dx = 0 \quad \forall \varphi \in W_0^{1,p}(\Omega)\,,$$

whose coefficients $A^i(z)$ satisfy relations (8.9)–(8.11). Then, u has second derivatives D^2u such that for every $\Sigma \subset\subset \Omega$

$$\int_\Sigma V^{p-2} |D^2 u|^2 \, dx \le c(\Sigma) \int_\Omega V^p \, dx\,. \tag{8.14}$$

Proof. Let $6R < \operatorname{dist}(\Sigma, \partial\Omega)$, and let $x_0 \in \Sigma$. Setting $Q_t = Q(x_0, t)$, let $\zeta \in C_0^\infty(Q_{2R})$, $0 \le \zeta \le 1$, $\zeta = 1$ in Q_R and $|D\zeta|^2 + |D^2\zeta| \le cR^{-2}$. Finally, let $|h| < R$. Writing $\varphi = \Delta_{s,-h}(\zeta^2 \Delta_{s,h} u)$ in (8.8), and integrating by parts by means of (8.5), we get:

$$\int \Delta_h A^i (\zeta^2 D_i \Delta_h u + 2\Delta_h u \, \zeta D_i \zeta) \, dx = 0\,, \tag{8.15}$$

where as usual we have written Δ_h instead of $\Delta_{s,h}$.

We now have

$$\begin{aligned} \Delta_h A^i &= \frac{1}{h} \int_0^1 \frac{d}{dt} A^i(Du + th\Delta_h Du) \, dt \\ &= \int_0^1 A^{ij}(Du + thD\Delta_h u) D_j \Delta_h u \, dt =: \alpha^{ij} D_j \Delta_h u\,, \end{aligned} \tag{8.16}$$

with

$$\alpha^{ij} = \int_0^1 A^{ij}(Du + th\Delta_h Du) \, dt\,.$$

[1] Of course, the inequalities (8.10) and (8.11), both expressing the ellipticity of the equation, are not independent, and actually (8.10) is a consequence of (8.11).

It is possible to estimate the coefficients α^{ij} by means of the assumptions (8.9)–(8.11) and of Lemma 8.3. We get:

$$|\alpha^{ij}| \leq c_1 W^{p-2},$$

$$\alpha^{ij}\xi_i\xi_j \geq c_2 W^{p-2}|\xi|^2,$$

where

$$W^2 = 1 + |Du(x)|^2 + |Du(x+he_s)|^2.$$

With these relations we can estimate the first term of (8.15):

$$\int \Delta_h A^i \zeta^2 D_i \Delta_h u\, dx \geq c \int W^{p-2}\zeta^2 |D\Delta_h u|^2\, dx. \tag{8.17}$$

In order to estimate the second term, we distinguish the two cases $p \geq 2$ and $1 < p < 2$. In the first case we can use again (8.16), obtaining:

$$\begin{aligned} 2\left|\int \Delta_h A^i\, \Delta_h u\, \zeta\, D_i\zeta\, dx\right| &= 2\left|\int \alpha^{ij} D_j(\Delta_h u)\, \zeta\, D_i\zeta\, \Delta_h u\, dx\right| \\ &\leq c\int W^{p-2}|\Delta_h u||D\Delta_h u|\,\zeta\,|D\zeta|\, dx \\ &\leq \epsilon \int W^{p-2}|D\Delta_h u|^2\,\zeta^2\, dx \\ &\quad + c\epsilon^{-1}\int W^{p-2}|\Delta_h u|^2\,|D\zeta|^2\, dx, \end{aligned}$$

and in conclusion:

$$\int_{Q_R} W^{p-2}|D\Delta_h u|^2\, dx \leq cR^{-2}\int_{Q_{2R}} W^{p-2}|\Delta_h u|^2\, dx. \tag{8.18}$$

On the other hand

$$W^{p-2}|\Delta_h u|^2 \leq c(W^p + |\Delta_h u|^p) \tag{8.19}$$

and, since $|h| < R$,

$$\int_{Q_{2R}} W^p\, dx \leq c\int_{Q_{3R}} V^p\, dx.$$

Recalling Lemma 8.1, we get

$$\int_{Q_R} W^{p-2}|D\Delta_h u|^2\, dx \leq cR^{-2}\int_{Q_{3R}} V^p\, dx. \tag{8.20}$$

If instead $1 < p < 2$, (8.19) does not hold, and we must take a different path, which consists of expressing the quantity $\Delta_h A^i$ in a different way. We have:

$$\Delta_h A^i = \frac{1}{h}\int_0^1 \frac{d}{dt} A^i(Du(x+the_s))\,dt$$

$$= \int_0^1 D_s A^i(Du(x+the_s))\,dt =: D_s\alpha^i\,,$$

where the functions α^i verify the inequality

$$|\alpha^i| \le Y =: \int_0^1 (1+|Du(x+the_s)|^2)^{\frac{p-1}{2}}\,dt\,.$$

We have then

$$2\int \Delta_h A^i\,\Delta_h u\,\zeta\,D_i\zeta\,dx = -2\int \alpha^i D_s(\Delta_h u\,\zeta\,D_i\zeta)\,dx$$

$$= -2\int \alpha^i D_s(\Delta_h u)\,\zeta\,D_i\zeta\,dx$$

$$-2\int \alpha^i\,\Delta_h u(\zeta\,D_{is}\zeta + D_i\zeta\,D_s\zeta)\,dx$$

and hence

$$2\left|\int \Delta_h A^i\,\Delta_h u\,\zeta\,D_i\zeta\,dx\right| \le cR^{-1}\int Y\zeta|D\Delta_h u|\,dx$$

$$+\,cR^{-2}\int_{Q_{2R}} Y|\Delta_h u|\,dx\,. \tag{8.21}$$

Let us evaluate the first term on the right-hand side. We have

$$R^{-1}Y\zeta|D\Delta_h u| = R^{-1}YW^{\frac{2-p}{2}}W^{\frac{p-2}{2}}\,\zeta\,|D\Delta_h u|$$

$$\le \epsilon W^{p-2}\zeta^2|D\Delta_h u|^2 + c\epsilon^{-1}R^{-2}Y^2W^{2-p}$$

and therefore

$$\int_{Q_R} W^{p-2}|D\Delta_h u|^2\,dx \le cR^{-2}\int_{Q_{2R}} (Y^2W^{2-p} + Y|\Delta_h u|)\,dx\,. \tag{8.22}$$

We remark now that

$$Y^2W^{2-p} \le c(W^p + Y^{\frac{p}{p-1}})\,,$$

$$Y|\Delta_h u| \le c(|\Delta_h u|^p + Y^{\frac{p}{p-1}})$$

and moreover

$$\int_{Q_{2R}} Y^{\frac{p}{p-1}}\,dx = \int_{Q_{2R}} \left\{ \int_0^1 (1+|Du(x+the_s)|^2)^{\frac{p-1}{2}}\,dt \right\}^{\frac{p}{p-1}} dx$$
$$\leq \int_0^1 \left\{ \int_{Q_{2R}} (1+|Du(x+the_s)|^2)^{\frac{p}{2}}\,dx \right\} dt$$
$$\leq \int_{Q_{3R}} V^p\,dx\,,$$

so that in conclusion (8.20) holds also for $1 < p < 2$.

We remark that if $1 < p < 2$, setting $2\alpha = p(2-p)$, we have:

$$|D\Delta_h u|^p = W^\alpha W^{-\alpha}|D\Delta_h u|^p \leq c(W^p + W^{p-2}|D\Delta_h u|^2)\,, \tag{8.23}$$

whereas if $p \geq 2$ it holds that

$$|D\Delta_h u|^2 \leq W^{p-2}|D\Delta_h u|^2\,.$$

The sequence $D\Delta_h u$ is therefore bounded in $L^\mu(Q_R)$ ($\mu = \min(2,p)$); by Lemma 8.2 it converges in $L^\mu_{\text{loc}}(Q_R)$ to $DD_s u$, and hence $u \in W^{2,\mu}_{\text{loc}}(Q_R)$. Moreover, from that sequence we can extract a subsequence converging almost everywhere; since also W tends to $(1+2|Du|^2)^{\frac{1}{2}}$ almost everywhere, passing to the limit in (8.20) we obtain the estimate

$$\int_{Q_R} V^{p-2}|D^2u|^2\,dx \leq cR^{-2}\int_{Q_{3R}} V^p\,dx \tag{8.24}$$

from which, covering Σ with cubes of sufficiently small side, we get immediately (8.14). □

Remark 8.2 The preceding theorem holds also for elliptic systems:

$$\int A^i_\alpha(Du)D_i\varphi^\alpha\,dx = 0$$

with coefficients satisfying (8.9) and the strong ellipticity condition, analogous to (8.11),

$$A^{ij}_{\alpha\beta}(p)\xi^\alpha_i\xi^\beta_j \geq \nu V^{p-2}|\xi|^2\,.$$

Apart from the multiplicity of the indices, the proof is exactly the same. □

Remark 8.3 If $1 < p < 2$ it follows from (8.23) that $u \in W^{2,p}_{\text{loc}}(\Omega)$, with the estimate

$$\int_\Sigma |D^2u|^p \, dx \leq c(\Sigma) \int_\Omega V^p \, dx \,. \tag{8.25}$$

□

Remark 8.4 Since $|DV| \leq |D^2u|$, (8.14) implies that the function $Z =: V^{\frac{p}{2}}$ belongs to $W^{1,2}_{\text{loc}}(\Omega)$, and

$$\int_\Sigma |DZ|^2 \, dx \leq c \int_\Omega Z^2 \, dx \,. \tag{8.26}$$

□

The fact that u has second derivatives permits one to write two interesting equations, both obtained from (8.8). The first is deduced by means of a simple integration by parts; setting as before

$$A^{ij} = \frac{\partial A^i}{\partial z_j}$$

we have

$$\int A^{ij} D_{ij} u \, \varphi \, dx = 0$$

for every $\varphi \in C_0^\infty(\Omega)$, and hence the function $u(x)$ verifies the equation

$$A^{ij}(Du) D_{ij} u = 0 \quad \text{a.e. in } \Omega \,. \tag{8.27}$$

In contrast, the second one is an integral relation; writing $D_s\varphi$ instead of φ in (8.8) and integrating by parts, we get

$$\int A^{ij}(Du) D_{js} u D_i \varphi \, dx = 0 \tag{8.28}$$

for every $\varphi \in C_0^\infty(\Omega)$ and therefore for every φ with compact support for which the integral makes sense.

8.3 Gradient Estimates

We shall use Eq. (8.28) in order to prove the boundedness and the Hölder continuity of the first derivatives of u.

Proposition 8.1 *Let u be a solution of Eq. (8.28), and assume that conditions (8.9)–(8.11) are satisfied. Then, for every $k \geq \frac{p}{2}$ the function $Z = V^k$ belongs to $W^{1,2}_{\text{loc}}(\Omega)$ and for every $\Sigma \subset\subset \Lambda \subset\subset \Omega$ we have*

$$\int_\Sigma |DZ|^2 \, dx \leq c(k, \Sigma, \Lambda) \int_\Lambda Z^2 \, dx \,. \tag{8.29}$$

Proof. For $T > 0$ we set

$$V_T^2 = 1 + \min\{|Du|^2, T\},$$

and we take $\varphi = \zeta^2 V_T^{2\alpha} D_s u$ in (8.28), ζ being a function with compact support. We obtain

$$\int A^{ij} D_{js}u[V_T^{2\alpha} D_{is}u + 2\alpha V_T^{2\alpha-1} D_i V_T D_s u]\zeta^2\, dx$$

$$= -2\int A^{ij} D_{is}u V_T^{2\alpha} D_s u\, \zeta D_i\zeta\, dx.$$

We can estimate the different terms in the usual way, remarking only that

$$|Du||DV_T| \le V_T|DV_T|,$$

$$D_i V_T D_s u D_{js} u = V_T D_i V_T D_j V_T.$$

In this way we arrive to the inequality

$$\int V^{p-2} V_T^{2\alpha}(|D^2u|^2 + 2\alpha|DV_T|^2)\zeta^2\, dx$$

$$\le c\int V^p V_T^{2\alpha}|D\zeta|^2\, dx. \tag{8.30}$$

From that estimate we can begin an iterative procedure. Assume that $V \in L^{p+2\alpha}_{\mathrm{loc}}(\Omega)$ for some $\alpha \ge 0$, and let $\Sigma_1 \subset\subset \Sigma_2 \subset\subset \Omega$. Taking $\zeta \in C_0^\infty(\Sigma_2)$ with $\zeta \equiv 1$ in Σ_1 and passing to the limit for $T \to \infty$ we deduce from (8.30):

$$\int_{\Sigma_1} |DV^{\frac{p}{2}+\alpha}|^2\, dx \le c(\Sigma_1, \Sigma_2, \alpha)\int_{\Sigma_2} V^{p+2\alpha}\, dx.$$

By the SOBOLEV immersion theorem, $V^{\frac{p}{2}+\alpha} \in L^{2^*}(\Sigma_1)$ and

$$\|V^{\frac{p}{2}+\alpha}\|_{2^*,\Sigma_1} \le c(\Sigma_1, \Sigma_2, \alpha)\|V^{\frac{p}{2}+\alpha}\|_{2,\Sigma_2}.$$

Since $(\frac{p}{2} + \alpha)2^* \ge p + 2\alpha + \frac{2p}{n-2}$, at every step (starting from $\alpha = 0$) we gain a fixed exponent, so that after a finite number of steps we arrive at the required exponent k, with the estimate (8.29). □

In particular, the function $w_0 = V^p$ belongs to $W^{1,2}_{\mathrm{loc}}$, and we have

$$D_j w_0 = pV^{p-2} D_i u D_{ij} u.$$

Setting $a^{ij}(x) = V^{2-p} A^{ij}(Du(x))$, we deduce from (8.28)

$$\int a^{ij} D_j w_0 D_i \varphi \, dx + p \int A^{ij} D_{js} u D_{is} u \varphi \, dx = 0 ,$$

that we write for simplicity

$$\int a^{ij} D_j w_0 D_i \varphi \, dx + \int g \varphi \, dx = 0 \tag{8.31}$$

with

$$g = p A^{ij} D_{js} u D_{is} u .$$

In particular, since the function g is non-negative,

$$\int a^{ij} D_j w_0 D_i \varphi \, dx \leq 0$$

for every $\varphi \geq 0$. Remarking that the coefficients a^{ij} are bounded and satisfy the ellipticity condition

$$a^{ij} \xi_i \xi_j \geq |\xi|^2 , \tag{8.32}$$

we conclude that w_0 is a sub-quasi-minimum of the DIRICHLET integral

$$\int |Du|^2 \, dx$$

(see Remark 6.4), and therefore from Theorem 7.5 we get

$$\sup_{Q_\varrho} V^q \leq \frac{c(q)}{(R-\varrho)^n} \int_{Q_R} V^q \, dx \tag{8.33}$$

for every $\varrho < R$ and for every $q > 0$.

Moreover, if $\Sigma \subset\subset \Omega$, and if we take the cube $Q_{R/2}$ in such a way that $\sup_{Q_{R/2}} V = \sup_\Sigma V$, we obtain easily the following:

Theorem 8.2 *Under the assumptions of the preceding proposition, the gradient of u is locally bounded in Ω, and for every $\Sigma \subset\subset \Omega$ and every $q > 0$ we have*

$$\sup_\Sigma V \leq \left(\frac{c(q)}{\operatorname{dist}(\Sigma, \partial\Omega)^n} \int_\Omega V^q \, dx \right)^{\frac{1}{q}} . \tag{8.34}$$

Once the boundedness of the gradient has been proved, it is immediate to show that $u \in C^{1,\alpha}$ for some $\alpha > 0$. Actually, the function $w = D_s u$ is a solution of Eq. (8.28), with

$$|A^{ij}| \leq M \quad \text{and} \quad A^{ij} \xi_i \xi_j \geq |\xi|^2 .$$

It follows that $D_s u$ is a quasi-minimum of the functional $\int |Du|^2\,dx$, and hence it is Hölder continuous.

We have therefore the following:

Theorem 8.3 *Let* $u \in W^{1,p}(\Omega)$ *be a solution of the Eq.* (8.28), *with conditions* (8.9)–(8.11), *and let* $p > 1$.

Then, the derivatives of u *are Hölder-continuous in* Ω, *and for every compact set* $K \subset \Omega$, *the norm* $\|u\|_{C^{1,\alpha}(K)}$ *can be estimated by means of* $\|u\|_{W^{1,p}(\Omega)}$.

In particular, the preceding theorem applies to the minima of the functionals discussed in Chapter 1. In fact they verify Eq. (8.8) with $A^i = F_{z_i}$ and, being Lipschitz-continuous functions, conditions (8.9)–(8.11) are satisfied with $p = 2$.

8.4 Boundary Estimates

We want now to obtain an analogue of (8.33) for solutions of the Eq. (8.3) in a half-ball

$$B^+ = \{x \in \mathbf{R}^n : |x| < 1, x_n > 0\}$$

with zero boundary value on the flat part

$$P = \{x \in \partial B^+ : x_n = 0\}$$

of the boundary of B^+.

We can assume that u is continuous in B^+, and that almost everywhere it satisfies Eq. (8.27).

We shall begin by repeating, with due caution, the argument that lead to the proof of (8.33). In the first place, we remark that if $s \neq n$, the function $\Delta_{s,h} u$ has null trace on P, and hence we may take again $\varphi = \Delta_{-h}(\zeta^2 \Delta_h u)$, with $\zeta \in C_0^\infty(B)$ but generally speaking different from zero on P. We arrive thus as above at the estimate

$$\int_{Q_R^+} V^{p-2} |DD'u|^2\,dx \leq cR^{-2} \int_{Q_{3R}^+} V^p\,dx\,, \tag{8.35}$$

in which we have denoted $D'u$ any derivative $D_i u$, with $i = 1, 2, \ldots, n-1$.

In this way we can estimate every second derivative, except $D_{nn}u$. For it we use Eq. (8.27), that we rewrite in the form

$$A^{nn} D_{nn} u = -\Sigma' A^{ij} D_{ij} u\,,$$

where the apex indicates that in the sum we have excluded the term with i and j both equal to n.

For $T > 0$ we set $w = \min\{\max\{D_n u, -T\}, T\}$, and let ζ be a function in $C_0^\infty(B)$. The function $D_n w$ is zero if $|D_n u| > T$, and is equal to $D_{nn} u$ otherwise. Multiplying the preceding equation by $D_n w$, and making the usual estimates, we get

$$V^{p-2}|D_n w|^2 \leq c V^{p-2}|DD'u|^2$$

for almost every $x \in B^+$.

The last inequality can be integrated on every $Q_R^+ \subset B^+$, giving

$$\int_{Q_R^+} V^{p-2}|D_n w|^2\, dx \leq c \int_{Q_R^+} V^{p-2}|DD'u|^2\, dx\,.$$

If we pass to the limit for $T \to \infty$ and use (8.35), we get in conclusion

$$\int_{Q_R^+} V^{p-2}|D^2 u|^2\, dx \leq c R^{-2} \int_{Q_{3R}^+} V^p\, dx\,. \tag{8.36}$$

The second step consists of proving the boundedness in P of the gradient of u. We shall start from Eq. (8.27):

$$A^{ij}(Du) D_{ij} u = 0\,,$$

or else, setting as above $a^{ij} = A^{ij} V^{2-p}$:

$$a^{ij}(Du) D_{ij} u = 0\,. \tag{8.37}$$

We have already proved that the function u is continuous (or better, Hölder-continuous) in $B^+ \cup P$, and that it belongs to $C^{1,\alpha}(B^+)$. We shall prove in Chapter 10 that $u \in C^{2,\alpha}(B^+)$.

For $0 < x_n < \delta$ we set

$$w = e^u - 1 + \mu e^{-\lambda x_n}\,.$$

We have

$$D_{ij} w = e^u (D_{ij} u + D_i u D_j u) + \lambda^2 \mu \delta_{in} \delta_{jn} e^{-\lambda x_n}$$

and hence, recalling (8.32):

$$a^{ij} D_{ij} w \geq \mu \lambda^2 e^{-\lambda x_n}\,.$$

We choose now the constants μ and λ in such a way that

$$a^{ij} D_{ij} w \geq 1 \quad \text{if } 0 < x_n < \delta$$

and

$$w(x) < \mu \quad \text{if } x_n = \delta\,.$$

To that purpose, it is sufficient for instance to take λ in such a way that $e^{-\lambda\delta} = \frac{1}{2}$, and then, setting $M = \sup|u|$, to choose $\mu > 2(e^M - 1)$ in such a way that $\mu\lambda^2 > 2$.

This being done, the function w cannot have a maximum in a point x with $0 < x_n < \delta$, since at the maximum point one should have $a^{ij}D_{ij}w \leq 0$. Since $w < \mu$ if $x_n = \delta$, the maximum will be taken on $\partial\Omega$, where $w \equiv \mu$. Consequently,

$$u \leq \gamma(x_n) =: \log[1 + \mu(1 - e^{-\lambda x_n})]\,.$$

Since $-u$ is solution of an equation of the same sort, we can conclude that

$$|u(x)| \leq \gamma(x_n)$$

in the strip $0 < x_n < \delta$.

From that we conclude immediately that Du is bounded on $\partial\Omega$, thanks to Corollary 3.2 and to the fact that, by virtue of Theorem 8.1 and of (8.36) the first derivatives of u belong to $W^{1,2}(B^+)$ if $m \geq 2$, and to $W^{1,p}(B^+)$ if $p < 2$.

At this point we can prove the analogue of Proposition 8.1. Let $M = \max_P |Du|^2$, and for $T > M$ let

$$W^2 = W^2_{M,T} =: 1 + \min\{\max\{|Du|^2, M\}, T\}\,,$$
$$\mu^2 = 1 + M\,.$$

We have $W = \max\{V_T, \mu\}$, and hence

$$V_T \leq W \leq V_T + \mu\,, \tag{8.38}$$

whereas $W = \mu$ on P, so that $W^{2\alpha} - \mu^{2\alpha} = 0$ on P for every $\alpha > 0$.

Taking then $\varphi = \zeta^2(W^{2\alpha} - \mu^{2\alpha})D_s u$ in (8.28), where ζ is a test function not necessarily equal to zero on P, we get

$$\int A^{ij}D_{js}u[(W^{2\alpha} - \mu^{2\alpha})D_{is}u + 2\alpha W^{2\alpha-1}D_iW D_su]\zeta^2\,dx$$
$$= -2\int A^{ij}D_{js}uD_su(W^{2\alpha} - \mu^{2\alpha})\zeta D_i\zeta\,dx\,.$$

If we remark that

$$|Du||DW| \leq W|DW|,$$

$$D_iWD_suD_{js}u = WD_iWD_jW,$$

we get

$$\int V^{p-2}(W^{2\alpha} - \mu^{2\alpha})|D^2u|^2\zeta^2\,dx$$
$$\leq c\int V^p(W^{2\alpha} - \mu^{2\alpha})|D\zeta|^2\,dx.$$

From the above estimate, recalling the inequalities (8.36) and (8.38), and choosing ζ in the usual way, we obtain

$$\int_{Q_R^+} V^{p-2}V_T^{2\alpha}|D^2u|^2\,dx \leq c(\mu)R^{-2}\int_{Q_{3R}^+} V^pV_T^{2\alpha}\,dx, \tag{8.39}$$

which, thanks to (8.36), holds also for $\alpha = 0$.

We can now proceed as in the proof of Proposition 8.1, obtaining the estimate

$$\int_{B_{\frac{1}{2}}^+} |DZ|^2\,dx \leq c(k,\mu)\int_{B^+} Z^2\,dx \tag{8.40}$$

with $Z = V^k$.

Arguing as in Theorem 8.2, and taking into account the boundedness of Z on P, we conclude that the gradient of u is locally bounded in $B^+ \cup P$, and that for every $\varrho < R$ and every $q > 0$ we have

$$\sup_{Q_\varrho^+} V^q \leq \frac{c(q,\mu)}{(R-\varrho)^n}\int_{Q_R^+} V^q\,dx. \tag{8.41}$$

At this point we could prove that the function u has Hölder continuous derivatives up to P. On the other hand, the case under examination (functionals dependent only on the gradient, flat boundary) is too particular to be interesting in itself. We shall therefore postpone the regularity results at the boundary till the next sections, when we shall deal with the general case.

8.5 ω-Minima

Under suitable assumptions, most of the theorems proved till now can be extended to weak solutions of general elliptic equations

$$\int (A^i(x,u,Du)D_i\varphi + B(x,u,Du)\varphi)\, dx = 0$$

and therefore to the minima of general functionals

$$\mathcal{F}(u,\Omega) = \int_\Omega F(x,u,Du)\, dx\,. \tag{8.42}$$

We shall not follow the path that is discussed in detail in the treatise by LADYŽENSKAYA and URAL'CEVA [1]. On the contrary, we will rather attack the problem of the regularity of the first derivatives by direct methods, treating the ω-minima of general functionals. In this way, we shall also recover most of the classical results, under substantially more general assumptions.

Postponing to the next chapter the discussion of the vector case, we shall prove now the regularity of the first derivatives of scalar ω-minima.

For what concerns the integrand $F(x,u,z)$, we shall assume that it is of class C^2 in z and that it verifies the inequalities

$$V^p \le F(x,u,z) \le LV^p\,, \tag{8.43}$$

$$|F_{zz}(x,u,z)| \le LV^{p-2}\,, \tag{8.44}$$

$$F_{z_i z_j}\xi_i\xi_j \ge V^{p-2}|\xi|^2\,, \tag{8.45}$$

where we have set

$$V^2 = V^2(z) = 1 + |z|^2\,.$$

Moreover we shall assume that the function $V^{-p}F(x,u,z)$ is continuous in $(x,u) \in \bar{\Omega} \times \mathbf{R}$, uniformly for $z \in \mathbf{R}^n$; in other words, that there exists a continuous, bounded, increasing and concave[2] function $\vartheta(t)$, with $\vartheta(0) = 0$,

[2]This last condition is not restrictive. Actually, if (8.46) is satisfied by a function σ continuous, bounded and increasing, it will suffice to take as ϑ the smallest concave function not smaller than σ. Such a function ϑ is obviously increasing, is bounded by the same constant M giving the bound for σ, and is continuous. We have finally $\vartheta(0) = 0$, since if $\vartheta(0) = 2l > 0$, taking a $d > 0$ such that $\sigma(t) < l$ in $[0,d]$, the function $\min\{\vartheta, l + (M-l)x/d\}$ would be itself concave, and would lie between σ and ϑ, against the definition of ϑ.

such that

$$|F(x,u,z) - F(y,v,z)| \le \vartheta(|x-y| + |u-v|)V^p . \tag{8.46}$$

Our goal is to prove the Hölder continuity of the first derivatives of the ω-minima, and therefore in particular of the minima of general functionals (8.1).

Let us begin with some preliminary remarks. Setting as above

$$\Omega_R = \Omega(x_0, R) = \Omega \cap Q(x_0, R) ,$$

we consider the frozen functional

$$\mathcal{F}_0(v, \Omega_R) = \int_{\Omega_R} F(x_0, u(x_0), Dv)\, dx .$$

Lemma 8.4 *Let u be a bounded function and let v minimize the functional $\mathcal{F}_0$ with* DIRICHLET *datum $v = u$ on $\partial\Omega_R$.*[3] *We have*

$$\operatorname{osc}(v, \Omega_R) \le \operatorname{osc}(u, \partial\Omega_R) + cR . \tag{8.47}$$

Proof. The function v is a quasi-minimum of the integral

$$\int_{\Omega_R} (1 + |Dv|^2)^{\frac{p}{2}}\, dx .$$

If $k \ge k_0 =: \sup_{\partial\Omega_R} u$, comparing v with $w = \min\{v, k\}$ we get easily

$$\int_{A(k)} |Dv|^p\, dx \le c|A(k)| ,$$

where

$$A(k) = \{x \in \Omega_R : v(x) > k\} .$$

It follows for $h > k$

$$(h-k)^p |A(h)| \le \int_{A(k)} (v-k)^p\, dx \le \left(\int_{A(k)} (v-k)^{p^*}\, dx \right)^{\frac{p}{p^*}} |A(k)|^{\frac{p}{n}}$$
$$\le c \int_{A(k)} |Dv|^p\, dx |A(k)|^{\frac{p}{n}} \le c|A(k)|^{1+\frac{p}{n}} .$$

[3]The existence of a minimizing function is guaranteed by Theorem 4.6.

Setting now $k_i = k_0 + d - d2^{-i}$, and writing briefly $a_i = |A(k_i)|$, we have

$$a_{i+1} \le cd^{-p} 2^{ip} a_i^{1+\frac{p}{n}} .$$

By Lemma 7.1, if $a_0 \le cd^n$ (a condition that will be satisfied taking $d = cR$) we have $\lim_{i\to\infty} a_i = 0$, and hence

$$v(x) \le k_0 + cR .$$

The conclusion then follows by remarking that $-v$ is a quasi-minimum, with boundary datum $-u$, of the integral

$$\int_{\Omega_R} F(x_0, u(x_0), -Dv)\, dx ,$$

which satisfies the same inequalities as $\mathcal{F}_0$. □

Now let $u(x)$ be an ω-minimum for the functional (8.1), with boundary datum[4] U. We have shown in the previous chapter that u is Hölder continuous in $\bar{\Omega}$; let $\delta > 0$ be such that $\operatorname{osc}_{\Omega_R} u \le cR^\delta$. We have then, taking into account the lemma just proved:

$$\begin{aligned}\mathcal{F}_0(u, \Omega_R) &= \mathcal{F}(u, \Omega_R) + \int_{\Omega_R} [F(x_0, u(x_0), Du) - F(x, u, Du)]\, dx \\ &\le (1 + \omega(R))\mathcal{F}(v, \Omega_R) + \vartheta(cR^\delta) \int_{\Omega_R} (1 + |Du|^2)^{\frac{p}{2}}\, dx \\ &\le \mathcal{F}_0(v, \Omega_R) + \sigma(R) \int_{\Omega_R} [(1 + |Du|^2)^{\frac{p}{2}} + (1 + |Dv|^2)^{\frac{p}{2}}]\, dx ,\end{aligned}$$

where we have set $\sigma(R) = c\omega(R) + \vartheta(cR^\delta)$.

On the other hand v minimizes $\mathcal{F}_0$, and u is a ω-minimum of $\mathcal{F}$; whence if $w = u = v$ on $\partial\Omega_R$, we have $\mathcal{F}_0(v, \Omega_R) \le \mathcal{F}_0(w, \Omega_R)$ and $\mathcal{F}(u, \Omega_R) \le c\mathcal{F}(w, \Omega_R)$. Consequently, recalling condition (8.43):

$$\mathcal{F}_0(u, \Omega_R) \le [1 + c\sigma(R)]\mathcal{F}_0(w, \Omega_R) \tag{8.48}$$

for every $w \in W^{1,p}(\Omega_R)$ with $w = u$ on $\partial\Omega_R$.

This property will simplify the following proofs.

We shall begin by proving the regularity of the first derivatives of the ω-minima in the MORREY spaces $L^{p,\lambda}$. For that purpose, the two following lemmas will help.

[4] Of course, as long as we are interested in local results, the boundary value of u is irrelevant.

Lemma 8.5 *Let $g(\xi) = V(\xi)^s = (1 + |\xi|^2)^{\frac{s}{2}}$, with $s > 0$. Then, for every $\xi, \pi \in \mathbf{R}^\nu$ and for every $\epsilon > 0$ we have*

$$g(\pi) \leq (1 + \epsilon)g(\xi) + \frac{c}{\epsilon}(1 + |\xi|^2 + |\pi|^2)^{\frac{s-2}{2}}|\xi - \pi|^2 . \tag{8.49}$$

Proof. We can assume that $0 \leq |\xi| < |\pi|$, and hence $g(\xi) < g(\pi)$. Setting $\xi_t = t\pi + (1 - t)\xi$, we have

$$g(\pi) - g(\xi) = \int_0^1 \frac{d}{dt} g(\xi_t)\, dt = s \int_0^1 \langle \pi - \xi, \xi_t \rangle V(\xi_t)^{s-2}\, dt .$$

From it, taking into account Lemma 8.3, we get

$$\begin{aligned} g(\pi) - g(\xi) &\leq s|\pi - \xi| \int_0^1 V(\xi_t)^{s-1}\, dt \\ &\leq cs|\pi - \xi|(1 + |\xi|^2 + |\pi|^2)^{\frac{s-1}{2}} \\ &\leq 2^{-s-1}\epsilon(1 + |\xi|^2 + |\pi|^2)^{\frac{s}{2}} \\ &\quad + \frac{c2^{s-1}s^2}{\epsilon}|\pi - \xi|^2(1 + |\xi|^2 + |\pi|^2)^{\frac{s-2}{2}} . \end{aligned}$$

It follows immediately that

$$\left(1 - \frac{\epsilon}{2}\right) g(\pi) \leq g(\xi) + \frac{c2^{s-1}s^2}{\epsilon}|\pi - \xi|^2(1 + |\xi|^2 + |\pi|^2)^{\frac{s-2}{2}}$$

whence (8.49). □

Lemma 8.6 *Let $x_0 \in \Omega$, $R < \operatorname{dist}(x_0, \partial\Omega)$ and*

$$F^0(z) = F(x_0, u(x_0), z) .$$

Let $w \in W^{1,p}(Q_R)$, and let $v(x)$ be the function minimizing the frozen functional

$$\mathcal{F}_0(v, Q_R) =: \int_{Q(x_0,R)} F^0(Dv)\, dx$$

among all the functions coinciding with w on ∂Q_R. Then,

$$\begin{aligned} &\int_{Q_R} (1 + |Dw|^2 + |Dv|^2)^{\frac{p-2}{2}} |D(w - v)|^2\, dx \\ &\qquad \leq c[\mathcal{F}_0(w, Q_R) - \mathcal{F}_0(v, Q_R)] . \end{aligned} \tag{8.50}$$

Proof. We have

$$F^0(\pi) - F^0(\xi) - \langle F^0_z(\xi), \pi - \xi\rangle$$
$$= \int_0^1 (1-t) F^0_{ij}(\xi + t(\pi - \xi))(\pi - \xi)_i(\pi - \xi)_j \, dt ,$$

and therefore

$$F^0(\pi) - F^0(\xi) - \langle F^0_z(\xi), \pi - \xi\rangle \geq A|\xi - \pi|^2$$

where, due to the Lemma 8.3,

$$A =: \int_0^1 (1-t)(1 + |\xi + t(\pi - \xi)|^2)^{\frac{p-2}{2}} \, dt \geq c(1 + |\xi|^2 + |\pi|^2)^{\frac{p-2}{2}} .$$

The conclusion follows immediately, taking $\xi = Dv$, $\pi = Dw$, integrating on Q_R, and remarking that, since v minimizes $\mathcal{F}_0$ and $w - v = 0$ on ∂Q_R, we have

$$\int_{Q_R} \langle F^0_z(Dv), Dw - Dv\rangle \, dx = 0 .$$

□

Theorem 8.4 *Let $u \in W^{1,p}$ be an ω-minimum of the functional*

$$\mathcal{F}(u, \Omega) = \int_\Omega F(x, u, Du) \, dx ,$$

and let conditions (8.43)–(8.46) *be satisfied. Then, the derivatives Du belong to $L^{p,\lambda}_{\text{loc}}(\Omega)$ for every $\lambda < n$, and for every $\Sigma \subset\subset \Omega$ we have the estimate*

$$\|Du\|_{p,\lambda;\Sigma} \leq c(\lambda, \Sigma)\|V(Du)\|_{p,\Omega} . \tag{8.51}$$

Proof. Let $x_0 \in \Sigma$ and let $R < R_0 =: \operatorname{dist}(\Sigma, \partial\Omega)$. As above, let $v(x)$ be the function minimizing the frozen functional $\mathcal{F}_0(v, Q_R)$ among all the functions coinciding with $u(x)$ on ∂Q_R.

The function v is a solution of the elliptic equation

$$\int F^0_{p_i}(Dv) D_i\varphi \, dx = 0 \quad \forall \varphi \in C^\infty_0(Q_R) .$$

From what we have shown in the preceding section (see in particular (8.33) with $q = p$), the gradient of v is locally bounded, and we have

$$\sup_{Q_{\frac{R}{2}}} V(Dv)^p \leq c \fint_{Q_R} V(Dv)^p \, dx , \tag{8.52}$$

from which we get at once

$$\int_{Q_\varrho} V(Dv)^p\,dx \le c\left(\frac{\varrho}{R}\right)^n \int_{Q_R} V(Dv)^p\,dx \tag{8.53}$$

for every $\varrho \le \frac{R}{2}$.

By Lemma 8.5 we have

$$\int_{Q_\varrho} V(Du)^p\,dx \le 2\int_{Q_\varrho} V(Dv)^p\,dx + c\int_{Q_\varrho}(1+|Du|^2+|Dv|^2)^{\frac{p-2}{2}}|D(u-v)|^2\,dx$$

and hence

$$\int_{Q_\varrho} V(Du)^p\,dx \le c\left(\frac{\varrho}{R}\right)^n \int_{Q_R} V(Dv)^p\,dx + c\int_{Q_R}(1+|Du|^2+|Dv|^2)^{\frac{p-2}{2}}|D(u-v)|^2\,dx\,.$$

On the other hand

$$\int_{Q_R} V(Dv)^p\,dx \le \mathcal{F}_0(v,Q_R) \le \mathcal{F}_0(u,Q_R) \le c\int_{Q_R} V(Du)^p\,dx\,,$$

from which, using (8.50) and (8.48) with $w = v$, we arrive to the estimate

$$\int_{Q_\varrho} V^p\,dx \le c\left\{\left(\frac{\varrho}{R}\right)^n + \sigma(R)\right\}\int_{Q_R} V^p\,dx\,, \tag{8.54}$$

where for the sake of brevity we have written V instead of $V(Du)$.

Let now $\tau^\epsilon = \frac{1}{2c}$, and let R_0 be such that for every $R < R_0$ we have $\sigma(R) < \tau^n$. Taking $\varrho = \tau R$ in (8.54), we obtain

$$\int_{Q_{\tau R}} V^p\,dx \le 2c\tau^n\int_{Q_R} V^p\,dx \le \tau^{n-\epsilon}\int_{Q_R} V^p\,dx\,.$$

Setting therefore $\varphi(\varrho) = \int_{Q_\varrho} V^p\,dx$, we can apply Lemma 7.3, getting

$$\int_{Q_\varrho} V^p\,dx \le c\left(\frac{\varrho}{R_0}\right)^{n-2\epsilon}\int_{Q_{R_0}} V^p\,dx$$

from which the conclusion follows at once. □

The same result holds up to the boundary for solutions of the DIRICHLET problem.

We remark in the first place that it is possible to assume $U = 0$. In fact the function $v(x) = u(x) - U(x)$ is a ω-minimum of the functional

$$\mathcal{G}(v) = \int F(x, v + U(x), Dv + DU(x))\, dx$$

and the function $f(x,u,z) = F(x, u + U(x), z + DU(x))$ verifies the same conditions (8.43)–(8.45) of F. This is quite simple to show if only we remark that, setting

$$X^2 = 1 + |z + DU|^2$$

and $\chi = \sup |DU|$, we have obviously $X \leq c(\chi)V$, and moreover $X \geq 1$ if $|z| < 2\chi$, whereas $X \geq \frac{1}{2}V$ if $|z| \geq 2\chi$, so that in conclusion

$$X \geq \frac{V}{2\sqrt{1+4\chi^2}}\,.$$

Assume now that $\partial\Omega$ is a regular manifold in $\mathbf{R}^n$. More precisely, let us assume that for every $x_0 \in \partial\Omega$ there exists a diffeomorphism γ between the unit ball B and a neighborhood W of x_0, mapping the upper half-ball $B^+ = B \cap \mathbf{R}^n_+$ onto $W \cap \Omega$ and the flat part P of ∂B^+ on $\partial\Omega \cap W$. Setting as usual $v(x) =: u \circ \gamma(x)$ and denoting by H the inverse of the matrix

$$\Gamma^i_j = \frac{\partial \gamma_j}{\partial x_i}\,,$$

and with J the Jacobian determinant, $J = \det \Gamma$, we have $Dv = \Gamma Du \circ \gamma$, and the function v minimizes the functional

$$\int G(x, v, Dv)\, dx$$

in B^+, where

$$G(x, v, z) = |J(x)| F(\gamma(x), v, Hz)\,.$$

It is not difficult to prove that if the map $\gamma(x)$ is of class C^1, the new integrand G verifies conditions (8.43)–(8.45). For instance, we have

$$E^{hk} =: \frac{\partial^2 G}{\partial z_h \partial z_k} = |J| H^h_i H^k_j F_{z_i z_j}(\gamma(x), z, Hz)\,.$$

Since γ is a diffeomorphism, we have $|J| \geq c > 0$ and $a|\xi| \leq |H\xi| \leq b|\xi|$, with $a > 0$, and therefore

$$\begin{aligned} E^{hk}\xi_h\xi_k &= |J| F_{z_i z_j}(\gamma(x), z, Hz) H_i^h \xi_h H_j^k \xi_k \\ &\geq \inf |J| V(Hz)^{p-2} |H\xi|^2 \geq \nu V(z)^{p-2} |\xi|^2 . \end{aligned}$$

We can now repeat the proof of Theorem 8.4, writing simply Q_R^+ instead of Q_R, and using (8.41) at the place of (8.33).

In particular, the function $u(x)$ is Hölder continuous with every exponent $\alpha < 1$.

8.6 Hölder Continuity of the Derivatives ($p = 2$)

Our program continues now with the proof of the Hölder continuity of the first derivatives of the ω-minima. We shall treat first the case $p = 2$, for which the proofs are simpler by far.

Theorem 8.5 *Let u be an ω-minimum of the functional $\mathcal{F}$, and let the conditions* (8.43)–(8.46) *with $p = 2$ be satisfied. Assume moreover that $\sigma(t) =: \omega(t) + \vartheta(ct^\delta) \leq At^\tau$ for some $\tau > 0$. Then, the derivatives of u are Hölder-continuous with some exponent α in Ω, and for every open set $\Sigma \subset\subset \Omega$ we have:*

$$\|u\|_{C^{1,\alpha}(\Sigma)} \leq c \|u\|_{W^{1,2}(\Omega)} . \tag{8.55}$$

Proof. As above, let $v(x)$ be the function minimizing the frozen functional $\mathcal{F}_0$ with $v = u$ on ∂Q_R. The derivatives $D_s v$ verify the EULER equation

$$\int F_{ij}^0(Dv) D_j(D_s v) D_i \varphi \, dx = 0 , \tag{8.56}$$

and hence they are quasi-minima of the DIRICHLET functional

$$\int_{Q_R} |Dz|^2 \, dx .$$

From Theorem 7.7 we have

$$\int_{Q_\varrho} |Dv - (Dv)_\varrho|^2 \, dx \leq c \left(\frac{\varrho}{R}\right)^{n+2\delta} \int_{Q_R} |Dv - (Dv)_R|^2 \, dx \tag{8.57}$$

and therefore

$$\int_{Q_\varrho} |Du - (Du)_\varrho|^2\,dx \le c\left(\frac{\varrho}{R}\right)^{n+2\delta} \int_{Q_R} |Du - (Du)_R|^2\,dx + c\int_{Q_R} |D(u-v)|^2\,dx\,.$$

The last integral can be estimated by means of Lemma 8.6; recalling the inequality (8.48) we get

$$\int_{Q_\varrho} |Du - (Du)_\varrho|^2\,dx \le c\left(\frac{\varrho}{R}\right)^{n+2\delta} \int_{Q_R} |Du - (Du)_R|^2\,dx + cAR^\tau \int_{Q_R} V^2\,dx\,. \tag{8.58}$$

We can use now Theorem 8.4 with $4\epsilon = \tau$. We have

$$\int_{Q_R} V^2\,dx \le cR^{n-2\epsilon}\|u\|_{1,2}$$

and in conclusion

$$\int_{Q_\varrho} |Du - (Du)_\varrho|^2\,dx \le c\left(\frac{\varrho}{R}\right)^{n+2\delta} \int_{Q_R} |Du - (Du)_R|^2\,dx + BR^{n+2\epsilon}$$

with $B = c\|u\|_{1,2}$.

Applying once again Lemma 7.3, we obtain the desired conclusion. □

The same result holds for ω-minima with zero DIRICHLET boundary value (or more generally with regular boundary datum on $\partial\Omega$); in this case one obtains the Hölder continuity of the derivatives up to the boundary of Ω. The proof proceeds as usual, considering first the derivatives $D_s u$ with $s \neq n$, and then using the equation to estimate the derivative with respect to x_n.

Let us consider the details. In the first place, we can assume that Ω is the half-ball B^+, and that u is zero on the flat part P of ∂B^+. If $Q(x_0, R)$ is a cube with $x_0 \in B^+_{\frac{1}{2}}$ and $R < \frac{1}{2}$, let v be the minimum of the frozen functional on $\Omega_R(x_0)$.

The function v is a solution of the equation

$$F^0_{ij}(Dv)D_{ij}v = 0\,, \tag{8.59}$$

and its derivatives $D_s v$ are quasi-minima of the DIRICHLET integral in $\Omega_R(x_0)$.

If $s \neq n$ the functions $D_s v$ have zero trace on the flat part P of the boundary, and hence the estimate (7.55) holds with $p = 2$ and $\chi = 0$. We thus have

$$\int_{Q_\varrho^+} |DD'v|^2 \, dx \le c \left(\frac{\varrho}{R}\right)^{n-2+2\delta} \int_{Q_R^+} |DD'v|^2 \, dx$$

for some $\delta > 0$, where D' indicates any of the derivatives, except that with respect to x_n.

For what concerns the derivative $D_{nn}u$, we use Eq. (8.59), from which we obtain

$$D_{nn}v = -\frac{1}{F_{nn}^0} \Sigma' F_{ij}^0 D_{ij} v \,,$$

where the apex indicates that in the sum we have excluded the term with $i = j = n$. Remarking that $F_{nn}^0 \ge 1$, we get

$$\int_{Q_\varrho^+} |D^2 v|^2 \, dx \le c \left(\frac{\varrho}{R}\right)^{n-2+2\delta} \int_{Q_R^+} |D^2 v|^2 \, dx$$

and hence

$$\int_{Q_\varrho^+} |Dv - (Dv)_\varrho|^2 \, dx \le c \left(\frac{\varrho}{R}\right)^{n+2\delta} \int_{Q_R^+} |Dv - (Dv)_R|^2 \, dx \,.$$

From that estimate, proceeding as above, we get

$$\int_{Q_\varrho^+} |Du - (Du)_\varrho|^2 \, dx \le c \left(\frac{\varrho}{R}\right)^{n+2\delta} \int_{Q_R^+} |Du - (Du)_R|^2 \, dx + BR^{n+2\epsilon} \,.$$

An additional application of Lemma 7.3 gives in conclusion the following:

Theorem 8.6 *Let u be an ω-minimum of the functional $\mathcal{F}$, and let conditions (8.43)–(8.46) be satisfied with $p = 2$. Assume that $\sigma(t) = At^\tau$, ($\tau > 0$), and that u has a boundary value U of class $C^{1,\delta}$ for some $\delta > 0$. Then, u belongs to $C^{1,\alpha}(\bar{\Omega})$ for some $\alpha > 0$, and we have*

$$\|u\|_{C^{1,\alpha}(\bar{\Omega})} \le c \|u\|_{W^{1,2}(\Omega)} \,. \tag{8.60}$$

Remark 8.5 The argument leading to the proof of Theorem 8.5 cannot be extended immediately to the case $p \neq 2$. In fact in this situation the coefficients $A_{ij} = F_{ij}^0$ of (8.56) are not bounded functions, and therefore it is not possible to deduce an estimate such as (8.57) with the above method.

More precisely, the derivatives of the function v, the solution of (8.56), are indeed bounded in $Q_{\frac{R}{2}}$, but by (8.52) we have

$$\sup_{Q_{\frac{R}{2}}}(1+|Dv|^2)^{\frac{p}{2}} \leq c \fint_{Q_R} (1+|Dv|^2)^{\frac{p}{2}}\, dx\,.$$

Consequently, the function Dv is always a quasi-minimum of the DIRICHLET integral, but the constant Q, and hence the constant c and the exponent δ in (8.57), depend on the quantity $\fint_{Q_R} V(Dv)^p\, dx$, which is of the same order than $\fint_{Q_R} V(Du)^p\, dx$, and hence depend on R. □

8.7 Other Gradient Estimates

We shall deal now with the more complex case in which the exponent p is different from 2. By Remark 8.5, we cannot use directly the inequality (8.57), and therefore we must look for new estimates, independent of R, for the minima of the frozen functional $\mathcal{F}_0$ in Q_R, or more generally for the weak solutions of the elliptic equation

$$D_i A^i(Dv) = 0\,. \tag{8.61}$$

As we have shown, the function v verifies the equation

$$\int A^{ij}(Dv) D_{js}v D_i\varphi\, dx = 0 \tag{8.62}$$

for every φ with compact support.

Let us begin with some simple consequences of (8.62). We have already remarked in Sec. 8.3 that, setting $w_0 = V^p$ ($V^2 = 1+|Dv|^2$) and $a^{ij}(x) = V^{2-p}A^{ij}(Dv(x))$, we have

$$\int a^{ij} D_j w_0 D_i\varphi\, dx + \int g\varphi\, dx = 0\,, \tag{8.63}$$

where

$$g = pA^{ij} D_{js}v D_{is}v\,,$$

The function g is obviously non-negative, and satisfies the inequalities

$$c_1 V^{p-2}|D^2v|^2 \leq g \leq c_2 V^{p-2}|D^2v|^2\,. \tag{8.64}$$

For $1 \leq k \leq n$, we define

$$w_k = w_k(Dv) =: V(Dv)^{p-1} D_k v\,.$$

If we indicate by w the vector of components w_k $(0 \leq k \leq n)$, we have

$$c_1 V^{2p-2}|D^2v|^2 \leq |Dw|^2 \leq c_2 V^{2p-2}|D^2v|^2 \tag{8.65}$$

and therefore

$$c_1|Dw|^2 \leq g w_0 \leq c_2|Dw|^2 . \tag{8.66}$$

Moreover, taking (8.62) into account, we have for every $k = 1, \ldots, n$:

$$\int a^{ij} D_j w_k D_i \varphi \, dx + \int g_k \varphi \, dx = 0$$

with

$$g_k = A^{ij}(Dv) D_{sj} v D_i \left(V \delta_{ks} + (p-1) \frac{D_k v D_s v}{V} \right)$$

and hence

$$|g_k| \leq c V^{p-2} |D^2 v|^2 \leq c g . \tag{8.67}$$

Let now ν be a unit vector, and let $v_\nu = \langle \nu, Dv \rangle$. The function v_ν is solution of the equation

$$\int A^{ij} D_j v_\nu D_i \varphi \, dx = 0$$

and hence, taking $\varphi = \zeta^2 v_\nu$ and estimating in the usual way, we get

$$\int V^{p-2} |Dv_\nu|^2 \zeta^2 \, dx \leq c \int V^{p-2} |v_\nu|^2 |D\zeta|^2 \, dx .$$

With the usual choice of ζ, the last expression becomes

$$\int_{B_t} V^{p-2} |Dv_\nu|^2 \, dx \leq \frac{c}{t^2} \int_{B_{2t}} V^{p-2} |v_\nu|^2 \, dx . \tag{8.68}$$

Lemma 8.7 *For every $\psi \in W^{1,2}(B_r)$ with zero average on B_r, we have*

$$r^{-2} \fint_{B_r} \psi^2 \, dx \leq c \left(\fint_{B_r} |D\psi|^2 \, dx \right)^{\frac{2}{n}} \fint_{B_r} |D\psi|^{2-\frac{4}{n}} \, dx . \tag{8.69}$$

Proof. It is an immediate consequence of the SOBOLEV–POINCARÉ inequality. We have in fact

$$\int \psi^2\,dx \le c\left(\int |D\psi|^{2_*}\,dx\right)^{\frac{2}{2_*}} = c\left(\int |D\psi|^{\alpha}|D\psi|^{2_*-\alpha}\,dx\right)^{\frac{2}{2_*}}$$

$$\le c\left(\int |D\psi|^2\,dx\right)^{\frac{\alpha}{2_*}}\left(\int |D\psi|^{\frac{2(2_*-\alpha)}{2-\alpha}}\,dx\right)^{\frac{2-\alpha}{2_*}}$$

from which (8.69) follows taking $\alpha = 2 - 2_* = \frac{4}{n+2}$. □

Let us define now the vector

$$\kappa = \int_{B_{2t}} V^{\frac{p-2}{2}} Dv\,dx\,.$$

The space orthogonal to κ has dimension $n-1$, and hence there exists $n-1$ orthonormal vectors ν_i such that $\langle \kappa, \nu_i\rangle = 0$. Since a rotation does not change the structure of Eq. (8.62), we can assume that $\kappa_i = 0$ for $i = 1,\ldots,n-1$.

The functions $\psi_i = V^{\frac{p-2}{2}} D_i v$ have therefore null average on B_{2t}, and we can apply the preceding lemma. Remarking that

$$|D\psi_i|^2 \le cV^{p-2}|D^2v|^2 \le cg$$

and taking (8.68) into account, we can conclude that

$$\Sigma' \fint_{B_t} V^{p-2}|D_{ij}v|^2\,dx \le c\left(\fint_{B_{2t}} g\,dx\right)^{\frac{2}{n}} \fint_{B_{2t}} g^{1-\frac{2}{n}}\,dx\,,$$

where as usual the apex indicates that in the sum we have excluded the term $i = j = n$. We can deal with this term by remarking that the function v is a solution of the equation

$$a^{ij}(x)D_{ij}v = 0\,,$$

and that the ellipticity condition implies that $a^{nn} \ge 1$. We have therefore

$$|D_{nn}v|^2 \le c\Sigma'|D_{ij}v|^2\,.$$

Introducing the last estimate in the previous one, and recalling (8.64), we obtain

$$\fint_{B_t} g\,dx \le c\left(\fint_{B_{2t}} g\,dx\right)^{\frac{2}{n}} \fint_{B_{2t}} g^{1-\frac{2}{n}}\,dx\,. \tag{8.70}$$

From that inequality we easily infer the following:

Lemma 8.8 *The function g belongs to $L^{1+\sigma}(Q_s)$ for some $\sigma > 0$, and we have*

$$\fint_{Q_s} g^{1+\sigma}\,dx \le c\left(\fint_{Q_{2s}} g\,dx\right)^{1+\sigma}. \tag{8.71}$$

Proof. It follows from (8.70) that there exists a constant $\lambda = \lambda(n)$ such that for any cube $Q(y,\Lambda)$ contained in Q_{2s}

$$\begin{aligned}\fint_{Q(y,t)} g\,dx &\le c\left(\fint_{Q(y,\lambda t)} g\,dx\right)^{\frac{2}{n}} \fint_{Q(y,\lambda t)} g^{1-\frac{2}{n}}\,dx \\ &\le \epsilon\fint_{Q(y,\lambda t)} g\,dx + c(\epsilon)\left(\fint_{Q(y,\lambda t)} g^{1-\frac{2}{n}}\,dx\right)^{\frac{n}{n-2}}.\end{aligned}$$

The conclusion of the lemma follows then from Corollary 6.1. □

The next result is a lemma of technical character, that will be useful later.

Lemma 8.9 *Let $\vartheta \in W_0^{1,2}(Q_{2s})$ be the solution of the equation*[5]

$$\int a^{ij} D_j\vartheta D_i\varphi\,dx = \frac{1}{s^2}\int \varphi\,dx\,. \tag{8.72}$$

There exists two positive constants c_1 and c_2, independent of s, such that $\vartheta \le c_2$ in Q_{2s} and $\vartheta \ge c_1$ in Q_s.

Proof. The function $\Theta(x) = \vartheta(sx)$ belongs to $W_0^{1,2}(Q_2)$ and it is a solution of the equation

$$\int \alpha^{ij} D_j\Theta D_i\varphi\,dx = \int \varphi\,dx \tag{8.73}$$

for every $\varphi \in W_0^{1,2}(Q_2)$, with $\alpha^{ij}(x) = a^{ij}(sx)$. In particular, Θ is a quasi-minimum of the functional $\int(|Du|^2 + |u|^2)\,dx$, and a positive super-quasi-minimum of the DIRICHLET integral.

---

[5] Or equivalently the minimum, necessarily unique by the strict convexity, of the functional

$$\mathcal{R}(\vartheta) =: \int \left(a^{ij} D_i\vartheta D_j\vartheta - 2\frac{\vartheta}{s^2}\right) dx\,.$$

We have consequently $\Theta \leq c_2$ in Q_2, and by Theorem 7.9

$$\inf_{Q_1} \Theta \geq c \left(\fint_{Q_2} \Theta^r \, dx \right)^{\frac{1}{r}}$$

for some $r > 0$.

The last integral cannot be zero, since otherwise Θ would be identically zero, and could not be a solution of (8.73). We have therefore $\Theta \geq c_1$ in Q_1, and coming back to the function ϑ we get the desired conclusion. □

The next result is again instrumental, but it is of some interest in itself.

Lemma 8.10 *Let $f \in W^{1,2}(Q_{2s})$ be a non-negative function, such that*

$$\int a^{ij} D_i f D_j \varphi \, dx \leq 0 \tag{8.74}$$

for every $\varphi \in W_0^{1,2}(Q_{2s})$, $\varphi \geq 0$. Then, for some $r > 0$ we have:

$$\fint_{Q_s} |Df| \, dx \leq \frac{c}{s} M(2s)^{1-r} [M(2s) - M(s)]^r \, , \tag{8.75}$$

where

$$M(s) = \sup_{Q_s} f \, .$$

Proof. Taking $\varphi = \zeta f$ ($\zeta \geq 0$ with compact support) in (8.74), we have

$$\int a^{ij} D_i f^2 D_j \zeta \, dx + 2 \int a^{ij} D_i f D_j f \zeta \, dx \leq 0 \tag{8.76}$$

and hence also f^2 satisfies the differential inequality (8.74).

Consequently, f^2 is a sub-quasi-minimum of the DIRICHLET functional $\int |Dw|^2 \, dx$, and hence inequality (7.21) holds with $\kappa_0 = \chi = 0$:

$$M(s)^2 = \sup_{Q_s} f^2 \leq c \fint_{Q_{2s}} f^2 \, dx \, . \tag{8.77}$$

In a similar way, the function $y = M(2s)^2 - f^2$ satisfies the inequality

$$\int a^{ij} D_i y D_j \zeta \, dx = 2 \int a^{ij} D_i f D_j f \zeta \, dx \geq 0$$

for every test function $\zeta \geq 0$, and hence we have HARNACK's inequality (7.61) with $\chi = 0$:

$$\left(\fint_{Q_{2s}} y^{2r}\, dx \right)^{\frac{1}{2r}} \leq c \inf_{Q_s} y \leq c(M(2s)^2 - M(s)^2)$$
$$\leq cM(2s)[M(2s) - M(s)]\,. \qquad (8.78)$$

Let now ϑ be the function introduced in the preceding lemma. We can choose $\varphi = \vartheta y$ in (8.72), obtaining

$$\int a^{ij} D_j \vartheta^2 D_i y\, dx = \frac{2}{s^2} \int \vartheta y\, dx - 2 \int a^{ij} D_j \vartheta D_i \vartheta y\, dx \leq \frac{2}{s^2} \int \vartheta y\, dx\,.$$

Taking now $\zeta = \vartheta^2$ in (8.76), we obtain

$$c \int |Df|^2 \vartheta^2\, dx - \int a^{ij} D_i y D_j \vartheta^2\, dx \leq 0$$

and therefore

$$s^2 \fint_{Q_s} |Df|^2\, dx \leq c \fint_{Q_{2s}} y\, dx \leq cM(2s)^{2-4r} \fint_{Q_{2s}} y^{2r}\, dx$$
$$\leq cM(2s)^{2-2r}[M(2s) - M(s)]^{2r}$$

from which (8.75) follows at once, recalling that

$$\fint |Df|\, dx \leq \left(\fint |Df|^2\, dx \right)^{\frac{1}{2}}\,. \qquad \square$$

Let us consider now, for $h = 0, 1, \ldots, n$, the function y_h minimizing the functional

$$\mathcal{Q}(y) = \int_{Q_{2t}} a^{ij}(x) D_i y D_j y\, dx$$

among all the functions taking the value w_h on ∂Q_{2t}; or in other words the weak solution of the DIRICHLET problem:

$$\begin{cases} \displaystyle\int a^{ij} D_j y_h D_i \varphi\, dx = 0 \quad \forall \varphi \in W_0^{1,2}(Q_{2t}) \\ y_h = w_h \text{ on } \partial Q_{2t} \end{cases}$$

Lemma 8.11 *We have $y_0 \geq w_0$. $\sup y_0 = \sup w_0$, and moreover*

$$\int_{Q_{2t}} |Dy_h|^2\, dx \leq c \int_{Q_{2t}} |Dw_h|^2\, dx\,.$$

Proof. The pointwise inequalities are simple consequences of the maximum principle. Indeed, setting $\varphi = \max(w_0 - y_0, 0)$, we have $\varphi \geq 0$, $\varphi = 0$ on ∂Q_{2t}, and hence

$$\int a^{ij} D_j(w_0 - y_0) D_i \varphi \, dx \leq 0 .$$

From that it follows immediately

$$\int |D\varphi|^2 \, dx \leq \int a^{ij} D_j \varphi D_i \varphi \, dx \leq 0$$

and therefore $\varphi = 0$, that is $w_0 \leq y_0$.

Moreover, since y_0 is a quasi-minimum of the DIRICHLET integral, it will take its maximum on the boundary, where it coincides with w_0.

Finally, the integral inequality can be proved taking $\varphi = y_h - w_h$ in the equation and estimating. □

Lemma 8.12 *Setting* $M(s) = \sup_{Q_s} w_0$, *we have*

$$M(t) \fint_{Q_t} g \, dx \leq c \fint_{Q_{2t}} |Dw|^2 \, dx . \tag{8.79}$$

Proof. Since $y_0 > 0$, we have by HARNACK's inequality:

$$\inf_{Q_t} y_0 \geq c \sup_{Q_t} y_0 \geq c \sup_{Q_t} w_0 = cM(t) .$$

Let us now define

$$E = \left\{ x \in Q_t : w_0(x) < \frac{1}{2} cM(t) \right\} ,$$

and let $f_0 = y_0 - w_0$. We have $f_0 \geq 0$, $f_0 = 0$ on ∂Q_{2t}, and $f_0 > \frac{c}{2} M(t)$ in E. Consequently,

$$\begin{aligned} M(t) \int_E g \, dx &\leq c \int_{Q_{2t}} g f_0 \, dx = -c \int_{Q_{2t}} a^{ij} D_j w_0 D_i f_0 \, dx \\ &\leq c \left(\int_{Q_{2t}} |Dw_0|^2 \, dx \right)^{\frac{1}{2}} \left(\int_{Q_{2t}} |Df_0|^2 \, dx \right)^{\frac{1}{2}} \\ &\leq c \int_{Q_{2t}} |Dw_0|^2 \, dx . \end{aligned}$$

On the other hand $w_0 \geq \frac{c}{2} M(t)$ in $Q_t - E$, and since $g w_0 \leq c|Dw|^2$, we have $M(t) g \leq c|Dw|^2$. That inequality, together with the preceding one, gives the lemma at once. □

At this point we come back to the functions y_h defined above, and we set $f_h = y_h - w_h$. The functions y_h minimize the quadratic integral $\mathcal{Q}$, and hence we have (7.46) with $m = 2$ and $\chi = 0$:

$$\int_{Q_\varrho} |Dy_h|^2\,dx \le c\left(\frac{\varrho}{t}\right)^{n-2+2\delta} \int_{Q_t} |Dy_h|^2\,dx$$
$$\le c\left(\frac{\varrho}{t}\right)^{n-2+2\delta} \int_{Q_t} |Dw_h|^2\,dx\,.$$

It follows that

$$\int_{Q_\varrho} |Dw_h|^2\,dx \le c\left(\frac{\varrho}{t}\right)^{n-2+2\delta} \int_{Q_t} |Dw_h|^2\,dx + \int_{Q_t} |Df_h|^2\,dx\,. \tag{8.80}$$

It remains to estimate the last integral. We have in the first place

$$\int_{Q_{2t}} |Df_h|^2\,dx \le c\int_{Q_{2t}} a^{ij} D_j f_h D_i f_h\,dx$$
$$= -c\int_{Q_{2t}} a^{ij} D_j w_h D_i f_h\,dx$$
$$= c\int_{Q_{2t}} g_h f_h\,dx\,.$$

From that inequality, using (8.67) and (8.71), we get

$$⨍_{Q_{2t}} |Df_h|^2\,dx \le c\left(⨍_{Q_{2t}} g^{1+\sigma}\,dx\right)^{\frac{1}{1+\sigma}} \left(⨍_{Q_{2t}} |f_h|^q\,dx\right)^{\frac{1}{q}}$$
$$\le c⨍_{Q_{4t}} g\,dx \left(⨍_{Q_{2t}} |f_h|^q\,dx\right)^{\frac{1}{q}},$$

where we have set $q = 1 + \frac{1}{\sigma}$.

The last integral can be estimated as follows:

$$⨍_{Q_{2t}} |f_h|^q\,dx \le cM(2t)^{q-2}⨍_{Q_{2t}} |f_h|^2\,dx \le ct^2M(2t)^{q-2}⨍_{Q_{2t}} |Df_h|^2\,dx$$
$$\le ct^2M(2t)^{q-2}⨍_{Q_{2t}} |Dw_h|^2\,dx \le ct^2M(2t)^{q-2}⨍_{Q_{2t}} gw_0\,dx$$
$$\le ct^2M(2t)^{q-1}⨍_{Q_{2t}} g\,dx\,.$$

On the other hand, if ζ is a function with support in Q_{4t} and such that $\zeta = 1$ in Q_{2t} and $|D\zeta| \le ct^{-1}$, we have

$$\begin{aligned} \fint_{Q_{2t}} g\,dx &\le c\fint_{Q_{4t}} g\zeta\,dx = -c\fint_{Q_{4t}} a^{ij} D_j w_0 D_i\zeta\,dx \\ &\le \frac{c}{t}\fint_{Q_{4t}} |Dw_0|\,dx \\ &\le \frac{c}{t^2} M(8t)^{1-r}[M(8t) - M(4t)]^r\,, \end{aligned} \tag{8.81}$$

where we have applied Lemma 8.10 to the function w_0.

From the above relation it follows that

$$\left(\fint_{Q_{2t}} |f_h|^q\,dx\right)^{\frac{1}{q}} \le cM(8t)\left\{1 - \frac{M(4t)}{M(8t)}\right\}^{\frac{r}{q}},$$

and introducing Lemma 8.12,

$$\int_{Q_{2t}} |Df_h|^2\,dx \le c\left\{1 - \frac{M(4t)}{M(8t)}\right\}^{\frac{r}{q}} \int_{Q_{16t}} |Dw|^2\,dx\,. \tag{8.82}$$

We have in conclusion:

$$\int_{Q_\varrho} |Dw|^2\,dx \le c\left\{\left(\frac{\varrho}{t}\right)^{n-2+2\delta} + \left[1 - \frac{M(\varrho)}{M(16t)}\right]^{\frac{r}{q}}\right\}\int_{Q_{16t}} |Dw|^2\,dx$$

for $\varrho < t$.

On the other hand, for $t \le \varrho < 16t$ we have

$$\int_{Q_\varrho} |Dw|^2\,dx \le c\left(\frac{\varrho}{t}\right)^{n-2+2\delta}\int_{Q_{16t}} |Dw|^2\,dx$$

so that the preceding relation holds for every $\varrho < 16t$. Writing t instead of $16t$, we obtain

$$\int_{Q_\varrho} |Dw|^2\,dx \le c\left\{\left(\frac{\varrho}{t}\right)^{n-2+2\delta} + \left[1 - \frac{M(\varrho)}{M(t)}\right]^{\frac{r}{q}}\right\}\int_{Q_t} |Dw|^2\,dx \tag{8.83}$$

for every $\varrho < t$.

From that inequality we deduce the required estimate for the gradient of v:

Theorem 8.7 *Let v be a weak solution of* (8.61). *There exists a constant $\mu > 0$ such that for every $x_0 \in \Omega$ and for every $\varrho < R < c\,\mathrm{dist}\,(x_0, \partial\Omega)$ we* *have*

$$\fint_{Q_\varrho} |w - w_\varrho|^2 \, dx \le c \left(\frac{\varrho}{R} \right)^{4\mu} M(R)^2 . \tag{8.84}$$

Proof. We define

$$\sigma(t) = t^{2-n} \int_{Q_t} |Dw|^2 \, dx .$$

For $0 < \tau < 1$, we have obviously

$$\sigma(\tau t) \le \tau^{2-n} \sigma(t) . \tag{8.85}$$

Moreover, taking $\varrho = \tau t$ in (8.83), we have

$$\sigma(\tau t) \le c \left\{ \tau^{2\delta} + \left[1 - \frac{M(\tau t)}{M(t)} \right]^{\frac{r}{q}} \tau^{2-n} \right\} \sigma(t) .$$

Choosing $\tau < \frac{1}{4}$ in such a way that $2c\tau^\delta < 1$, if

$$\left[1 - \frac{M(\tau t)}{M(t)} \right]^{\frac{r}{q}} \le \tau^{n-2+2\delta} \tag{8.86}$$

then

$$\sigma(\tau t) \le \tau^\delta \sigma(t) . \tag{8.87}$$

If instead (8.86) is false, we have

$$M(\tau t) < \left(1 - \tau^{\frac{q(n-2+2\delta)}{r}} \right) M(t) =: \gamma M(t) \tag{8.88}$$

with $\gamma < 1$.

We choose now λ in such a way that $\delta + \lambda(2 - n - 2\delta) = \frac{\delta}{2}$. Let k be an integer, and let us consider the preceding relations for $t = R, \tau R, \dots, \tau^{k-1} R$.

If (8.86) is false at most for $j \le \lambda k$ indices between 0 and $k - 1$, we can use either (8.85) or (8.87) according to circumstances, obtaining

$$\begin{aligned} \sigma(\tau^k R) &\le \tau^{j(2-n)+(k-j)\delta} \sigma(R) \\ &\le \tau^{k(\delta + \lambda(2-n-\delta))} \sigma(R) = \tau^{\frac{k\delta}{2}} \sigma(R) . \end{aligned} \tag{8.89}$$

If instead (8.86) is false for more than $j > \lambda k$ indices between 0 and $k - 1$, we have

$$M(\tau^k R) \le \gamma^j M(R) \le \gamma^{\lambda k} M(R) .$$

We remark now that by (8.66) and (8.81) we have

$$\sigma(t) \leq ct^2 \fint_{Q_t} g w_0 \, dx \leq ct^2 M(t) \fint_{Q_t} g \, dx \leq cM(4t)^2$$

and hence, recalling that $\tau < \frac{1}{4}$:

$$\sigma(\tau^k R) \leq cM(\tau^{k-1}R) \leq c\gamma^{2\lambda(k-1)} M(R)^2 \leq c\gamma^{2\lambda k} M(R)^2 .$$

In the same way, we get from (8.89)

$$\sigma(\tau^k R) \leq c\tau^{\frac{k\delta}{2}} M(R)^2 ,$$

and hence, setting $4\mu = \min\{\frac{\delta}{2}, 2\lambda \frac{\log \lambda}{\log \tau}\}$, we have in any case

$$\sigma(\tau^k R) \leq c\tau^{4k\mu} M(R)^2 ,$$

whence

$$\sigma(\varrho) \leq c \left(\frac{\varrho}{R}\right)^{4\mu} M(R)^2$$

from which (8.84) follows at once thanks to POINCARÉ's inequality. □

In particular, taking into account the inequality

$$\fint_{Q_\varrho} |w - w_\varrho| \, dx \leq \left(\fint_{Q_\varrho} |w - w_\varrho|^2 \, dx \right)^{\frac{1}{2}}$$

we deduce without difficulty from (8.84) the estimate

$$\fint_{Q_\varrho} |w - w_\varrho| \, dx \leq c \left(\frac{\varrho}{R}\right)^{2\mu} M(R) , \tag{8.90}$$

which will be the starting point for the proof of the regularity of the ω-minima.

8.8 Hölder Continuity of the Derivatives ($p \neq 2$)

Let us consider a ω-minimum u of the functional

$$\mathcal{F}(u) = \int_\Omega F(x, u, Du) \, dx \tag{8.91}$$

with the function $F(x, u, z)$ satisfying conditions (8.43)–(8.45) with $p > 1$. Assume moreover that (8.46) is satisfied with $\vartheta(t) = At^\delta$ for some $\delta > 0$.

Let $\Sigma \subset\subset \Omega$, and let $x_0 \in \Sigma$ and $Q_R = Q(x_0, R) \subset\subset \Omega$. Setting $u_0 = u_{x_0,R}$, let v be the minimum of the frozen functional

$$\mathcal{F}_0(v) = \int_{Q_R} F(x_0, u_0, Dv)\, dx$$

among all the functions coinciding with u on ∂Q_R. Recalling (8.90) and (8.33), we obtain easily the estimate

$$\fint_{Q_\varrho} |w(Dv) - \{w(Dv)\}_\varrho|\, dx \le c\left(\frac{\varrho}{R}\right)^{2\mu} \fint_{Q_R} w_0(Dv)\, dx$$

and setting $\lambda = \{w(Du)\}_\varrho$:

$$\begin{aligned} \int_{Q_\varrho} |w(Du) - \lambda|\, dx \le c\left(\frac{\varrho}{R}\right)^{n+2\mu} \int_{Q_R} V^p\, dx \\ + c\int_{Q_R} |w(Du) - w(Dv)|\, dx\,, \qquad (8.92) \end{aligned}$$

where as usual we have set $V^2 = 1 + |Du|^2$.

We can estimate the last term thanks to Lemma 8.6. Setting

$$W^2 = 1 + |Dv|^2 + |Du|^2\,,$$

we have

$$|w(Du) - w(Dv)| \le cW^{p-1}|Du - Dv|$$

so that, taking (8.48) into account,

$$\begin{aligned} \int_{Q_R} |w(Du) - w(Dv)|\, dx &\le c\left(\int_{Q_R} W^{p-2}|Du - Dv|^2\, dx\right)^{\frac{1}{2}}\left(\int_{Q_R} W^p\, dx\right)^{\frac{1}{2}} \\ &\le c\left(\int_{Q_R} V^p\, dx\right)^{\frac{1}{2}} [\mathcal{F}_0(u, Q_R) - \mathcal{F}_0(v, Q_R)]^{\frac{1}{2}} \\ &\le cR^{2\tau} \int_{Q_R} V^p\, dx \qquad (8.93) \end{aligned}$$

for some $\tau > 0$.

If we insert the last inequality into (8.92), we get

$$\int_{Q_\varrho} |w(Du) - \lambda|\, dx \le c\left\{\left(\frac{\varrho}{R}\right)^{n+2\mu} + R^{2\tau}\right\} \int_{Q_R} V^p\, dx\,.$$

We can estimate the last integral by means of Theorem 8.4; we have

$$\int_{Q_R} V^p\,dx \leq cR^{n-\epsilon}\|V\|^p_{p,\Omega}$$

so that in conclusion

$$\int_{Q_\varrho} |w(Du)-\lambda|\,dx \leq c\|V\|^p_{p,\Omega}\left\{\left(\frac{\varrho}{R}\right)^{n+2\mu} R^{n-\epsilon} + R^{n+2\tau-\epsilon}\right\}.$$

Let now $\alpha > 0$, and let $\varrho < \varrho_0 < 1$. Choosing R in the preceding formula in such a way that $\varrho = R^{1+\alpha}$, we get

$$\int_{Q_\varrho} |w(Du)-\lambda|\,dx \leq c\|V\|^p_{p,\Omega}(R^{\alpha(n+2\mu)+n-\epsilon} + R^{n+2\tau-\epsilon}).$$

If we choose $\alpha = \frac{\tau}{n+\mu}$ and $\epsilon = \alpha\mu$, we get easily

$$\begin{aligned}\int_{Q_\varrho} |w(Du)-\lambda|\,dx &\leq c\|V\|^p_{p,\Omega}R^{n(1+\alpha)+\alpha\mu}\\ &\leq c\|V\|^p_{p,\Omega}\varrho^{n+\gamma} \qquad (8.94)\end{aligned}$$

with $\gamma = \frac{\alpha\mu}{1+\alpha} > 0$.

From the above relation it follows that the functions $w_i = V^{p-1}D_iu$ are Hölder-continuous with exponent γ. The same can be said of the derivatives D_iu, since for every ξ, $\xi_0 \in \mathbf{R}^n$ we have

$$|w(\xi) - w(\xi_0)| \geq |\xi - \xi_0|,$$

and therefore, choosing ξ_0 in such a way that $w(\xi_0) = \lambda$,

$$\int_{Q_\varrho} |D_iu - \xi_0|\,dx \leq c\|V\|^p_{p,\Omega}\varrho^{n+\gamma}.$$

We have thus proved the following:

Theorem 8.8 *Let $u(x)$ be an ω-minimum of the functional* (8.91), *with the integrand $F(x,u,z)$ satisfying* (8.43)–(8.45), *and* (8.46) *with $\vartheta(t) = At^\delta$. Then, the first derivatives of u are locally Hölder-continuous in Ω, and for every $\Sigma \subset\subset \Omega$ we have*

$$\|w(Du)\|_{C^{0,\gamma}(\Sigma)} \leq c(\Sigma)\|V(Du)\|^p_{p,\Omega}. \qquad (8.95)$$

8.9 Elliptic Equations

The results of the preceding section can be extended to bounded solutions of the equation

$$\int \{A^i(x,u,Du)D_i\varphi + B(x,u,Du)\varphi\}\,dx = 0 \tag{8.96}$$

with the conditions

$$|z||A| + |z|^2|A_z| + |B| \le cV^p\,, \tag{8.97}$$

$$A^i(x,u,z)z_i \ge |z|^p - c\,, \tag{8.98}$$

$$A^i_{z_j}\xi_i\xi_j \ge V^{p-2}|\xi|^2 \tag{8.99}$$

and with the coefficients $A(x,u,z)$ continuous in (x,u); or more precisely such that

$$|A(x,u,z) - A(y,v,z)| \le V^{p-1}\vartheta(|x-y| + |u-v|)\,. \tag{8.100}$$

We begin by the remark that under these hypotheses every solution u is a quasi-minimum of the functional $\int(1+|Du|^2)^{\frac{p}{2}}\,dx$ (Theorem 6.2), and therefore it is Hölder-continuous in Ω.

Let now $x_0 \in \Omega$, $u_0 = u_{x_0,R}$ and let v be a weak solution in Q_R of the equation[6]

$$D_iA^i(x_0,u_0,Dv) = 0 \tag{8.101}$$

taking the value $u(x)$ on ∂Q_R.

The function v is a quasi-minimum of the same integral, and hence we have

$$\begin{aligned} c_1\int_{Q_R}(1+|Du|^2)^{\frac{p}{2}}\,dx &\le \int_{Q_R}(1+|Dv|^2)^{\frac{p}{2}}\,dx \\ &\le c_2\int_{Q_R}(1+|Du|^2)^{\frac{p}{2}}\,dx\,. \end{aligned}$$

Moreover, there hold for v all those properties dependent only on its quality of quasi-minimum, such as

$$\operatorname{osc}(v,Q_R) \le \operatorname{osc}(u,\partial Q_R) + cR$$

---

[6] The existence and the uniqueness of the solution are well known from the theory of elliptic equations; see for instance MORREY [3].

(Lemma 8.4), and all the results proved in Secs. 8.3 and 8.7; in particular the estimates

$$\sup_{Q_{\frac{R}{2}}}(1+|Dv|^2)^{\frac{p}{2}} \leq c \fint_{Q_r} (1+|Dv|^2)^{\frac{p}{2}}\, dx\,, \tag{8.102}$$

$$\int_{Q_\varrho} (1+|Dv|^2)^{\frac{p}{2}}\, dx \leq c\left(\frac{\varrho}{R}\right)^n \int_{Q_R} (1+|Dv|^2)^{\frac{p}{2}}\, dx\,, \tag{8.103}$$

$$\fint_{Q_\varrho} |w(Dv)-\{w(Dv)\}_\varrho|^2\, dx \leq c\left(\frac{\varrho}{R}\right)^{n+2\delta} \fint_{Q_R} (1+|Dv|^2)^{\frac{p}{2}}\, dx\,, \tag{8.104}$$

that we have proved in Theorems 8.4 and 8.7.

It is now question to pass from (8.103) and (8.104) to their analogous for the function u. For that, we must estimate the difference

$$\int_{Q_R} |w(Du)-w(Dv)|\, dx \leq c \int_{Q_R} W^{p-1}|Du-Dv|\, dx\,.$$

Recalling that u and v are solutions respectively of (8.96) and (8.101), we have

$$\begin{aligned}
&\int_{Q_R} [A^i(x,u,Du)-A^i(x_0,u_0,Du)]D(u-v)\, dx \\
&\qquad + \int_{Q_R} [A^i(x_0,u_0,Du)-A^i(x_0,u_0,Dv)]D_i(u-v)\, dx \\
&\qquad + \int_{Q_R} B(x,u,Du)(u-v)\, dx = 0\,.
\end{aligned}$$

On the other hand

$$\begin{aligned}
&A^i(x_0,u_0,Du)-A^i(x_0,u_0,Dv) \\
&\qquad = \int_0^1 A^i_{z_j}(x_0,u_0,Dv+t(Du-Dv))\, dt\, D_j(u-v)
\end{aligned}$$

so that, recalling (8.99) and Lemma 8.3:

$$\begin{aligned}
&\int_{Q_R} [A^i(x_0,u_0,Du)-A^i(x_0,u_0,Dv)]D_i(u-v)\, dx \\
&\qquad \geq \int_{Q_R} W^{p-2}|Du-Dv|^2\, dx\,.
\end{aligned}$$

Moreover, keeping into account the estimate $|u - u_0| \leq cR^\delta$, we have:

$$\begin{aligned} |A^i(x,u,Du) - A^i(x_0,u_0,Du)||Du - Dv| &\leq \vartheta(cR^\delta)V^{p-1}|Du - Dv| \\ &\leq \frac{1}{2}V^{p-2}|Du - Dv|^2 + \vartheta(cR^\delta)^2 V^p \end{aligned}$$

and

$$\begin{aligned} |B(x,u,Du)(u-v)| &\leq \{\operatorname{osc}(u,Q_R) + \operatorname{osc}(v,Q_R)\}V^p \\ &\leq cR^\delta V^p . \end{aligned}$$

From these inequalities we obtain, arguing as in (8.93),

$$\int_{Q_R} W^{p-1}|Du - Dv|\,dx \leq c\omega(R)\int_{Q_R} V^p\,dx\,,$$

so that, using (8.103) and (8.104):

$$\int_{Q_\varrho} V^p\,dx \leq c\left\{\left(\frac{\varrho}{R}\right)^n + \omega(R)\right\}\int_{Q_R} V^p\,dx\,, \tag{8.105}$$

$$\int_{Q_\varrho} |w(Du) - \lambda|\,dx \leq c\left\{\left(\frac{\varrho}{R}\right)^{n+2\delta} + \omega(R)\right\}\int_{Q_R} V^p\,dx\,, \tag{8.106}$$

with $\omega(R) = cR^\delta + \vartheta(cR^\delta)$.

From this point on we continue as in the proof of Theorems 8.4 and 8.8, and we arrive at the following:

Theorem 8.9 *Let $u \in W^{1,2}$ be a weak solution of the equation*

$$D_i A^i(x,u,Du) = B(x,u,Du)$$

with conditions (8.97)–(8.100). *Then:*

(i) *if the function $\vartheta(t)$ goes to zero with t, the derivatives Du belong to $L^{p,\lambda}_{\mathrm{loc}}(\Omega)$ for every $\lambda < n$ and for every open set $\Sigma \subset\subset \Omega$ we have*

$$\|Du\|_{p,\lambda,\Sigma} \leq c(\lambda,\Sigma)\|V(Du)\|_{p,\Omega}\,.$$

(ii) *if $\vartheta(t) = At^\delta$ for some $\delta > 0$, the derivatives are Hölder-continuous in Ω, with the estimate*

$$\|w(Du)\|_{C^{0,\gamma}(\Sigma)} \leq c(\Sigma)\|V(Du)\|_{p,\Omega}\,.$$

8.10 Notes and Comments

The methods of this chapter are taken mostly from the regularity theory for solutions of quasi-linear elliptic equations in divergence form

$$\int_\Omega [A_i(x, u, Du)D_i\varphi + B(x, u, Du)\varphi]\, dx = 0 \tag{8.107}$$

and from their applications to the minima of the integrals of the calculus of variations, through their EULER equation.

The extension of these methods to minima of functionals, even those without the EULER equation, is due to GIAQUINTA and GIUSTI [3, 4] in the case $p = 2$, and later, after a paper by GIAQUINTA and G. MODICA [3], relative to the case $p > 2$ under stronger assumptions, to LEWIS [1] and MANFREDI [2] (see also DI BENEDETTO [1] and TOLKSDORF [1]) for $p \neq 2$. We have followed here the method introduced by LEWIS [1] for the generalized DIRICHLET functional

$$\mathcal{D}_p(u) = \int_\Omega |Du|^p\, dx\,.$$

We remark that some of the above regularity results hold even for degenerate functionals (in which the quantity V^p is replaced by $|Du|^p$), of which $\mathcal{D}_p$ is the typical representative. In this case, we cannot expect that the derivatives of u are Hölder-continuous with every exponent, even less so that u is of class C^2 (see for instance GIAQUINTA and MODICA [3]).

The interest of that extension, as well as of that to ω-minima, introduced by ANZELLOTTI [1] in a slightly different context, is two-fold. In the first place, it shows how the $C^{1,\alpha}$ regularity is governed by the notion of ω-minimum, much in the same way in which that of quasi-minimum superintends the Hölder regularity.

Secondly, the results so obtained do not require the differentiability of the function $F(x, u, z)$ with respect to u, being sufficient a sort of uniform Hölder continuity. Moreover, even in the case of regular functions F, Theorems 8.5 and 8.8 represent a generalization of those already known, since no behavior of the derivatives of F with respect to u is required.

It must be noted that, whereas in the case $p = 2$ we obtain the regularity up to the boundary, a similar result is not known for $p \neq 2$, although it seems plausible.

The method of difference quotients was used first in the proof of the regularity of solutions to linear elliptic equations and systems (see

Chapter 10). Its extension to non-linear equations was the object of some discussion, relative to the case $p < 2$. We have followed here a technique introduced by ACERBI and FUSCO [4], that seems free from the difficulties that have troubled some of the former methods.

Chapter 9

Partial Regularity

9.1 Preliminaries

In this chapter we discuss the problem of the regularity of the minima of the functional

$$\mathcal{F}(u,\Omega) = \int_\Omega F(x,u(x),Du(x))dx \tag{9.1}$$

in the vector case, that is when $u(x)$ is a function with values in $\mathbf{R}^N$, and $N > 1$.

It is easily seen that, opposite to what happens if $N = 1$, in this case we cannot expect that the minima are regular everywhere in Ω, even if the function F is quadratic in the gradient:

$$F(x,u,Du) = A_{\alpha\beta}^{ij}(x,u)D_i u^\alpha D_j u^\beta$$

and $A_{\alpha\beta}^{ij}$ are analytic functions of x and u. We actually have:

Example 9.1 (GIUSTI and MIRANDA [1]) For $N = n$ sufficiently large, the function $u(x) = x|x|^{-1}$ minimizes the functional

$$\mathcal{A}(u,B) = \int_B A_{\alpha\beta}^{ij}(u)D_i u^\alpha D_j u^\beta \, dx\,,$$

with

$$A_{\alpha\beta}^{ij}(u) = \delta_{\alpha\beta}\delta_{ij} + \left[\delta_{i\alpha} + \frac{4}{n-2}\cdot\frac{u_i u_\alpha}{1+|u|^2}\right]\left[\delta_{j\beta} + \frac{4}{n-2}\cdot\frac{u_j u_\beta}{1+|u|^2}\right].$$

The matrix $A^{ij}_{\alpha\beta}$ is bounded, and satisfies the ellipticity condition

$$A^{ij}_{\alpha\beta}(u)\xi^\alpha_i\xi^\beta_j \geq |\xi|^2$$

so that the integrand is a convex function of z. Increasing possibly the dimension n, the above function is the unique minimum — and even the unique extremum — of the functional $\mathcal{A}$, among the functions of $W^{1,2}(B, \mathbf{R}^n)$ taking the value x on ∂B.

On the other hand it will be possible to prove the *partial regularity* of the minima, that is the regularity in an open set $\Omega_0 \subset \Omega$, with the singular set $\Omega - \Omega_0$ of zero measure, or better of dimension smaller than n.

Unlike the preceding chapter, we shall consider only the case of growth $p \geq 2$,

$$|F(x,u,z)| \leq c(\lambda + |z|^2)^{\frac{p}{2}}, \quad \lambda \geq 0\,,^{1} \tag{9.2}$$

since no result is known when $p < 2$.

For what concerns estimates from below, we shall make two assumptions. In the first place we shall assume that the function F is *strictly quasi-convex*; or more precisely that there exists a constant $\nu > 0$ such that for every $(x_0, u_0, z_0) \in \Omega \times \mathbf{R}^N \times \mathbf{R}^{nN}$ and every $\varphi \in W^{1,2}_0(\Omega, \mathbf{R}^N)$

$$\int_\Omega [F(x_0, u_0, z_0 + D\varphi(x)) - F(x_0, u_0, z_0)]dx$$

$$\geq \nu \int_\Omega (V_0^{p-2}|D\varphi|^2 + |D\varphi|^p)dx \tag{9.3}$$

with $V_0^2 = 1 + |z_0|^2$.

Moreover we shall assume that there exists a function $\tilde{F} = \tilde{F}(z)$, strictly quasi-convex in 0, such that

$$\tilde{F}(z) \leq F(x,u,z) \tag{9.4}$$

for every x, u and z.

As we have shown in Lemma 5.2, from the quasi-convexity of F and from inequality (9.2) we get the estimate

$$|F(x,u,z) - F(x,u,w)| \leq c(\lambda + |z| + |w|)^{p-1}|z - w|\,, \tag{9.5}$$

---

[1] Of course, the most important cases are $\lambda = 0$ and $\lambda = 1$.

and if F is differentiable with respect to z:

$$|F_z(x,u,z)| \le c(\lambda + |z|)^{p-1}. \tag{9.6}$$

Moreover, under the above assumptions we have proved in Theorem 6.8 the higher summability of the gradient:

Proposition 9.1 *Let u be a cubical Q-minimum of the functional*

$$\mathcal{F}(u) = \int F(x, u(x), Du(x))dx,$$

with F satisfying the assumptions (9.2) *and* (9.4). *Then, u belongs to $W^{1,pr}$ for some $r > 1$, and*

$$\fint_{Q_{R/2}} (\lambda + |Du|)^{pr}\, dx \le c\left(\fint_{Q_R} (\lambda + |Du|)^p\, dx\right)^r \tag{9.7}$$

Remark 9.1 More generally, the estimate (9.7) holds if u satisfies the inequality

$$\int_{Q_s} F(x,u,Du)dx \le Q \int_{Q_s} (\lambda + |Dv|^2)dx$$

for every v with $\operatorname{supp}(u-v) \subset Q_s$. □

9.2 Quadratic Functionals

We shall begin our research by the study of a special yet meaningful class of functionals, that we call *quadratic*:

$$\mathcal{Q}(u,\Omega) = \int_\Omega A^{ij}_{\alpha\beta}(x,u) D_i u^\alpha D_j u^\beta\, dx \tag{9.8}$$

As we have already remarked, these functionals are strictly quasi-convex if and only if they satisfy the Legendre-Hadamard condition:

$$A^{ij}_{\alpha\beta}(x,u)\xi_i\xi_j\eta^\alpha\eta^\beta \ge \nu|\xi|^2|\eta|^2, \quad \nu > 0. \tag{9.9}$$

Moreover we shall assume that are verified the conditions leading to Caccioppoli's inequality; in particular that there exists a strictly quasi-convex function $\tilde{F}(z)$ such that

$$A^{ij}_{\alpha\beta}(x,u) z^\alpha_i z^\beta_j \ge \tilde{F}(z). \tag{9.10}$$

Finally, we will assume that the coefficients $A^{ij}_{\alpha\beta}$ are bounded and uniformly continuous;[2] or in other words that there exists an increasing, continuous and concave function $\gamma(t)$, with $0 \leq \gamma \leq 1$ and $\gamma(0) = 0$, such that

$$|A^{ij}_{\alpha\beta}(x,u) - A^{ij}_{\alpha\beta}(y,v)| \leq c\gamma(|x-y|^2 + |u-v|^2) \tag{9.11}$$

for every x, $y \in \Omega$ e for every u, $v \in \mathbf{R}^N$.

Now let $\omega(t)$ be a continuous increasing function, with $\omega(0) = 0$, and let $u(x)$ be a ω-minimum of the functional $\mathcal{Q}$; that is such that in any $Q_R \subset\subset \Omega$ we have

$$\mathcal{Q}(u, Q_R) \leq (1 + \omega(R))\mathcal{Q}(v, Q_R)$$

for every $v \in W^{1,2}(Q_R)$ with $v = u$ on ∂Q_R.

For $x_0 \in \Omega$ we set $u_0 = u_{x_0,R}$, and we call $v(x)$ the function minimizing the "frozen" functional

$$\mathcal{Q}^0(v, B_R) = \int_{B_R} A^{ij}_{\alpha\beta}(x_0, u_0) D_i v^\alpha D_j v^\beta \, dx$$

in the ball B_R, among all the functions taking the value u on ∂B_R.[3]

The function v is a solution in B_R of the equation

$$A^{ij}_{\alpha\beta}(x_0, u_0) D_i D_j v^\beta = 0 \tag{9.12}$$

and therefore, taking $r = \frac{R}{\sqrt{n}}$ (so that $Q_r \subset B_R$), we have, as we shall prove in the next chapter, (Theorem 10.7):

$$\int_{Q_\varrho} |Dv|^2 \, dx \leq c \left(\frac{\varrho}{r}\right)^n \int_{Q_r} |Dv|^2 \, dx \,, \tag{9.13}$$

$$\int_{Q_\varrho} |Dv - (Dv)_\varrho|^2 \, dx \leq c \left(\frac{\varrho}{r}\right)^{n+2} \int_{Q_r} |Dv - \xi|^2 \, dx \tag{9.14}$$

for every $\varrho < r$ and for every $\xi \in \mathbf{R}^{nN}$.

---

[2]Strictly speaking, the uniform continuity is not necessary, and it is sufficient to assume that the coefficients are simply continuous. The reader can easily make the changes necessary to conclude the proof in the general case, following the ideas of Proposition 9.4.

[3]The existence of the minimum is guaranteed by the coercivity and the weak semicontinuity of $\mathcal{Q}^0$ in the class $u + W^{1,2}_0(B_R)$. The coercivity is a consequence of Lemma 5.1; the semicontinuity of Theorem 4.3.

From (9.13), setting $w = u - v$, we get

$$\begin{aligned}\int_{Q_\varrho} |Du|^2\,dx &\le c\left(\frac{\varrho}{r}\right)^n \int_{Q_r} |Du|^2\,dx + c\int_{Q_r} |Dw|^2\,dx \\ &\le c\left(\frac{\varrho}{R}\right)^n \int_{Q_R} |Du|^2\,dx + c\int_{B_R} |Dw|^2\,dx \qquad (9.15)\end{aligned}$$

and it remains to estimate only the last term.

We have

$$\begin{aligned}\int_{B_R} |Dw|^2\,dx &\le c\int_{B_R} A^{ij}_{\alpha\beta}(x_0,u_0) D_j w^\beta D_i w^\alpha\,dx \\ &= c[\mathcal{Q}^0(u,B_R) - \mathcal{Q}^0(v,B_R)] \\ &= c[\mathcal{Q}(u,B_R) - \mathcal{Q}(v,B_R)] \\ &\quad + c\int_{B_R} [A^{ij}_{\alpha\beta}(x_0,u_0) - A^{ij}_{\alpha\beta}(x,u)] D_j u^\beta D_i u^\alpha\,dx \\ &\quad - c\int_{B_R} [A^{ij}_{\alpha\beta}(x_0,u_0) - A^{ij}_{\alpha\beta}(x,v)] D_j v^\beta D_i v^\alpha\,dx\,. \qquad (9.16)\end{aligned}$$

The first term on the right-hand side can be estimated using the fact that u is an ω-minimum for $\mathcal{Q}$. Extending $v = u$ outside B_R we have

$$\begin{aligned}\mathcal{Q}(u,B_R) &= \mathcal{Q}(u,Q_R) - \mathcal{Q}(u,Q_R - B_R) \\ &\le [1+\omega(R)]\mathcal{Q}(v,Q_R) - \mathcal{Q}(u,Q_R - B_R) \\ &= [1+\omega(R)]\mathcal{Q}(v,B_R) + \omega(R)\mathcal{Q}(u,Q_R)\,,\end{aligned}$$

and hence

$$\mathcal{Q}(u,B_R) - \mathcal{Q}(v,B_R) \le c\omega(R)\int_{Q_R} |Du|^2\,dx \qquad (9.17)$$

since

$$\mathcal{Q}(v,B_R) \le c\int_{B_R} |Du|^2\,dx\,.$$

The remaining part of (9.16) is bounded by

$$c\int_{B_R} [\gamma(R^2 + |u-u_0|^2)|Du|^2 + \gamma(R^2 + |v-u_0|^2)|Dv|^2]dx\,. \qquad (9.18)$$

In order to evaluate these quantities, we shall use the inequality[4]

$$\int_{B_R} |Dv - \lambda|^{2r}\,dx \le c\int_{B_R} |Du - \lambda|^{2r}\,dx \tag{9.19}$$

valid for every constant vector λ.

Let us estimate for instance the second integral in (9.18). Recalling that $\gamma \le 1$, we have

$$\begin{aligned}
\fint_{B_R} \gamma |Dv|^2\,dx &\le c\left(\fint_{B_R} \gamma(R^2 + |v - u_0|^2)dx\right)^{\frac{r-1}{r}} \left(\fint_{B_R} |Dv|^{2r}\,dx\right)^{\frac{1}{r}} \\
&\le c\gamma(R^2 + \fint_{B_R} |v - u_0|^2\,dx)^{\frac{r-1}{r}} \left(\fint_{B_R} |Du|^{2r}\,dx\right)^{\frac{1}{r}} \\
&\le c\gamma(R^2 + \fint_{Q_R} |v - u_0|^2\,dx)^{\frac{r-1}{r}} \left(\fint_{Q_R} |Du|^{2r}\,dx\right)^{\frac{1}{r}} \\
&\le c\gamma(R^2 + \fint_{Q_R} |v - u_0|^2\,dx)^{\frac{r-1}{r}} \fint_{Q_{2R}} |Du|^2\,dx\,,
\end{aligned}$$

where we have used Proposition 9.1 with $\lambda = 0$.

On the other hand

$$\begin{aligned}
\fint_{Q_R} |v - u_0|^2\,dx &\le c\fint_{Q_R} (|u - u_0|^2 + |w|^2)dx \\
&\le cR^2 \fint_{Q_R} (|Du|^2 + |Dw|^2)dx \le cR^2 \fint_{Q_R} |Du|^2\,dx
\end{aligned}$$

and therefore

$$\fint_{Q_R} \gamma |Dv|^2\,dx \le c\gamma\left(R^2 + cR^{2-n}\int_{Q_R} |Du|^2\,dx\right)^{\frac{r-1}{r}} \fint_{Q_{2R}} Du|^2\,dx\,.$$

The other term in (9.18) can be estimated in the same way, so that, setting

$$E(s) = E(x_0, s) = s^{2-n}\int_{Q(x_0,s)} |Du|^2\,dx$$

we arrive to the relation:

$$\int_{Q_\varrho} |Du|^2\,dx \le c\left\{\left(\frac{\varrho}{R}\right)^n + \zeta(2R)\right\}\int_{Q_{2R}} |Du|^2\,dx\,, \tag{9.20}$$

[4] The proof will be given in the following chapter (Theorem 10.15).

where

$$\zeta(R) = \omega(R) + \gamma(R^2 + cE(R))^{\frac{r-1}{r}}.$$

The above inequality will be the starting point in the proof of the partial regularity of the function u.

We remark in the first place that it holds for every $\varrho < 2R$, since it is trivially true (possibly with a different constant c) when $\frac{R}{\sqrt{n}} \leq \varrho < 2R$. It follows that we can write R instead of $2R$. Setting $\varrho = \tau R$, we have

$$E(\tau R) \leq c\tau^2[1 + \tau^{-n}\zeta(R)]E(R).$$

Let $\alpha < 1$ and let τ be such that $c\tau^{2-2\alpha} \leq \frac{1}{3}$. Let ϵ_0 a positive number, such that $\tau^{-n}\gamma((1+c)\epsilon_0)^{\frac{r-1}{r}} \leq 1$ and let R_0, $0 < R_0 < \sqrt{\epsilon_0}$ be such that $\tau^{-n}\omega(R) < 1$ for every $R < R_0$. Assume finally that for some $R < R_0$ we have $E(R) < \epsilon_0$. Then $\tau^{-n}\zeta(R) \leq 2$, and hence

$$E(\tau R) \leq \tau^{2\alpha} E(R).$$

Repeating the procedure, we obtain for every integer k

$$E(\tau^k R) \leq \tau^{2k\alpha} E(R).$$

Let now $\varrho < R$, and let k be such that $\tau^{k+1}R \leq \varrho < \tau^k R$. We have

$$E(\varrho) \leq cE(\tau^k R) \leq c\tau^{2k\alpha} E(R) \leq c\left(\frac{\varrho}{R}\right)^{2\alpha} E(R). \qquad (9.21)$$

From the above estimate we deduce easily the following:

Proposition 9.2 *Let $\omega(t)$ be a continuous increasing function, with $\omega(0) = 0$, and let u be an ω-minimum of the quadratic functional $\mathcal{Q}$. There exists $\epsilon_0 > 0$ and $R_0 > 0$ such that if for some x_0 and for $R < R_0$ we have $E(x_0, R) < \epsilon_0$, then the derivatives Du belong to $L^{2,\lambda}$ in a neighborhood I of x_0, for every $\lambda < n$.*

Proof. Since $E(y, R)$ is a continuous function of y, if $E(x_0, R) < \epsilon_0$, there will be $E(y, R) < \epsilon_0$ for every y in a neighborhood I of x_0. We can therefore write the inequality (9.21) for every $y \in I$:

$$E(y, \varrho) \leq c\left(\frac{\varrho}{R}\right)^{2\alpha} E(y, R),$$

and hence, setting as usual $I_\varrho = I \cap Q(y,\varrho)$,

$$\int_{I_\varrho} |Du|^2\,dx \le \int_{Q_\varrho} |Du|^2\,dx \le c\left(\frac{\varrho}{R}\right)^{n-2+2\alpha} \int_{Q_R} |Du|^2\,dx$$

$$\le c\varrho^{n-2+2\alpha} \int_\Omega |Du|^2\,dx\,, \tag{9.22}$$

so that the function Du belongs to $L^{2,n-2+2\alpha}(I)$. Since $\alpha < 1$ is arbitrary, we get the conclusion of the proposition. □

In particular, by Proposition 3.7, we have $u \in \mathcal{L}^{2,n+2\alpha}(I)$, and hence u is Hölder-continuous with exponent α in I. As a consequence, u is of class $C^{0,\alpha}$ in an open set $\Omega_0 \subset \Omega$.

A point y belongs to the singular set $\Sigma = \Omega - \Omega_0$ if and only if

$$\liminf_{R\to 0} R^{2-n} \int_{Q(y,R)} |Du|^2\,dx > 0\,, \tag{9.23}$$

since if u is Hölder-continuous with exponent α in a neighborhood of a point y, we have by Caccioppoli's inequality:

$$\int_{Q(y,\frac{R}{2})} |Du|^2\,dx \le \frac{c}{R^2} \int_{Q(y,R)} |u-u_R|^2\,dx \le cR^{n-2+2\alpha}$$

and the lim inf in (9.23) is zero.

We want to evaluate the dimension of the singular set Σ. For that, we remark that $u \in W^{1,2r}(\Omega)$ with $r > 1$ (Theorem 6.8), and

$$E(x_0,R) = R^2 \fint_{Q_R} |Du|^2\,dx \le cR^2\left(\fint_{Q_R} |Du|^{2r}\,dx\right)^{\frac{1}{r}}$$

$$\le c\left(R^{2r-n}\int_{Q_R}(1+|Du|^{2r})dx\right)^{\frac{1}{r}}.$$

We can therefore apply Proposition 2.8, with

$$\mu(Q_\varrho) = \int_{Q_\varrho}(1+|Du|^{2r})dx$$

and we get $H^{n-2r}(\Sigma) = 0$ for some $r > 1$. We have thus proved our first partial regularity theorem:

Theorem 9.1 *Let $\omega(t)$ be a continuous increasing function, with $\omega(0) = 0$, and let $u(x)$ be a ω-minimum of the functional*

$$\mathcal{Q}(u,\Omega) = \int_\Omega A^{ij}_{\alpha\beta}(x,u)D_iu^\alpha D_ju^\beta\,dx$$

with the coefficients $A^{ij}_{\alpha\beta}(x,u)$ *satisfying the Legendre–Hadamard condition* (9.9) *and the inequality* (9.10).

Then, u is Hölder-continuous with any exponent $\alpha < 1$ *in an open set* $\Omega_0 \subset \Omega$, *and* $\dim_H(\Omega - \Omega_0) < n - 2$.

We have moreover:

Theorem 9.2 *If* $\omega(R) \le cR^{2\sigma}$ *for some* $\sigma > 0$, *and if the coefficients* $A^{ij}_{\alpha\beta}$ *are Hölder-continuous functions of their arguments, then every* ω*-minimum* u *of the functional* $\mathcal{Q}$ *has Hölder-continuous first derivatives in* Ω_0.

Proof. Let $K \subset \Omega_0$ be a compact set, and let Q_R be a cube with center in K, contained in Ω_0. From (9.14) we get:

$$\int_{Q_\varrho} |Du - (Du)_\varrho|^2\, dx \le c\left(\frac{\varrho}{r}\right)^{n+2} \int_{Q_r} |Du - \xi|^2\, dx + c\int_{Q_r} |Dw|^2\, dx\,, \tag{9.24}$$

where as usual $w = u - v$.

The last term can be estimated as above:

$$\int_{Q_t} |Dw|^2\, dx \le c\zeta(R)\int_{Q_R} |Du|^2\, dx$$

with

$$\zeta(R) = \omega(R) + \gamma(R^2 + cE(R))^{\frac{r-1}{r}}\,.$$

If we remark that from the Hölder continuity of the coefficients we get $\gamma(s) \le cs^\delta$ for some $\delta > 0$, and that by Proposition 9.2 we have $E(R) \le cR^\mu$ for every $\mu < 2$, we conclude easily that

$$\int_{Q_\varrho} |Du - (Du)_\varrho|^2\, dx \le c\left(\frac{\varrho}{R}\right)^{n+2} \int_{Q_R} |Du - (Du)_R|^2\, dx + cR^{n+2\alpha} \tag{9.25}$$

for some $\alpha > 0$, and for every couple of concentric cubes $Q_\varrho \subset Q_R \subset \Omega_0$.

Applying Lemma 7.3 to the function $\varphi(\varrho) = \int_{Q_\varrho} |Du - (Du)_\varrho|^2\, dx$, we arrive at once to the inequality

$$\int_{Q_\varrho} |Du - (Du)_\varrho|^2\, dx \le c\varrho^{n+2\alpha}\,,$$

which implies that the first derivatives of u are Hölder-continuous in K, and hence in Ω_0. □

The fact that the dimension of the singular set is smaller than that of $\partial\Omega$ suggests the possibility of proving partial regularity up to the boundary for the ω-minima with Dirichlet boundary data, provided the boundary of Ω and the datum $U(x)$ are regular.

We can always reduce locally to the case of flat boundary, by means of a diffeomorphism which does not change the quadratic structure of the functional in question;[5] moreover, writing $y = u - U$, we can restrict to the case of zero boundary data. If (once the boundary has been flattened) u is a ω-minimum of the quadratic functional $\mathcal{Q}$, the new function y will be a ω-minimum of the functional

$$\mathcal{P}(y, Q_R^+) = \int_{Q_R^+} \{B_{\alpha\beta}^{ij}(x,y) D_j y^\beta D_i y^\alpha + B_{\alpha\beta}^{ij}(x,y)(2D_j y^\beta + D_j U^\beta) D_i U^\alpha\} dx \,, \tag{9.26}$$

where we have set

$$B(x,y) = A(x, y + U(x)) \,.$$

Let now x_0 be a point lying on the flat portion of $\partial\Omega$, and let R be such that $Q(x_0, R) \cap \Omega = Q^+(x_0, R)$. We shall replace the cubes Q_R^+ with sets Λ_R with regular boundary, and such that $Q_{R/2}^+ \subset \Lambda_R \subset Q_R^+$. In order to avoid artificial dependence on R, we start from a regular set Λ, with $Q_{1/2}^+ \subset \Lambda \subset Q^+$, and we shall take Λ_R homothetic to Λ:

$$\Lambda_R = \Lambda_R(x_0) =: \{x \in \mathbf{R}^n : R^{-1}(x - x_0) \in \Lambda\} \,.$$

Let v be the function minimizing the functional

$$\mathcal{P}_0(v, \Lambda_R) = \int_{\Lambda_R} \langle B_0 Dv, Dv + 2DU \rangle dx$$

among all the functions assuming the value y on $\partial\Lambda_R$, where as usual $B_0 = B(x_0, y_0)$. The function v is a solution of the Euler equation:

$$\int_{\Lambda_R} \langle B_0(Dv + DU), D\varphi \rangle dx = 0 \quad \forall \varphi \in W_0^{1,2}(\Lambda_R) \,.$$

[5] For that, it is sufficient to assume that $\partial\Omega$ is of class C^1.

From Theorem 10.7 of the next chapter we get for every $\varrho < R/2$ the estimate

$$\int_{Q_\varrho^+} |Dv|^2\,dx \le c\left\{\left(\frac{\varrho}{R}\right)^n \int_{Q_{R/2}^+} |Dv|^2\,dx + \int_{Q_{R/2}^+} |DU|^2\,dx\right\}$$

from which we obtain at once

$$\int_{Q_\varrho^+} |Dy|^2\,dx \le c\left\{\left(\frac{\varrho}{R}\right)^n \int_{Q_R^+} |Dy|^2\,dx \right.$$

$$\left. + \int_{Q_R^+} |DU|^2\,dx + \int_{\Lambda_R^+} |D(y-v)|^2\,dx\right\}. \qquad (9.27)$$

The corrective term represented by the last integral can be estimated as above, using the continuity of the coefficients, the fact that the function y is a ω-minimum, the L^{2r} estimates for the gradient found in Proposition 9.1, and the inequality

$$\int_{\Lambda_R} |Dv|^{2r}\,dx \le c\int_{\Lambda_R} |Dy|^{2r}\,dx$$

that we shall prove in the next chapter (Theorem 10.17; see Remark 10.4). Without entering in the details, that the reader can easily check, if we assume that DU belongs to the space $L^{2,n-2+2\sigma}$, and hence that

$$\int_{Q_R^+} |DU|^2\,dx \le M_0 R^{n-2+2\sigma}\,,$$

we arrive as above to the estimate

$$E(\tau R) \le c\tau^2\{1+\tau^{-n}\zeta(R)\}E(R) + \tau^{2-n}M_0R^{2\sigma}\,, \qquad (9.28)$$

where we have set

$$\zeta(R) = \omega(R) + \gamma(R^2 + cE(R))^{\frac{r-1}{r}}$$

and

$$E(s) = E(x_0,s) =: s^{2-n}\int_{Q_s^+} |Dy|^2\,dx\,.$$

We can now proceed as above, choosing first $\alpha > \sigma$, secondly τ such that $c\tau^{2-2\alpha} = \frac{1}{3}$, then ϵ_0 in such a way that $\tau^{-n}\gamma((1+c)\epsilon_0)^{\frac{r-1}{r}} \le 1$, and finally $R_0 < \sqrt{\epsilon_0}$ such that $\tau^{2\alpha}\epsilon_0 + M_1R_0^{2\sigma} \le \epsilon_0$, and $\tau^{-n}\omega(R_0) \le 1$.

Assume now that for some $R < R_0$ we have $E(R) < \epsilon_0$. Then,

$$E(\tau R) \le \tau^{2\alpha} E(R) + M_1 R^{2\sigma} ,$$

and therefore in particular $E(\tau R) < \epsilon_0$. By induction, if $E(\tau^k R) < \epsilon_0$ we have

$$E(\tau^{k+1} R) \le \tau^{2\alpha} E(\tau^k R) + M_1 (\tau^k R)^{2\sigma} < \varepsilon_0 ,$$

and therefore

$$E(\tau^k R) \le \tau^{2k\alpha} E(R) + M_1 (\tau^{k-1} R)^{2\sigma} \sum_{j=0}^{k-1} \tau^{2j(\alpha-\sigma)} . \qquad (9.29)$$

In particular

$$E(\tau^k R) \le \tau^{2k\alpha} E(R) + M_2 (\tau^k R)^{2\sigma}$$

and hence

$$E(x_0, \varrho) \le M_1 \left(\frac{\varrho}{R}\right)^{2\sigma} E(x_0, R) + M_2 \varrho^{2\sigma} .$$

Let now $x_1 \in \Omega$ be such that $Q(x_1, R)$ intersects only the flat part of $\partial\Omega$, and assume that $R \le R_1$ and

$$E(x_1, R) < \epsilon_1 ,$$

where $R_1 < R_0$ and $\epsilon_1 < \epsilon_0$ are such that

$$2^{n-2}(M_1 \epsilon_1 + M_2 R_1^{2\sigma}) < \epsilon_0 .$$

We distinguish two cases:

(i) $d = \mathrm{dist}(x_1, \partial\Omega) \ge \frac{R}{2}$.
We have

$$E\left(x_1, \frac{R}{2}\right) \le 2^{n-2} E(x_1, R) < \epsilon_0$$

and therefore by (9.21):

$$E(x_1, \varrho) \le c \left(\frac{\varrho}{R}\right)^{2\alpha} E\left(x_1, \frac{R}{2}\right) \le c \left(\frac{\varrho}{R}\right)^{2\sigma} E(x_1, R)$$

since $\sigma < \alpha$.

(ii) $d < \frac{R}{2}$.

If $x_0 \in \partial\Omega$ is the projection of x_1 on $\partial\Omega$, we have $Q^+(x_0, R) \subset Q^+(x_1, R)$, and hence $E(x_0, R) \leq E(x_1, R) < \epsilon_0$, so that, if $2d \leq r < R$:

$$E(x_0, r) \leq M_1 \left(\frac{r}{R}\right)^{2\sigma} E(x_0, R) + M_2 r^{2\sigma} .$$

Remarking that $Q(x_1, d) \subset Q^+(x_0, 2d)$, we get from this estimate

$$E(x_1, d) \leq 2^{n-2} E(x_0, 2d) < 2^{n-2}(M_1\epsilon_1 + M_2 R_1^{2\sigma}) < \epsilon_0 ,$$

and hence, if $\varrho < d$:

$$E(x_1, \varrho) \leq c \left(\frac{\varrho}{d}\right)^{2\sigma} E(x_1, d) \leq c \left(\frac{\varrho}{R}\right)^{2\sigma} E(x_1, R) + c\varrho^{2\sigma} . \tag{9.30}$$

Changing possibly the constant c, the last inequality holds in any case, and for every $\varrho < R$.

Taking into account the continuity of $E(x, R)$ with respect to x, we can then conclude that if $E(x_1, R)$ is small enough, there exists a neighborhood V of x_1 such that the function Dy belongs to the space $L^{2,n-2+2\sigma}(V)$, and therefore y is Hölder-continuous with exponent σ. With the same argument as above we can now prove the Hölder continuity of the derivatives. We thus have the following:

Theorem 9.3 *Let $\omega(t)$ be a continuous increasing function, with $\omega(0) = 0$, let u be a ω-minimum of the quadratic functional $\mathcal{Q}$ in Ω, and assume that the boundary datum U has derivatives in $L^{2,n-2+\sigma}(\Omega)$. Then u is a Hölder-continuous function in $\bar{\Omega} - \Sigma$, where Σ is a closed set of dimension less than $n - 2$.*

If moreover we have $\omega(R) \leq cR^{2\sigma}$ for some $\sigma > 0$, if the coefficients $A^{ij}_{\alpha\beta}$ are Hölder-continuous functions of their arguments, and if U has Hölder-continuous derivatives in $\bar{\Omega}$, then every ω-minimum u of the functional $\mathcal{Q}$ has Hölder-continuous derivatives in $\bar{\Omega} - \Sigma$.

9.3 The Second Caccioppoli Inequality

When we pass from quadratic functionals to the general situation, the inequality (9.7) alone does not suffice to get the regularity, and we need a second Caccioppoli inequality, in which u is replaced by $u - P$, P being an arbitrary polynomial of the first degree.

To prove it, we shall make the following assumptions:

(i) $F(x,u,z)$ is a continuous strictly quasi-convex function, of class C^2 in z, with growth $p \geq 2$:

$$|F(x,u,z)| \leq c_1 V^p \quad (V^2 = 1 + |z|^2)\,. \tag{9.31}$$

We note that from the above inequality it follows that

$$|F_z(x,u,z)| \leq c_2 V^{p-1}\,. \tag{9.32}$$

(ii) The function $V^{-p}F(x,u,z)$ is Hölder-continuous (with exponent 2δ) in $(x,u) \in \bar{\Omega} \times \mathbf{R}^N$, uniformly with respect to z. That means that there exists an increasing function $\varrho(s) \geq 2c_1$ such that

$$|F(x,u,z) - F(y,v,z)| \leq \vartheta(|v|, |x-y|^2 + |u-v|^2)V^p \tag{9.33}$$

with $\vartheta(s,t) = \min\{2c_1, \varrho(s)t^\delta\}$.
We remark that the function ϑ is concave in its second argument.

(iii) There exists a function $\tilde{F}(z)$, strictly quasi-convex in 0, such that

$$\tilde{F}(z) \leq F(x,u,z)\,. \tag{9.34}$$

Lemma 9.1 *Let $F(z)$ be a function of class C^2, with $|F(z)| \leq cV^p$ and $|F_z(z)| \leq cV^{p-1}$. Then, setting*

$$\bar{F}(z) = F(z_0 + z) - F(z_0) - \langle F_z(z_0), z\rangle\,, \tag{9.35}$$

we have

$$|\bar{F}(z)| \leq c(z_0)V^{p-2}|z|^2\,, \tag{9.36}$$

$$|\bar{F}_z(z)| \leq c(z_0)V^{p-2}|z|\,, \tag{9.37}$$

$$\frac{|\bar{F}(z) - \bar{F}(w)|}{|z-w|} \leq c(z_0)(1 + |z|^{p-2} + |w|^{p-2})(|z| + |w|)\,. \tag{9.38}$$

Proof. The inequality (9.38) follows at once from (9.37). We shall prove (9.36); (9.37) well be proved in a similar way.

Let $k(z_0) = \sup_{|w| \leq 1 + |z_0|} |F_{zz}(w)|$. If $|z| \leq 1$ we have

$$\bar{F}(z) = \frac{1}{2}|\langle F_{zz}(z_0 + tz)z, z\rangle| \leq \frac{1}{2}k(z_0)|z|^2 \leq c(z_0)V^{p-2}|z|^2\,.$$

If instead $|z| > 1$:

$$\begin{aligned}\bar{F}(z) &\leq c(1 + |z|^2 + |z_0|^2)^{\frac{p}{2}} + c(z_0)|z| \\ &\leq c(z_0) + |z|^p \leq c(z_0)|z|^p \leq c(z_0)V^{p-2}|z|^2\,. \qquad \square\end{aligned}$$

Remark 9.2 If the function F depends on a parameter u_0, the preceding estimates hold with the constant c depending on u_0 and z_0. □

Remark 9.3 If $p = 2$ and the function F has bounded second derivatives, the constant c in (9.36)–(9.38) can be taken independent of z_0. □

We van now prove a result central in our regularity program.

Theorem 9.4 (CACCIOPPOLI's inequality II) *Let $u \in W^{1,p}$ be an ω-minimum of the functional*

$$\mathcal{F}(u,\Omega) = \int_\Omega F(x,u,Du)dx\,,$$

and let the function F verify (9.31)–(9.34) with $p \geq 2$.

Then, for every $x_0 \in \Omega$, for every $\varrho < R \leq \frac{1}{2\sqrt{n}}\mathrm{dist}(x_0,\partial\Omega)$, for every $u_0 \in \mathbf{R}^N$ and for every polynomial $P(x) = a + \langle z_0, x - x_0\rangle$ of the first degree, we have

$$\begin{aligned}\int_{Q_\varrho} &(V_0^{p-2}|Du - z_0|^2 + |Du - z_0|^p)dx \\ &\leq \frac{cV_0^{p-2}}{(R-\varrho)^2}\int_{Q_R}|u-P|^2\,dx \\ &\quad + \frac{c}{(R-\varrho)^p}\int_{Q_R}|u-P|^p\,dx + c\omega(R)R^n \\ &\quad + c\int_{Q_R}\vartheta(|u_0|, R^2 + |u-u_0|^2 + |u-P|^2)(V^p + V_0^p)dx\,. \end{aligned} \tag{9.39}$$

Proof. Let $t < s < R$, let η be the usual test function, $0 \leq \eta \leq 1$, $\mathrm{supp}(\eta) \subset Q_s$, $\eta = 1$ in Q_t, $|D\eta| \leq \frac{2}{s-t}$, and let $\varphi = \eta(u-P)$, $\psi = (1-\eta)(u-P)$, so that $\varphi + \psi = u - P$ and $D\varphi + D\psi = Du - z_0$.

Set now

$$\bar{F}(z) =: F(x_0,u_0,z_0+z) - F(x_0,u_0,z_0) - \langle F_z(x_0,u_0,z_0), z\rangle\,.$$

Remarking that $\int\langle F_z(z_0), D\varphi\rangle dx = 0$, we get from the strict quasi-convexity of F:

$$\begin{aligned}\int_{Q_s} &(V_0^{p-2}|D\varphi|^2 + |D\varphi|^p)dx \\ &\leq \int_{Q_s}\bar{F}(D\varphi)dx\end{aligned}$$

$$
\begin{aligned}
&= \int_{Q_s} \bar{F}(Du - z_0 - D\psi)dx \\
&= \int_{Q_s} \bar{F}(Du - z_0)dx \\
&\quad + \int_{Q_s} [\bar{F}(Du - z_0 - D\psi) - \bar{F}(Du - z_0)]dx \\
&\leq \int_{Q_s} \bar{F}(Du - z_0)dx \\
&\quad + c\int_{Q_s} (1 + |D\psi| + |Du - z_0|)^{p-2}(|Du - z_0| + |D\psi|)|D\psi|dx\,.
\end{aligned}
\tag{9.40}
$$

We estimate now the first term on the right-hand side. If we set for simplicity $F^0(z) = F(x_0, u_0, z)$, we have:

$$
\begin{aligned}
&\int_{Q_s} \bar{F}(Du - z_0)dx \\
&\quad = \int_{Q_s} F^0(Du)dx - \int_{Q_s} F^0(z_0)dx \\
&\quad\quad - \int_{Q_s} \langle F_z^0(z_0), Du - z_0\rangle dx \\
&\quad = \int_{Q_s} F(x, u, Du)dx \\
&\quad\quad + \int_{Q_s} [F(x_0, u_0, Du) - F(x, u, Du)]dx \\
&\quad\quad - \int_{Q_s} F^0(z_0)dx - \int_{Q_s} \langle F_z^0(z_0), Du - z_0\rangle dx\,.
\end{aligned}
\tag{9.41}
$$

On the other hand

$$
\int_{Q_s} F(x, u, Du)dx \leq [1 + \omega(s)] \int_{Q_s} F(x, u - \varphi, Du - D\varphi)dx \tag{9.42}
$$

and moreover

$$
\begin{aligned}
&\int_{Q_s} F(x, u - \varphi, Du - D\varphi)dx \\
&\quad = \int_{Q_s} F(x, \psi + P, D\psi + z_0)dx
\end{aligned}
$$

$$= \int_{Q_s} F^0(D\psi + z_0)dx$$

$$+ \int_{Q_s} [F(x, \psi + P, D\psi + z_0) - F(x_0, u_0, D\psi + z_0)]dx$$

$$= \int_{Q_s} \bar{F}(D\psi)dx + \int_{Q_s} F^0(z_0)dx + \int_{Q_s} \langle F_z^0(z_0), D\psi\rangle dx$$

$$+ \int_{Q_s} [F(x, \psi + P, D\psi + z_0) - F(x_0, u_0, D\psi + z_0)]dx\,. \tag{9.43}$$

Combining (9.40), (9.41) and (9.43), we get:

$$\int_{Q_s} (V_0^{p-2}|D\varphi|^2 + |D\varphi|^p)dx$$

$$\leq c\int_{Q_s} \bar{F}(D\psi)dx$$

$$+ c\int_{Q_s - Q_t} (1 + |Du - z_0| + |D\psi|)^{p-2}(|D\psi| + |Du - z_0|)|D\psi|dx$$

$$+ \int_{Q_s} [F(x_0, u_0, Du) - F(x, u, Du)]dx$$

$$+ \omega(s)\left(\int_{Q_s} F^0(z_0)dx + \int_{Q_s} \langle F_z^0(z_0), D\psi\rangle dx\right)$$

$$+ \int_{Q_s} [F(x, \psi + P, D\psi + z_0) - F(x_0, u_0, D\psi + z_0)]dx\,. \tag{9.44}$$

From that inequality, taking (9.33) into account, and recalling that $\psi = 0$ in Q_t, we obtain

$$\int_{Q_s} (V_0^{p-2}|D\varphi|^2 + |D\varphi|^p)dx$$

$$\leq c\int_{Q_s - Q_t} V_0^{p-2}(|D\psi|^2 + |Du - z_0|^2)dx$$

$$+ c\int_{Q_s - Q_t} (|D\psi|^p + |Du - z_0|^p)dx + c\omega(s)s^n$$

$$+ \int_{Q_s} \vartheta(|u_0|, |x - x_0|^2 + |u - u_0|^2 + |u - P|^2)(V^p + V_0^p)dx \tag{9.45}$$

Finally, introducing the expression of ψ, we get

$$\begin{aligned}\int_{Q_t} & (V_0^{p-2}|Du - z_0|^2 + |Du - z_0|^p)dx \\ & \leq c_1 \int_{Q_s - Q_t} (V_0^{p-2}|Du - z_0|^2 + |Du - z_0|^p)dx \\ & \quad + \frac{c}{(s-t)^2} \int_{Q_s} V_0^{p-2}|u - P|^2\, dx \\ & \quad + \frac{c}{(s-t)^p} \int_{Q_s} |u - P|^p\, dx + c\omega(R)R^n \\ & \quad + c\int_{Q_R} \vartheta(|u_0|, |x - x_0|^2 + |u - u_0|^2 + |u - P|^2)(V^p + V_0^p)dx\,.\end{aligned} \tag{9.46}$$

The conclusion follows as usual, summing to both sides the first integral multiplied by c_1 and applying Lemma 6.1. □

Remark 9.4 We note that $\vartheta = 0$ if the function F depends only on z. □

Starting from (9.39) we can prove the higher summability of $Du - z_0$. For that, we need the following:

Lemma 9.2 *For $z \in \mathbf{R}^n$ let $w(z) = zV(z)^{2\sigma}$. For every $\sigma > -\frac{1}{2}$, there exist two constants c_1 and c_2 such that for every z, $z_0 \in \mathbf{R}^n$*

$$\begin{aligned}c_1(1 + |z|^2 + |z_0|^2)^\sigma |z - z_0| &\leq |w(z) - w(z_0)| \\ &\leq c_2(1 + |z|^2 + |z_0|^2)^\sigma |z - z_0|\,.\end{aligned} \tag{9.47}$$

Proof. The second inequality follows at once from the formula

$$\begin{aligned}|w(z) - w(z_0)| &= \left|\int_0^1 \frac{d}{dt} w(z_0 + t(z - z_0))dt\right| \\ &\leq c(1 + |2\sigma|)|z - z_0| \int_0^1 (1 + |z_0 + t(z - z_0)|^2)^\sigma\, dt\end{aligned}$$

and from Lemma 8.3.

In order to prove the first one, let $s \in \mathbf{R}$ and $\eta(s) = s(1 + s^2)^\sigma$. We have

$$\eta'(s) = (1 + s^2)^{\sigma - 1}(1 + (1 + 2\sigma)s^2) \geq \min(1, 1 + 2\sigma)(1 + s^2)^\sigma\,.$$

We remark that we can assume $|z| \geq |z_0|$, and we distinguish the two cases $|z| - |z_0| \geq \epsilon|z - z_0|$ and $|z| - |z_0| < \epsilon|z - z_0|$, with $0 < \epsilon \leq \frac{1}{2}$. In the first case, we have

$$\begin{aligned} |w(z) - w(z_0)| &\geq |w(z)| - |w(z_0)| = \eta(|z|) - \eta(|z_0|) \\ &= (|z| - |z_0|) \int_0^1 (1 + |z_0 + t(z - z_0)|^2)^\sigma \, dt \\ &\geq \epsilon|z - z_0|(1 + |z|^2 + |z_0|^2)^\sigma , \end{aligned}$$

thanks to Lemma 8.3.

If instead $|z| - |z_0| < \epsilon|z - z_0|$, and hence $|z| < 3|z_0|$, we set $\xi = z\frac{|z_0|}{|z|}$, so that $|\xi| = |z_0|$ and $|z - \xi| = |z| - |z_0| < \epsilon|z - z_0|$ and $|z_0 - \xi| \geq |z - z_0| - |z - \xi| \geq (1 - \epsilon)|z - z_0|$. Then

$$\begin{aligned} |w(z) - w(z_0)| &\geq |w(\xi) - w(z_0)| - |w(z) - w(\xi)| \\ &\geq V(z_0)^{2\sigma}|\xi - z_0| - c(1 + |z|^2 + |z_0|^2)^\sigma|z - \xi| \\ &\geq (1 + |z|^2 + |z_0|^2)^\sigma|z - z_0|(10^{-2\sigma}(1 - \epsilon) - c\epsilon) . \end{aligned}$$

With a suitable choice of $\varepsilon > 0$, we get the required inequality. □

We remark that in (9.47) we can replace $1 + |z|^2 + |z_0|^2$ with the equivalent quantity $1 + |z - z_0|^2 + |z_0|^2$.

Theorem 9.5 *Let the hypotheses of Theorem 9.4 hold, and let $\omega(R) = cR^{2\sigma}$. There exist $s > 1$ and $\mu > 0$ such that*

$$\left(⨍_{Q_R} |w(Du) - w(z_0)|^{2s} \, dx\right)^{\frac{1}{s}} \leq c⨍_{Q_{2R}} |w(Du) - w(z_0)|^2 \, dx + cR^\mu \left(⨍_{Q_{2R}} (V^p + V_0^p) dx\right)^{1+\frac{\mu}{p}} , \tag{9.48}$$

where $w(z) = zV^{\frac{p-2}{2}}(z)$.

Proof. Let $Q_{R_0} \subset\subset \Omega$, and let $Q_R \subset Q_{R_0}$. Setting

$$G = \omega(R_0) + \vartheta(|u_0|, R_0^2 + |u - u_0|^2 + |u - P|^2)(V^p + V_0^p)$$

we get from (9.39):

$$\int_{Q_{\frac{R}{2}}} (V_0^{p-2}|Du - z_0|^2 + |Du - z_0|^p)dx$$

$$\leq c \int_{Q_R} (R^{-2}V_0^{p-2}|u - P|^2 + R^{-p}|u - P|^p + G)dx\,. \tag{9.49}$$

The polynomial P in (9.49) is completely arbitrary. If $\bar{x}$ is the center of Q_R, and if we choose $P(x) = u_{\bar{x},R} + \langle z_0, x - \bar{x}\rangle$, the function $u - P$ has zero average in Q_R, and hence from the Sobolev–Poincaré inequality:

$$\int_{Q_R} |u - P|^2 \, dx \leq c \left(\int_{Q_R} |Du - z_0|^{2_*} \, dx \right)^{\frac{2}{2_*}},$$

where $s_* = \frac{sn}{s+n}$. Remarking that $2p_* \leq 2_* p$ we conclude that

$$\int_{Q_R} |u - P|^p \, dx \leq c \left(\int_{Q_R} |Du - z_0|^{p_*} \, dx \right)^{\frac{p}{p_*}}$$

$$\leq cR^{p-2} \left(\int_{Q_R} |Du - z_0|^{\frac{p2_*}{2}} \, dx \right)^{\frac{2}{2_*}}$$

and therefore, taking into account Lemma 9.2:

$$\fint_{Q_{\frac{R}{2}}} |w(Du) - w(z_0)|^2 \, dx \leq c \left(\fint_{Q_R} |w(Du) - w(z_0)|^{2_*} \, dx \right)^{\frac{2}{2_*}}$$

$$+ c \fint_{Q_R} G \, dx\,. \tag{9.50}$$

With our choice of P the function G in (9.50) depends on R. In order to apply Theorem 6.6 we must estimate the last integral with the integral of an analogous quantity independent of R. We have

$$\fint_{Q_R} G \, dx = \omega(R_0) + \fint_{Q_R} \vartheta(V^p + V_0^p)dx\,.$$

We can estimate the right-hand side by remarking that since ϑ is an increasing function,[6] we have $\vartheta(a + b) \leq \vartheta(2a) + \vartheta(2b)$ and therefore

$$\vartheta(R^2 + |u - u_0|^2 + |u - P|^2) \leq \vartheta(2R_0^2 + 2|u - u_0|^2) + \vartheta(2|u - P|^2)\,.$$

[6]For the sake of simplicity, we shall write $\vartheta(s)$ instead of $\vartheta(r, s)$.

By Proposition 9.1, we have $V^p \in L^r(Q_{R_0})$ for some $r > 1$, and hence

$$\begin{aligned}
&\fint_{Q_R} \vartheta(2|u-P|^2)(V^p + V_0^p)dx \\
&\le \left(\fint_{Q_R} \vartheta(2|u-P|^2)^{\frac{r}{r-1}}\,dx\right)^{1-\frac{1}{r}} \left(\fint_{Q_R} (V^p + V_0^p)^r\,dx\right)^{\frac{1}{r}} \\
&\le c\left(\fint_{Q_R} \vartheta(2|u-P|^2)dx\right)^{1-\frac{1}{r}} \left(\fint_{Q_R} (V^p + V_0^p)^r\,dx\right)^{\frac{1}{r}},
\end{aligned}$$

where we have taken into account the boundedness of ϑ. Moreover, since ϑ is a concave function, we have

$$\fint_{Q_R} \vartheta(2|u-P|^2)dx \le \vartheta\left(2\fint_{Q_R} |u-P|^2\,dx\right),$$

which with our choice of P gives

$$\begin{aligned}
\fint_{Q_R} \vartheta(2|u-P|^2)dx &\le \vartheta\left(cR^2\fint_{Q_R} |Du-z_0|^2\,dx\right) \\
&\le \vartheta\left(cR_0^2\fint_{Q_{R_0}} |Du-z_0|^2\,dx\right).
\end{aligned}$$

Using now Proposition 9.1, we obtain:

$$\begin{aligned}
&\fint_{Q_R} \vartheta(2|u-P|^2)(V^p + V_0^p)dx \\
&\le \vartheta\left(cR_0^2\fint_{Q_{R_0}} |Du-z_0|^2\,dx\right)^{1-\frac{1}{r}} \fint_{Q_{2R}} (V^p + V_0^p)dx
\end{aligned}$$

and in conclusion:

$$\fint_{Q_R} G\,dx \le \fint_{Q_{2R}} (\alpha(x) + B)(V^p + V_0^p)dx$$

where

$$\alpha(x) = cR_0^{2\sigma} + \vartheta(2R_0^2 + 2|u(x)-u_0|^2)$$

and

$$B = \vartheta\left(cR_0^2\fint_{Q_{R_0}} |Du-z_0|^2\,dx\right)^{1-\frac{1}{r}},$$

are both independent of R.

With the above estimate, (9.50) becomes

$$\fint_{Q_{\frac{R}{2}}} |w(Du) - w(z_0)|^2 \, dx \le c \left(\fint_{Q_R} |w(Du) - w(z_0)|^{2_*} \, dx \right)^{\frac{2}{2_*}}$$

$$+ c \fint_{Q_{2R}} (\alpha(x) + B)(V^p + V_0^p) dx \, . \qquad (9.51)$$

Covering the cube $Q_{R/2}$ with 4^n cubes $Q(x_i, \frac{R}{8})$, we can write R instead of $2R$ in the last integral.

Applying now Theorem 6.6, we conclude that for some $r > 1$ we have

$$\left(\fint_{Q_{\frac{R_0}{2}}} |w(Du) - w(z_0)|^{2r} \, dx \right)^{\frac{1}{r}}$$

$$\le c \fint_{Q_{R_0}} |w(Du) - w(z_0)|^2 \, dx + c \left(\fint_{Q_{R_0}} (\alpha + B)^r (V^p + V_0^p)^r \, dx \right)^{\frac{1}{r}} . \qquad (9.52)$$

Let us write now R instead of R_0, and let us estimate the last integral. We have

$$\left(\fint_{Q_R} B^r (V^p + V_0^p)^r \, dx \right)^{\frac{1}{r}} \le B \fint_{Q_{2R}} (V^p + V_0^p) dx \, .$$

The other term can be estimated as above, choosing $u_0 = u_R$. We obtain

$$\left(\fint_{Q_R} \alpha^r (V^p + V_0^p)^r \, dx \right)^{\frac{1}{r}} \le A \fint_{Q_{2R}} (V^p + V_0^p) dx$$

with

$$A = cR^{2\sigma} + \vartheta \left(cR^2 + cR^2 \fint_{Q_R} |Du - z_0|^2 \, dx \right)^{1 - \frac{1}{r}} .$$

Recalling that $\vartheta(t) \le ct^\delta$, we get in conclusion:

$$A + B \le cR^\mu \left(\fint_{Q_{2R}} (V^p + V_0^p) dx \right)^{1 + \frac{\mu}{p}} ,$$

where we have set

$$\mu = 2\delta \left(1 - \frac{1}{r} \right)$$

and we have assumed $2\sigma \le \mu$, as we are allowed to do.

From the above inequality we get at once (9.48) with $Q_{\frac{R}{2}}$ on the left-hand side. Covering the cube Q_R with a finite number of cubes $Q_{\frac{R}{4}}$, we arrive finally at the required inequality. □

Remark 9.5 The inequality (9.48) is a key tool in the proof of the regularity of the ω-minima of quasi-convex functionals. If F depends only on z, we have $\vartheta = 0$, and the whole proof above can be extremely simplified. We remark that in this case the constants do not depend on z_0 when $p = 2$ and the second derivatives F_{zz} are bounded. □

9.4 The Case $F = F(z)$ $(p = 2)$

In order to clarify the main idea of the proof, before discussing the general case we shall treat the simpler problem of the regularity of the minima (i.e. $\omega = 0$) of functionals dependent only on the gradient, and with growth $p = 2$, under suitable assumptions of uniformity. We shall deal with a function $F(z)$ of class C^2 in $\mathbf{R}^{nN}$, and possessing bounded and *uniformly continuous* second order derivatives; in other words, we shall assume the existence of an increasing, concave and continuous function $\gamma(t)$, with $0 \leq \gamma \leq 1$ and $\gamma(0) = 0$, such that

$$|F_{zz}(z) - F_{zz}(w)| \leq c\gamma(|z - w|^2). \tag{9.53}$$

In this situation, the inequality (9.48) becomes

$$\left(\fint_{Q_R} |Du - z_0|^{2s}\, dx\right)^{\frac{1}{s}} \leq c \fint_{Q_{2R}} |Du - z_0|^2\, dx \tag{9.54}$$

with a constant c independent of z_0.

Let $Q_R = Q(x_0, R)$ be a cube contained in Ω, and let

$$(Du)_R = (Du)_{x_0,R} = \fint_{Q(x_0,R)} Du\, dx\,, \tag{9.55}$$

$$E(x_0, R) = \fint_{Q_R} |Du - (Du)_R|^2\, dx\,. \tag{9.56}$$

We have the following:

Theorem 9.6 *With the above assumptions on the function F, let u be a minimum of the functional*

$$\mathcal{F}(u) = \int_\Omega F(Du) dx\,.$$

Then, for every $\varrho < R$:

$$E(x_0, \varrho) \le c\left\{\left(\frac{\varrho}{R}\right)^2 + \left(\frac{R}{\varrho}\right)^n \gamma(cE(x_0, R))^{\frac{r-1}{r}}\right\} E(x_0, R)\,. \tag{9.57}$$

Proof. Setting $z_0 = (Du)_R$, let

$$g(z) = F(z_0) + \langle F_z(z_0), z - z_0\rangle + \frac{1}{2}\langle F_{zz}(z_0)(z - z_0), z - z_0\rangle\,, \tag{9.58}$$

and let $v(x)$ be the function minimizing the functional[7]

$$\mathcal{G}(u, B_R) = \int_{B_R} g(Dv)dx$$

among all the functions assuming the value u on ∂B_R.

The function v is a solution of the Dirichlet problem

$$\begin{cases} F_{z_i^\alpha z_j^\beta}(z_0)D_iD_jv = 0 & \text{in } B_R \\ v = u & \text{on } \partial B_R \end{cases}$$

and therefore satisfies (9.13) and (9.19). Moreover, for every $\varrho < t = R/\sqrt{n}$ and for every $\pi \in \mathbf{R}^{nN}$ we have

$$\fint_{Q_\varrho} |Dv - (Dv)_\varrho|^2\, dx \le c\left(\frac{\varrho}{t}\right)^2 \fint_{Q_t} |Dv - \pi|^2\, dx\,. \tag{9.59}$$

From (9.59) with $\pi = (Du)_{x_0,R}$, setting $w = u - v$, we deduce

$$E(x_0, \varrho) \le c\left(\frac{\varrho}{t}\right)^2 E(x_0, t) + c\left(\frac{t}{\varrho}\right)^n \fint_{B_R} |Dw|^2\, dx \tag{9.60}$$

so that we only need an estimate of the last term.

By the strict quasi-convexity of $F(z)$ we get

$$\begin{aligned} \fint_{B_R} |Dw|^2\, dx &\le c\fint_{B_R} [F(z_0 + Dw) - F(z_0)]dx \\ &= c\fint_{B_R} [F(z_0 + Dw) - g(z_0 + Dw)]dx \\ &\quad + \frac{1}{2}\fint_{B_R} \langle F_{zz}(z_0)Dw, Dw\rangle dx\,. \end{aligned} \tag{9.61}$$

---

[7] The existence of a minimizing function is guaranteed by the coerciveness and the weak semicontinuity of $\mathcal{G}$ in the class $u + W_0^{1,2}(B_R)$.

Let us begin with the first term on the right-hand side. From the uniform continuity of the second derivatives of F, we obtain

$$\fint_{B_R} [F(z_0 + Dw) - g(z_0 + Dw)]dx \le c \fint_{B_R} \gamma(|Dw|^2)|Dw|^2 \, dx .$$

On the other hand the derivatives Dw belong to L^{2r}, for some $r > 1$, and hence, using the estimate (9.19), we get

$$\begin{aligned} \fint_{B_R} |Dw|^{2r} \, dx &\le c \fint_{B_R} (|Du - z_0|^{2r} + |Dv - z_0|^{2r}) dx \\ &\le c \fint_{B_R} |Du - z_0|^{2r} \, dx \le c \left(\fint_{Q_{2R}} |Du - z_0|^2 \, dx \right)^r \\ &\le c E(x_0, 2R)^r , \end{aligned}$$

where in the last passage we have made use of the inequality (3.36) (Remark 3.3).

We have therefore, taking into account the concavity and boundedness of the function $\gamma(t)$ and making use of Jensen's inequality (5.10):

$$\begin{aligned} \fint_{B_R} \gamma(|Dw|^2)|Dw|^2 \, dx &\le c \left(\fint_{B_R} \gamma(|Dw|^2) dx \right)^{\frac{r-1}{r}} E(x_0, 2R) \\ &\le \gamma \left(\fint_{Q_R} |Dw|^2 \, dx \right)^{\frac{r-1}{r}} E(x_0, 2R) . \end{aligned}$$

The integral of $|Dw|^2$ can be estimated by $cE(x_0, R) \le cE(x_0, 2R)$, and hence in conclusion:

$$\fint_{B_R} [F(z_0 + Dw) - g(z_0 + Dw)]dx \le c\gamma(cE(x_0, 2R))^{\frac{r-1}{r}} E(x_0, 2R) . \tag{9.62}$$

Let us now consider the second term of (9.61). We have

$$\begin{aligned} \frac{1}{2} \fint_{Q_R} \langle F_{zz}(z_0) Dw, Dw \rangle dx &= \fint_{Q_R} [g(Du) - g(Dv)] dx \\ &= \fint_{Q_R} [g(Du) - F(Du)] dx \\ &\quad + \fint_{Q_R} [F(Du) - F(Dv)] dx \\ &\quad + \fint_{Q_R} [F(Dv) - g(Dv)] dx . \end{aligned}$$

The second integral is negative, since u is a minimum of $\mathcal{F}$; the other two can be estimated as above. Writing R instead of $2R$ we get the conclusion (9.57) if $\varrho < R/2$. On the other hand if $\varrho \geq R/2$ we have $E(x_0, \varrho) \leq 2^n E(x_0, R) \leq 2^{n+2}(\varrho/R)^2 E(x_0, R)$, and hence (9.57) is valid for every $\varrho < R$. □

From the preceding theorem we deduce a first partial regularity result.

Theorem 9.7 *Let $u \in W^{1,2}(\Omega, \mathbf{R}^n)$ be a minimum of the functional*

$$\mathcal{F}(u, \Omega) = \int_\Omega F(Du)dx ,$$

in which the function $F(z)$ is strongly quasi-convex, and has second derivatives uniformly continuous and bounded.

There exists an open set $\Omega_0 \subset \Omega$, with $|\Omega - \Omega_0| = 0$, such that $u(x)$ is of class $C^{1,\alpha}(\Omega_0)$ for every $\alpha < 1$.

Proof. The starting point is inequality (9.57), in which we set $\varrho = \tau R$ and we write for simplicity $E(R)$ instead of $E(x_0, R)$:

$$E(\tau R) \leq c\tau^2\{1 + \tau^{-n-2}\gamma(cE(R))^{1-\frac{1}{r}}\}E(R) . \tag{9.63}$$

Let $\alpha < 1$, and let τ be such that $c\tau^{2-2\alpha} \leq \frac{1}{2}$. Let $\epsilon_0 > 0$ be such that $c\tau^{-n-2}\gamma(c\epsilon_0) \leq \frac{1}{2}$, and assume that

$$E(x_0, R) < \epsilon_0 . \tag{9.64}$$

From (9.63) we get

$$E(\tau R) \leq \tau^{2\alpha} E(R)$$

and by iteration

$$E(\tau^k R) \leq \tau^{2k\alpha} E(R) .$$

From the last inequality we deduce at once

$$E(\varrho) \leq c\left(\frac{\varrho}{R}\right)^{2\alpha} E(R) \tag{9.65}$$

for every $\varrho < R$. Assume now that for some $x_0 \in \Omega$ and $R < \frac{\mathrm{dist}(x_0, \partial\Omega)}{2\sqrt{n}}$ (9.64) holds. Since $E(y, R)$ is continuous in y, we shall have $E(y, R) < \epsilon_0$ for every y in a neighborhood I of x_0. For every $y \in I$ we then have:

$$\int_{I_\varrho(y)} |Du - (Du)_{y,\varrho}|^2 \, dx \leq c\left(\frac{\varrho}{R}\right)^{n+2\alpha} \int_{I_R(y)} |Du - (Du)_{y,R}|^2 \, dx .$$

The last inequality implies that u is of class $C^{1,\alpha}$ in I, so that in conclusion the function u has Hölder-continuous derivatives in an open set Ω_0 which contains all the points y such that[8]

$$\liminf_{r\to 0} E(y,r) = 0\,.$$

Since the last relation holds for almost every $y \in \Omega$, we have $|\Omega - \Omega_0| = 0$, and the theorem follows. □

9.5 Partial Regularity

Having seen the method at work in the simple case $F = F(z)$, let us come to the partial regularity of the ω-minima of the functional

$$\mathcal{F}(x,\Omega) = \int_\Omega F(x,u,Du)dx$$

under general assumption for the function F.

To be precise, we shall assume that $F(x,u,z)$ is a strictly quasi-convex function, satisfying the hypotheses (9.31)–(9.34)) of Sec. 9.3. Moreover, we shall assume that the second derivatives F_{zz} of the function F are continuous, or better that there exists a function $\gamma(s,t)$ defined for s, $t \leq 0$, increasing in both its arguments, bounded, continuous and concave in t for every s, with $\gamma(s,0) = 0$, such that

$$|F_{zz}(x,u,z) - F_{zz}(x,u,w)| \leq \gamma(|u| + |z| + |w|, |z-w|)\,. \tag{9.66}$$

We shall consider ω-minima u of the functional $\mathcal{F}$; namely functions $u \in W^{1,P}(\Omega)$ such that for every cube $Q_R \subset \Omega$ and for every $v \in W^{1,p}(Q_R)$, with $v = u$ on ∂Q_R, it holds that

$$\mathcal{F}(u,Q_R) \leq [1 + \omega(R)]\mathcal{F}(v,Q_R)\,,$$

and we shall assume that $\omega(t) \leq ct^{2\sigma}$.

Let now $u_0 = u_{x_0,R}$, $z_0 = (Du)_{x_0,R}$, and let us denote with F^0 the "frozen" function

$$F^0(z) = F(x_0,u_0,z)\,,$$

[8] Actually, Ω_0 coincides with the set of such points, since if Du is continuous in a neighborhood of y, $E(y,r)$ is infinitesimal with r.

and with $g(z)$ the quadratic approximation of F^0:

$$g(z) = F^0(z_0) + \langle F_z^0(z_0), z - z_0\rangle + \frac{1}{2}\langle F_{zz}^0(z_0)(z - z_0), z - z_0\rangle . \tag{9.67}$$

Let $\mathcal{G}(u, \Omega)$ be the functional correspondent to the function g, and let v be the minimum of $\mathcal{G}(v, B_R)$ among all the functions assuming the value u on ∂B_R. The function v satisfies (9.19) and (9.59) with every exponent $p > 2$ (see later, Remark 10.4, and in particular (10.59)). As a consequence for every $\varrho < r = R/\sqrt{n}$ we have

$$\begin{aligned}
&\fint_{Q_\varrho} |w(Dv) - w((Dv)_\varrho)|^2\, dx \\
&\qquad \le cV^{p-2}((Dv)_\varrho) \fint_{Q_\varrho} |Dv - (Dv)_\varrho|^2\, dx \\
&\qquad\quad + c\fint_{Q_\varrho} |Dv - (Dv)_\varrho|^p\, dx \\
&\qquad \le c\left(\frac{\varrho}{r}\right)^2 V^{p-2}((Dv)_\varrho) \fint_{Q_r} |Dv - (Dv)_r|^2\, dx \\
&\qquad\quad + c\left(\frac{\varrho}{r}\right)^p \fint_{Q_r} |Dv - (Dv)_r|^p\, dx .
\end{aligned} \tag{9.68}$$

On the other hand, we have from (9.13)

$$\begin{aligned}
|(Dv)_\varrho|^2 &\le \fint_{Q_\varrho} |Dv|^2\, dx \le c\fint_{Q_r} |Dv|^2\, dx \\
&\le c|(Dv)_r|^2 + c\fint_{B_r} |Dv - (Dv)_r|^2\, dx
\end{aligned}$$

and therefore

$$\begin{aligned}
&\fint_{Q_\varrho} |w(Dv) - w((Dv)_\varrho)|^2\, dx \\
&\quad \le c\left(\frac{\varrho}{r}\right)^2 V^{p-2}((Dv)_r) \fint_{Q_r} |Dv - (Dv)_r|^2\, dx \\
&\quad\quad + c\left(\frac{\varrho}{r}\right)^2 \left(\fint_{Q_r} |Dv - (Dv)_r|^2\, dx\right)^{\frac{p}{2}} + c\left(\frac{\varrho}{r}\right)^p \fint_{Q_r} |Dv - (Dv)_r|^p\, dx .
\end{aligned}$$

From the above formula we conclude easily that

$$\fint_{Q_\varrho} |w(Dv) - w((Dv)_\varrho)|^2\, dx \le c\left(\frac{\varrho}{r}\right)^2 \fint_{Q_r} |w(Dv) - w((Dv)_r)|^2\, dx$$

and therefore

$$\fint_{Q_\varrho} |w(Du) - w((Du)_\varrho|^2\, dx \le c\left(\frac{\varrho}{r}\right)^2 \fint_{Q_r} |w(Du) - w((Du)_r|^2\, dx$$
$$+ \left(\frac{R}{\varrho}\right)^n \fint_{B_R} |w(Du) - w(Dv)|^2\, dx\,. \tag{9.69}$$

We need an estimate of the last integral. For that, we begin with the remark that, setting $\zeta = u - v$, we have

$$\begin{aligned}|w(Du) - w(Dv)|^2 &\le c(V^{p-2}|D\zeta|^2 + |D\zeta|^p)\\ &\le c((V_0^{p-2} + |Du - z_0|^{p-2})|D\zeta|^2 + |D\zeta|^p)\\ &\le \epsilon|Du - z_0|^p + c(\epsilon)(V_0^{p-2}|D\zeta|^2 + |D\zeta|^p)\,.\end{aligned}$$

The quasi-convexity of F now gives

$$\begin{aligned}\fint_{B_R} (V_0^{p-2}|D\zeta|^2 + |D\zeta|^p)dx &\le \fint_{B_R} [F^0(z_0 + D\zeta) - F^0(z_0)]dx\\ &= \fint_{B_R} [F^0(z_0 + D\zeta) - g(z_0 + D\zeta)]dx\\ &\quad + \frac{1}{2}\fint_{B_R} \langle F^0_{zz}(z_0)D\zeta, D\zeta\rangle dx\,,\end{aligned} \tag{9.70}$$

and moreover

$$\begin{aligned}&\frac{1}{2}\fint_{B_R} \langle F^0_{zz}(z_0)D\zeta, D\zeta\rangle dx\\ &= \fint_{B_R} [g(Du) - g(Dv)]dx\\ &= \fint_{B_R} [g(Du) - F^0(Du)]dx + \fint_{B_R} [F^0(Du) - F(x,u,Du)]dx\\ &\quad + \fint_{B_R} [F(x,u,Du) - F(x,v,Dv)]dx + \fint_{B_R} [F(x,v,Dv) - F^0(Dv)]dx\\ &\quad + \fint_{B_R} [F^0(Dv) - g(Dv)]dx\\ &= (I) + (II) + (III) + (IV) + (V)\,.\end{aligned}$$

Recalling that u is an ω-minimum of $\mathcal{F}$, we get

$$(III) \leq R^{-n}\omega(R)\mathcal{F}(v, B_R) \leq cR^{2\sigma} \fint_{Q_R} V^p \, dx\,.$$

For what concerns the remaining terms, (I), (V), as well as remaining term on the right-hand side of (9.70), are of the type already treated in the preceding section, whereas (II) and (IV) are similar to those we have encountered in Sec. 9.3.

For these, we can argue in the same way, getting

$$(II) + (IV) \leq cR^{\mu} \left(\fint_{Q_{2R}} (V_0^p + V^p) dx \right)^{1+\frac{\mu}{p}}\,. \tag{9.71}$$

We come now to the terms (I) and (V), and to the residual term on the right-hand side of (9.70). In comparison with the estimates of the preceding section, here the situation is complicated by the fact that we do not assume that the second derivatives are bounded, let alone uniformly continuous. We shall consider (V) in detail, the estimates for the remaining terms being obtained in the same way.

Let β be a constant that we shall fix later, and let

$$K = K(R, \beta) =: \{x \in B_R : |Dv(x) - z_0| > \beta\}\,.$$

We have, for $|z - z_0| < \beta$,

$$|F^0(z) - g(z)| \leq \gamma(|u_0| + 2|z_0| + \beta, |z - z_0|^2)|z - z_0|^2$$

and hence

$$\begin{aligned}
R^n(V) &= \int_{B_R - K} [F^0(Dv) - g(Dv)] dx + \int_K [F^0(Dv) - g(Dv)] dx \\
&\leq \int_{B_R} \gamma(|u_0| + 2|z_0| + \beta, |Dv - z_0|^2)|Dv - z_0|^2 \, dx \\
&\quad + c \int_K (V(Dv)^p + V_0^p) dx \\
&\leq \int_{B_R} \gamma(|u_0| + 2|z_0| + \beta, |Dv - z_0|^2)|Dv - z_0|^2 \, dx \\
&\quad + c \int_K (V_0^p + |Dv - z_0|^p) dx\,.
\end{aligned}$$

We estimate now the first term on the right-hand side. Neglecting for the sake of simplicity the first argument of the function γ, we get:

$$
\begin{aligned}
&\fint_{B_R} \gamma(|Dv - z_0|^2)|Dv - z_0|^2\, dx \\
&\quad \le c\left(\fint_{B_R} \gamma(|Dv - z_0|^2) dx\right)^{\frac{r-1}{r}} \left(\fint_{B_R} |Dv - z_0|^{2r}\, dx\right)^{\frac{1}{r}} \\
&\quad \le c\gamma\left(\fint_{B_R} |Du - z_0|^2\, dx\right)^{\frac{r-1}{r}} \left(\fint_{B_R} |Du - z_0|^{2r}\, dx\right)^{\frac{1}{r}} \\
&\quad \le c\gamma\left(\fint_{Q_R} |w(Du) - w(z_0)|^2\, dx\right)^{\frac{r-1}{r}} \left(\fint_{Q_{2R}} |w(Du) - w(z_0)|^{2r}\, dx\right)^{\frac{1}{r}} \\
&\quad \le c\gamma\left(\fint_{Q_R} |w(Du) - w(z_0)|^2\, dx\right)^{\frac{r-1}{r}} \\
&\qquad \times \left\{\fint_{Q_{2R}} |w(Du) - w(z_0)|^2\, dx + R^{\mu}\left(\fint_{Q_{2R}} (V^p + V_0^p) dx\right)^{1+\frac{\mu}{p}}\right\}.
\end{aligned}
$$

On the other hand we have

$$
\begin{aligned}
&\int_K |Dv - z_0|^p\, dx \\
&\quad \le \left(\int_{B_R} |Dv - z_0|^{pr}\, dx\right)^{\frac{1}{r}} |K|^{\frac{r-1}{r}} \\
&\quad \le c\left(\int_{B_R} |Du - z_0|^{pr}\, dx\right)^{\frac{1}{r}} |K|^{\frac{r-1}{r}} \\
&\quad \le c\left(\int_{Q_R} |w(Du) - w(z_0)|^{2r}\, dx\right)^{\frac{1}{r}} |K|^{\frac{r-1}{r}} \\
&\quad \le c\left(\frac{|K|}{|Q_R|}\right)^{\frac{r-1}{r}} \left\{\int_{Q_{2R}} |w(Du) - w(z_0|^2\, dx\right. \\
&\qquad \left. + R^{n+\mu}\left(\fint_{Q_{2R}} (V^p + V_0^p) dx\right)^{1+\frac{\mu}{p}}\right\}.
\end{aligned}
$$

We estimate the measure of K. We have

$$
\int_K |Dv - z_0|^2\, dx \ge \beta^2 |K|
$$

and hence

$$\frac{|K|}{|Q_R|} \le c\beta^{-2} \fint_{Q_R} |Dv - z_0|^2 \, dx \le c\beta^{-2} \fint_{Q_R} V_0^{p-2} |Du - z_0|^2 \, dx$$

$$\le c\beta^{-2} \fint_{Q_{2R}} |w(Du) - w(z_0)|^2 \, dx \, .$$

From these inequalities we get at once

$$R^{-n} \int_K (V_0^p + |Du - z_0|^p) dx \le c[\beta^{-2} + (\beta^{-2} E_{2R})^{1-\frac{1}{r}})](E_{2R} + R^\mu P_{2R}) \, ,$$

where

$$E_t = E(x_0, t) = \fint_{Q_t} |w(Du) - w((Du)_t|)^2 \, dx \, , \tag{9.72}$$

$$P_t = P(x_0, t) = \left(\fint_{Q_t} (V^p + V_0^p) dx \right)^{1+\frac{\mu}{p}} \, . \tag{9.73}$$

In conclusion:

$$(V) \le c(\gamma(cE_{2R})^{\frac{r-1}{r}} + E_{2R}^{\frac{r-1}{r}} + \beta^{-2})(E_{2R} + R^\mu P_{2R}) \, .$$

The remaining terms can be estimated exactly in the same way, so that introducing all these inequalities in (9.69), and remarking that

$$P_t \le c(1 + E_t)^{1+\frac{\mu}{p}} \, ,$$

we get in conclusion the following:

Proposition 9.3 *Let u be a ω-minimum of the functional*

$$\mathcal{F}(u, \Omega) = \int_\Omega F(x, u, Du) dx$$

with the function F satisfying the assumptions stated at the beginning. Let $Q(x_0, R)$ be a cube contained in Ω, and let $\varrho < R$. Then,

$$E_\varrho \le A \left\{ \left(\frac{\varrho}{R} \right)^2 + \left(\frac{R}{\varrho} \right)^n [\epsilon + c(\epsilon)\beta^{-2} + c(\epsilon)\chi(E_{2R})] \right\} E(R)$$

$$+ \left(\frac{R}{\varrho} \right)^n H_2 R^\mu \, , \tag{9.74}$$

where A is an increasing function of $|u_0| + |z_0| = |u_{x_0,R}| + |(Du)_{x_0,R}|$, H_2 is increasing in $|u_0| + |z_0| + E(x_0, R)$, whereas

$$\chi(E) = \gamma^{1-\frac{1}{r}}(cE) + E^{1-\frac{1}{r}}$$

depends also on β and is infinitesimal with E, uniformly if $|u_0| + |z_0| + \beta$ remains bounded.

We remark that in (9.74) above we can write R instead of $2R$.

We choose now $\alpha > \mu$ and $M > 0$, and $\varrho = \tau R$, where $\tau = \tau(M) < 1$ is such that

$$A(M)\tau^{2-\alpha} \le \frac{1}{4} . \tag{9.75}$$

Let us take $\epsilon = \epsilon(M)$ in such a way that

$$A\tau^{-n}\epsilon < \frac{1}{4}\tau^{\alpha} \tag{9.76}$$

and $\beta = \beta(M)$ such that

$$A(M)c(\epsilon)\tau^{-n}\beta^{-2} \le \frac{1}{4}\tau^{\alpha} . \tag{9.77}$$

Finally, let $\kappa_0 = \kappa_0(M) < 1$ be such that

$$A(M)\tau^{-n}c(\epsilon)\chi(M + 1 + \beta(M), \kappa_0) \le \frac{1}{4}\tau^{\alpha} . \tag{9.78}$$

From the preceding proposition it follows that if for some r we have $E(x_0, r) < \kappa_0$ and $|u_{x_0,r}| + |(Du)_{x_0,r}| \le M$, then

$$E(x_0, \tau r) \le \tau^{\alpha} E(x_0, r) + H_3 r^{\mu} . \tag{9.79}$$

with $H_3 = H_2\tau^{-n}$.

Lemma 9.3 *For every s and every τ, $0 < \tau < 1$ we have*

$$||u_{x_0,s}| - |u_{x_0,\tau s}|| \le cs\tau^{-\frac{n}{2}} E(x_0, s)^{\frac{1}{2}} ,$$

$$||(Du)_{x_0,s}| - |(Du)_{x_0,\tau s}|| \le \tau^{-\frac{n}{2}} E(x_0, s)^{\frac{1}{2}} .$$

Proof. For every function v we have

$$||v_{x_0,s}| - |v_{x_0,\tau s}|| \le \fint_{Q_{\tau s}} |v - v_{x_0,s}| dx \le \left(\fint_{Q_{\tau s}} |v - v_{x_0,s}|^2 \, dx \right)^{\frac{1}{2}}$$

$$\le \tau^{-\frac{n}{2}} \left(\fint_{Q_s} |v - v_{x_0,s}|^2 \, dx \right)^{\frac{1}{2}} .$$

The second inequality follows at once taking $v = Du$; the first is proved choosing $v = u - \langle (Du)_{x_0,s}, x - x_0 \rangle$ (note that v and u have the same average both over Q_s and over $Q_{\tau s}$) and estimating the last integral by means of Poincaré inequality (3.33). □

Define now

$$E_k = E(x_0, \tau^k R),$$

$$T_k = |u_{x_0,\tau^k R}| + |(Du)_{x_0,\tau^k R}|,$$

and assume that

$$T_0 \le \frac{M}{2} - 1, \quad E_0 \le \frac{\kappa_1}{2} \tag{9.80}$$

in which $\kappa_1 < \kappa_0$ will be chosen later.

We want to prove that

$$E_k \le \tau^{k\alpha} E_0 + H_3(M)(\tau^{k-1}R)^\mu \sum_{j=0}^{k-1} \tau^{j(\alpha-\mu)}. \tag{9.81}$$

Inequality (9.81) holds for $k = 1$ by (9.79). Assuming that it holds for $k \le h$, we shall prove it for $h + 1$.

We remark in the first place that from (9.81) it follows

$$E_k \le \tau^{k\mu}\left(\frac{\kappa_1}{2} + \frac{H_3 R^\mu}{\tau^\mu - \tau^\alpha}\right) \le \kappa_1 \tau^{k\mu}$$

whenever

$$R^\mu \le R_1^\mu =: \frac{\kappa_1(\tau^\mu - \tau^\alpha)}{2H_3 + 1}.$$

Moreover, if we take

$$c\tau^{-\frac{n}{2}} \frac{\kappa_1^{\frac{1}{2}}}{1 - \tau^{\frac{\mu}{2}}} \le \frac{M}{2}. \tag{9.82}$$

we get from the preceding lemma

$$\begin{aligned} T_{h+1} &\le T_0 + c\tau^{-\frac{n}{2}} \sum_{k=0}^{h} E_k^{\frac{1}{2}} \\ &\le \frac{M}{2} - 1 + c\tau^{-\frac{n}{2}} \frac{\kappa_1^{\frac{1}{2}}}{1 - \tau^{\frac{\mu}{2}}} \\ &\le M - 1. \end{aligned}$$

With the above choice of κ_1 and of R_1 we can write (9.79) with $r = \tau^h R \le \tau^h R_1$, getting

$$
\begin{aligned}
E_{h+1} &\le \tau^\alpha E_h + (\tau^h R)^\mu H_3(M) \\
&\le \tau^{\alpha(h+1)} E_0 + H_3(M) \left\{ (\tau^h R)^\mu + \tau^\alpha (\tau^{h-1} R)^\mu \sum_{k=0}^{h-1} \tau^{k(\alpha-\mu)} \right\} \\
&= \tau^{\alpha(h+1)} E_0 + H_3(M)(\tau^h R)^\mu \left(1 + \sum_{k=0}^{h-1} \tau^{(k+1)(\alpha-\mu)} \right)
\end{aligned}
$$

from which we deduce at once the required inequality (9.81) for $h+1$.

From (9.81) it follows immediately

$$
E_h \le \tau^{h\mu} \left(E_0 + \frac{H_3(M) R_1^\mu}{\tau^\mu - \tau^\alpha} \right) .
$$

Finally, if $0 < \varrho < R$, choosing h in such a way that $\tau^{h+1} R \le \varrho < \tau^h R$, we get

$$
\begin{aligned}
E(x_0, \varrho) \le c E_h &\le c\tau^{h\mu} \left(E_0 + \frac{H_3(M) R_1^\mu}{\tau^\mu - \tau^\alpha} \right) \\
&\le c \left(\frac{\varrho}{R} \right)^\mu \left(E(x_0, R) + \frac{H_3(M) R_1^\mu}{\tau^\mu - \tau^\alpha} \right) .
\end{aligned}
$$

We have thus proved the following:

Proposition 9.4 *Let u be a ω-minimum for the functional $\mathcal{F}$, with $\omega(R) = cR^\mu$. For every $M > 0$ there exist $\kappa_1 > 0$ and $R_1 > 0$ such that if for some $x_0 \in \Omega$ and some $R < R_1$*

$$
|u_{x_0,R}| + |(Du)_{x_0,R}| \le \frac{M}{2} - 1 \text{ and } E(x_0, R) < \kappa_1 , \tag{9.83}
$$

then for every $\varrho < R$:

$$
E(x_0, \varrho) \le c \left(\frac{\varrho}{R} \right)^\mu \left(E(x_0, R) + \frac{H_3(M) R_1^\mu}{\tau^\mu - \tau^\alpha} \right) . \tag{9.84}
$$

At this point it is not difficult to prove the required result of partial regularity:

Theorem 9.8 *Let $u(x)$ be a ω-minimum of the functional*

$$
\mathcal{F}(u, \Omega) = \int_\Omega F(x, u, Du) dx
$$

with the function F satisfying the assumptions stated at the beginning of the section, and with $\omega(R) = cR^\mu$.

There exists an open set $\Omega_0 \subset \Omega$, with $|\Omega - \Omega_0| = 0$, in which the function u has Hölder-continuous derivatives.

Proof. If inequalities (9.83) hold for some x_0 and R, they will continue to hold for every x in a neighborhood I of x_0. It follows that inequality (9.84) will be satisfied for every $x \in I$ and $\varrho < R$, and hence u has Hölder-continuous derivatives in I.

On the other hand, for almost every $x \in \Omega$ we have

$$\lim_{R\to 0}(|u_{x,R}| + |(Du)_{x,R}|) < +\infty$$

$$\lim_{R\to 0} E(x,R) = 0$$

and therefore the singular set $\Omega - \Omega_0$ has zero measure. □

9.6 Notes and Comments

The first partial regularity results for variational problems in several dimensions were obtained by DE GIORGI [3] and REIFENBERG [1] in the framework of the theory of minimal surfaces of codimension 1, and were extended by FEDERER [2] and ALMGREN [1] to minimal currents and varifolds in any codimension.

The adaptation of these methods to the regularity theory for nonlinear elliptic systems was first achieved by MORREY [4], followed by GIUSTI and M. MIRANDA[9] [2] and GIUSTI [2]. The term "partial regularity" is a re-interpretation of the title of Morrey's paper, *Partial regularity results* .... Actually, when Morrey's paper appeared, hope was not yet given up to extend to linear elliptic systems De Giorgi's results for second order elliptic equations, described in Chapter 7. The appearance of Example 6.2 of De Giorgi, followed by Example 9.1 of GIUSTI and MIRANDA [1], showed the impossibility of such an extension, and more generally of proving the regularity of the solutions of nonlinear elliptic systems, since the function $u(x) = x|x|^{-1}$ of Example 9.1 is a weak solution of the elliptic system in divergence form

$$\int_\Omega A^{ij}_{\alpha\beta}(u) D_j u^\beta D_i \varphi^\alpha \, dx = 0 \, .$$

[9]Contrary to what EVANS says in [1], these papers, or at least the last two, were inspired by DE GIORGI [3].

About at the same time, similar examples were found by MAZ'YA [1]; later NEČAS [2] and NEČAS, JOHN and STARÀ [1] gave a large number of these examples, that STARÀ, JOHN and MALY [1] extended to the parabolic case.

The use of a theorem by GEHRING [1], or more precisely of its extension due to GIAQUINTA and MODICA [1] (Theorem 6.6), made possible an extension of the partial regularity results first to systems with quadratic right-hand side (GIAQUINTA and GIUSTI [1]), then to minima of quadratic functionals (GIAQUINTA and GIUSTI [6]) and in general to minima of quasi-convex functionals (EVANS [1], GIAQUINTA and MODICA [2]).

For what concerns quadratic functionals (9.8), we remark that the estimate of the dimension of the singular set Σ in Theorems 9.1 and 9.3 can be ameliorated in the case of *separated coefficients*:

$$A^{ij}_{\alpha\beta}(x,u) = G_{\alpha\beta}(x,u)g^{ij}(x)\,. \tag{9.85}$$

For $x_0 \in \Omega$, consider the functional

$$\mathcal{Q}^0(u,Q) = \int_Q G_{\alpha\beta}(x_0,u)g^{ij}(x_0)D_iuD_ju\,dx\,.$$

Since the coefficients of this functional depend only on u, one can assume that the independent variable x is in $\mathbf{R}^k$, $1 \le k \le n$ (and hence the indices i and j vary from 1 to k), and consider the relative problem of minimum in $\mathbf{R}^k$. It can be shown (GIAQUINTA and GIUSTI [6]) that if the minimum problem for $\mathcal{Q}^0$ in $\mathbf{R}^k$ possesses only regular solutions, then the minima of the functional $\mathcal{Q}$ have a singular set whose dimension does not exceed $n-k-1$. Moreover, if $k=n-1$, the minima of $\mathcal{Q}$ can have at most isolated singularities. In particular, since in dimension 2 all the minima are regular, if $n=3$ there are only isolated singularities, while generally speaking the dimension of the singularities does not exceed $n-3$.

Functionals of the above type occur in the theory of harmonic mappings between Riemannian manifolds. Actually these mappings are stationary points of the *energy*:

$$\mathcal{E}(u) = \int |du|^2\,,$$

which in local coordinates takes the form

$$\mathcal{E}(u) = \int g^{ij}(x)G_{\alpha\beta}(u)D_iuD_ju|\det g(x)|dx\,,$$

where g^{ij} and $G_{\alpha\beta}$ are the metric tensors respectively on the domain and on the target manifold.

Of course, the dimension of the singularities will depend on the structure of the target manifold only, since, being a local problem with respect to the domain manifold, the latter can be considered a n-dimensional ball B^n. For instance, whereas in general the minimizing harmonic mappings have singularities of dimension at most $n-3$, those from B^n into the sphere S^n are regular up to dimension 6 (GIAQUINTA and SOUČEK [1], SCHOEN and UHLENBECK [3]), and therefore, by virtue of what we have said above, their singular set has at most dimension $n-7$.

Actually, we can introduce local coordinates on the target manifold only after having proved at least the continuity of the mapping, a property which is essentially the problem in question. As a consequence, the above-mentioned theorem can be applied to harmonic mappings only if the target manifold can be covered with a single chart, as for instance in the case of $\mathbf{R}^n$ with an arbitrary metric, or else of the sphere S^n, if the mapping u omits at least a point.

The result remains nevertheless true in general, as was shown by SCHOEN and UHLENBECK [1].

SCHOEN and UHLENBECK [2] have also proved that in the case of the Dirichlet problem with regular data, the singularities of harmonic maps cannot reach the boundary. A similar result, in the case of separated coefficients (9.85), was proved independently by JOST and MEIER [1].

In the general case of quasi-convex functionals, the theorem of partial regularity is due to EVANS [1] for $p = 2$, and to GIAQUINTA and MODICA [2], for growth $p > 2$.[10] The artifice allowing one to avoid the uniform continuity of the second derivatives was introduced by ACERBI and FUSCO [2], whereas HONG M-C [1] has replaced condition (9.4) with the more stringent inequality

$$F(x, u, z) \geq |z|^p - \lambda .$$

In any case, even condition (9.4) is not simple to verify, and the extension of the partial regularity results to quadratic functionals satisfying for instance the condition of Legendre–Hadamard remains an open problem.

Except for very special situations (some systems with diagonal principal part, functionals "close" to the Dirichlet functional), we do not know any results of global regularity for minima of vector-valued functionals, even under assumptions of convexity in z. A case of some interest is that of

[10] It is not known whether the result holds also for $p < 2$.

functionals depending only on the modulus of the gradient:

$$\mathcal{F}(u) = \int_\Omega G(|Du|^2)dx \tag{9.86}$$

with the function $G(t)$ satisfying suitable conditions of growth and of convexity, as for instance $G(t) = (1+t)^s$, with $s > \frac{1}{2}$.

For the minima of these functionals one can prove global regularity results, in particular the Hölder continuity of the first derivatives, even in the degenerated case, for instance when $G(t) = t^s$, with $s > \frac{1}{2}$.

The first results in the above direction are due to K. UHLENBECK [2], and were obtained with a method that inspired the proof of Theorem 8.7. When the function G depends on x and u, we have results of partial regularity, with an estimate of the dimension of the singular set (see GIAQUINTA and MODICA [3] for $s \geq 1$, ACERBI and FUSCO [4] for $\frac{1}{2} < s < 1$).

Chapter 10

Higher Derivatives

In order to complete our program of gradual proof of the regularity of the minima of regular functionals of the calculus of variations

$$\mathcal{F}(u, \Omega) = \int_\Omega F(x, u, Du)\, dx\,,$$

it remains to discuss the regularity of the derivatives of higher order, beginning from the second. For that, we shall abandon in the first place all the various generalizations of the concept of minimum we have introduced so far (quasi-minima, ω-minima), and devote our attention to the true minima. Secondly, we shall assume from the beginning that the function $F(x, u, p)$ is regular, and satisfies suitable assumptions of convexity (when u is a scalar function) or of quasi-convexity (for vector-valued functions).

The results of the previous chapters tell us that every minimum $u(x)$ of the functional $\mathcal{F}$ is of class $C^{1,\alpha}$ in an open set $\Omega_0 \subset \Omega$ (which in the scalar case coincides with Ω), has second derivatives in L^2, and is a solution of the EULER equation

$$\frac{\partial}{\partial x_i} \frac{\partial F}{\partial z_i^\alpha}(x, u(x), Du(x)) = \frac{\partial F}{\partial u^\alpha}(x, u(x), Du(x))\,, \tag{10.1}$$

which can also be written in the form

$$A_{\alpha\beta}^{ij}(x, u, Du) D_{ij} u^\beta = B_\alpha(x, u, Du)\,, \tag{10.2}$$

where

$$A^{ij}_{\alpha\beta}(x,u,z) = \frac{\partial^2 F}{\partial z^\alpha_i \partial z^\beta_j}(x,u,z)\,,$$

$$B_\alpha(x,u,z) = \frac{\partial F}{\partial u^\alpha} - \frac{\partial^2 F}{\partial z^\alpha_i \partial u^\beta} z^\beta_i - \frac{\partial^2 F}{\partial z^\alpha_i \partial x_i}\,.$$

If the function F has Hölder-continuous second derivatives, the new coefficients $a^{ij}_{\alpha\beta}(x) =: A^{ij}_{\alpha\beta}(x, u(x), Du(x))$ and the right-hand side $g_\alpha(x) = B_\alpha(x, u(x), Du(x))$ are Hölder-continuous functions themselves. Moreover, we shall assume that there holds the LEGENDRE–HADAMARD ellipticity condition

$$a^{ij}_{\alpha\beta}(x)\xi_i\xi_j\eta^\alpha\eta^\beta \geq |\xi|^2|\eta|^2\,. \tag{10.3}$$

In conclusion our problem is reduced to that of the regularity of the solutions $u \in W^{2,2}$ of the elliptic equation

$$a^{ij}_{\alpha\beta}(x)D_{ij}u^\beta = g_\alpha(x)\,.$$

Once the desired regularity results are proved for these equations, it will not be difficult to deduce analogous results for the minima of the functional $\mathcal{F}$. In all these results, an essential role will be played by estimates for the solutions of elliptic linear equations and systems, in particular those with constant coefficients.

10.1 Hilbert Regularity

We shall begin by proving the regularity in the spaces $W^k =: W^{k,2}$ for weak solutions of linear elliptic systems

$$\int_\Omega a^{ij}_{\alpha\beta}(x)D_j u^\beta D_i\varphi^\alpha\,dx = \int_\Omega [f^i_\alpha(x)D_i\varphi^\alpha + g_\alpha(x)\varphi^\alpha]\,dx\,. \tag{10.4}$$

We shall prove internal regularity for weak solutions, and the boundary regularity for the solutions of the DIRICHLET problem, under suitable assumptions for the coefficients $a^{ij}_{\alpha\beta}$, for the functions f and g, and possibly for the boundary datum boundary U. We remark that writing $w = u - U$ instead of u, we can always assume that u is zero on $\partial\Omega$. In this case the functions f^i_α will be replaced by $f^i_\alpha - a^{ij}_{\alpha\beta}D_jU^\beta$, and therefore the assumptions on f will contain those on the boundary datum U.

For what concerns the coefficients $a^{ij}_{\alpha\beta}$, we shall always assume that they verify the LEGENDRE–HADAMARD condition (10.3).

We begin by proving a coerciveness result. Setting

$$\langle AD\varphi, D\psi\rangle = a^{ij}_{\alpha\beta} D_j\varphi^\beta D_i\psi^\alpha$$

we have the following:

Theorem 10.1 (GÅRDING inequality) *Let Ω be a bounded set, and let the coefficients $a^{ij}_{\alpha\beta}$ be uniformly continuous in $\bar{\Omega}$ and satisfy the condition* (10.3). *Let φ be any function in $W^1_0(\Omega)$. Then,*

(i) *If the coefficients are constant, we have*

$$\int_\Omega \langle AD\varphi, D\varphi\rangle\, dx \geq \mu \int_\Omega |D\varphi|^2\, dx\,. \tag{10.5}$$

with $\mu > 0$.

(ii) *There exists a constant R_0, depending only on the modulus of continuity of the coefficients, such that the preceding inequality holds if the diameter of the support of φ is less than R_0.*

(iii) *There exists two constants $\nu > 0$ and H such that*

$$\int_\Omega \langle AD\varphi, D\varphi\rangle\, dx \geq \nu \int_\Omega |D\varphi|^2\, dx - H\int_\Omega |\varphi|^2\, dx\,. \tag{10.6}$$

Proof. Part (i) was already proved in Lemma 5.1. To prove (ii), we remark that if $x_0 \in \operatorname{supp}\varphi$, setting $A_0 = A(x_0)$, we have

$$\langle AD\varphi, D\varphi\rangle = \langle A_0 D\varphi, D\varphi\rangle + \langle (A - A_0)D\varphi, D\varphi\rangle\,,$$

and hence, denoting by $\omega(t)$ the modulus of continuity of the coefficients, and by R the diameter of the support of φ, we get

$$\int \langle AD\varphi, D\varphi\rangle\, dx \geq (\mu - \omega(R)) \int |D\varphi|^2\, dx\,.$$

If now R is so small that $\omega(R) < \frac{\mu}{2}$, we have inequality (10.5) with $\frac{\mu}{2}$ instead of μ.

Let us turn now to (iii). Consider a covering of $\bar{\Omega}$ with balls of diameter smaller than R_0, and let α_h $(h = 1, \ldots, N)$ be the partition of the unit given by Theorem 3.2. Setting $\eta_h = \alpha_h^{\frac{1}{2}}$, we have

$$\sum_{h=1}^N \eta_h^2 = 1 \text{ in } \Omega\,.$$

Moreover,

$$\eta_h^2\langle AD\varphi, D\varphi\rangle = \langle AD(\eta_h\varphi), D(\eta_h\varphi)\rangle - \langle AD(\eta_h\varphi), \varphi D\eta_h\rangle$$
$$- \langle A\varphi D\eta_h, D(\eta_h\varphi)\rangle + \langle A\varphi D\eta_h, \varphi D\eta_h\rangle .$$

Let us now integrate over Ω. The first term on the right-hand side can be estimated from below by means of (ii); the last is positive due to (10.3), and the others can be estimated by

$$\left(\int |\varphi|^2|D\eta_h|^2\,dx\right)^{\frac{1}{2}}\left(\int |D(\eta_h\varphi)|^2\,dx\right)^{\frac{1}{2}} .$$

Summing on h, and using the inequality $ab \le \epsilon a^2 + \epsilon^{-1}b^2$, we get:

$$\int \langle AD\varphi, D\varphi\rangle\,dx \ge \left(\frac{\mu}{2} - \epsilon\right)\sum_h \int |D(\eta_h\varphi)|^2\,dx$$
$$- c(\epsilon)\sum_h \int |\varphi|^2|D\eta_h|^2\,dx .$$

On the other hand we have

$$\|D(\eta_h\varphi)\|_2^2 \ge \frac{1}{2}\|\eta_h D\varphi\|_2^2 - \|\varphi D\eta_h\|_2^2 ,$$

and inserting this inequality in the preceding one, in which we choose $\epsilon = \mu/4$, we arrive easily to (10.6). □

We can now prove a first theorem of internal regularity.

Proposition 10.1 *Assume that the coefficients $a^{ij}_{\alpha\beta}(x)$ are Lipschitz-continuous in Ω, and that $f \in W^1(\Omega)$ and $g \in L^2(\Omega)$. Let $u \in W^1(\Omega)$ be a weak solution of system (10.4). There exists a number $R_0 > 0$, depending only on the modulus of continuity of the coefficients, such that if $R < R_0$ and $Q_{2R} \subset\subset \Omega$, then u belongs to $W^2(Q_R)$ and for every $0 < t \le 1$ we have*

$$\int_{Q_R} |D^2u|^2\,dx \le c\int_{Q_{(1+t)R}} \left(\frac{|Du|^2}{(tR)^2} + |Df|^2 + |g|^2\right)dx . \qquad (10.7)$$

Proof. Let $\zeta \in C_0^\infty(Q_{(1+t)R})$, with $0 \le \zeta \le 1$, $\zeta = 1$ in Q_R and $|D\zeta| \le \frac{2}{tR}$. Choosing $\varphi = \Delta_{-h}(\zeta^2\Delta_h u)$ in (10.4), we have

$$\int a^{ij}_{\alpha\beta}D_ju^\beta D_i\varphi^\alpha\,dx = \int \Delta_h(a^{ij}_{\alpha\beta}D_ju^\beta)\zeta(D_i(\zeta\Delta_h u^\alpha) + \Delta_h u^\alpha D_i\zeta)\,dx .$$

We remark now that

$$\Delta_h(a^{ij}_{\alpha\beta}D_ju^\beta) = a^{ij}_{\alpha\beta}(x + he_s)D_j(\Delta_h u^\beta) + D_ju^\beta\Delta_h A^{ij}_{\alpha\beta} ,$$

and as usual

$$\zeta D_j(\Delta_h u^\beta) = D_j(\zeta\Delta_h u^\beta) - \Delta_h u^\beta D_j\zeta\,.$$

Inserting the above relations in the preceding one, isolating at the left-hand side the principal term and estimating the others, and choosing R small enough, we get

$$\begin{aligned}\int a_{\alpha\beta}^{ij}(x) D_j u^\beta D_i\varphi^\alpha\,dx \geq \frac{\mu}{2}\int |D(\zeta\Delta_h u)|^2\,dx\\ - c\int (\zeta^2|Du|^2 + |D\zeta|^2|\Delta_h u|^2)\,dx \qquad (10.8)\end{aligned}$$

We now estimate the terms on the right-hand side of (10.4):

$$\begin{aligned}\int f_\alpha^i D_i\varphi^\alpha\,dx &= \int \Delta_h f_\alpha^i D_i(\zeta^2\Delta_h u^\alpha)\,dx\\ &\leq \epsilon\int |D(\zeta\Delta_h u)|^2\,dx + c(\epsilon)\int \zeta^2|\Delta_h f|^2\,dx\\ &\quad + \int |D\zeta|^2|\Delta_h u|^2\,dx\,.\end{aligned}$$

On the other hand we get from Lemma 8.1:

$$\begin{aligned}\int g_\alpha\Delta_{-h}(\zeta^2\Delta_h u)\,dx &\leq \epsilon\int |D(\zeta^2\Delta_h u)|^2\,dx + c(\epsilon)\int_{Q_{2R}} |g|^2\,dx\\ &\leq \epsilon\int |D(\zeta\Delta_h u)|^2\,dx + \int |D\zeta|^2|\Delta_h u|^2\,dx\\ &\quad + c(\epsilon)\int_{Q_{2R}} |g|^2\,dx\,.\end{aligned}$$

and the conclusion follows at once. □

At this point it is not difficult to prove, with the assumptions of the preceding proposition, that the solution u belongs to $W^2_{\mathrm{loc}}(\Omega)$. It is sufficient to cover $\Sigma \subset\subset \Omega$ with a finite number of cubes Q_R of side sufficiently small, to write the estimate (10.7) for each of them, and to sum. In this way we get the following:

Theorem 10.2 *Let the coefficients $a_{\alpha\beta}^{ij}(x)$ be Lipschitz-continuous in Ω, and let $f \in W^1(\Omega)$ and $g \in L^2(\Omega)$. If $u \in W^1(\Omega)$ is a weak solution of*

(10.4), *then u belongs to $W^2_{\rm loc}(\Omega)$, and for every $\Sigma \subset\subset \Omega$ we have*

$$\int_\Sigma |D^2u|^2\,dx \le c(\Sigma)\int_\Omega (|Du|^2 + |Df|^2 + |g|^2)\,dx\,. \tag{10.9}$$

Integrating by parts in (10.4), we conclude moreover that u satisfies the equation

$$a^{ij}_{\alpha\beta}D_iD_ju^\beta = D_if^i_\alpha - g_\alpha - D_ju^\beta D_ia^{\alpha\beta}_{ij}\,, \quad (\alpha = 1,\dots,N) \tag{10.10}$$

almost everywhere in Ω.

The preceding result can be extended up to the boundary, for solutions of the DIRICHLET problem with zero boundary data. In this case, we begin by flattening the boundary of Ω by means of a diffeomorphism γ, which is described in detail in Sec. 8.5. Setting $v = u\circ\gamma$, the function v will be a weak solution of the equation

$$\int A^{ij}_{\alpha\beta}(x)D_jv^\beta D_i\psi^\alpha\,dx = \int [F^i_\alpha(x)D_i\psi^\alpha + G_\alpha(x)\psi^\alpha]\,dx$$

in the half-ball B^+, with

$$A^{ij}_{\alpha\beta}(x) = |J(x)|H^i_hH^j_ka^{hk}_{\alpha\beta}(\gamma(x))\,,$$

$$F^i_\alpha(x) = |J(x)|H^i_hf^h_\alpha(\gamma(x))\,,$$

$$G_\alpha(x) = |J(x)|g_\alpha(\gamma(x))\,.$$

It is easily seen that, under suitable assumptions for the function γ, the new coefficients and the new right-hand side have the same properties of the original coefficients and right-hand side. In particular, if γ is of class C^2, the coefficients $A^{ij}_{\alpha\beta}$ are Lipschitz-continuous, and the functions F^i_α belong to W^1, whereas the functions G_α are obviously in L^2. We can therefore consider Eq. (10.4) only in a half-ball B^+, with the solution u taking zero values on the flat part P of ∂B^+.

We can then proceed combining the methods of Sec. 8.4 with the proof of Proposition 10.1. If $s \ne n$, the function $\Delta_{h,s}u$ has zero trace on P, and hence we can take $\varphi = \Delta_{-h}(\zeta^2\Delta_hu)$ in (10.4), with a test function ζ with support in the cube Q_{2R} centered in the origin, but possibly different from zero in P. Arguing as in Proposition 10.1, we get easily the estimate

$$\int_{Q^+_R} |DD'u|^2\,dx \le c\int_{Q^+_{(1+t)R}} \left(\frac{|Du|^2}{(tR)^2} + |Df|^2 + |g|^2\right)dx \tag{10.11}$$

for every R small enough (depending only on the modulus of continuity of the coefficients), in which $D'u$ indicates any derivative $D_s u$, with $s \neq n$.

Finally, the estimate for the derivatives $D_{nn}u^\beta$ can be derived from Eq. (10.10), that can be written in the form

$$a^{nn}_{\alpha\beta} D_n D_n u^\beta = D_i f^i_\alpha - g_\alpha - D_j u^\beta D_i a^{\alpha\beta}_{ij} - \Sigma' a^{ij}_{\alpha\beta} D_i D_j u^\beta$$

with the usual meaning of the apex in the sum on the right-hand side.

The matrix $K_{\alpha\beta} =: a^{nn}_{\alpha\beta}$ can be inverted thanks to (10.3), and its inverse matrix is bounded. Taking (10.11) into account, we eventually get the estimate

$$\int_{Q^+_R} |D^2 u|^2 \, dx \leq c \int_{Q^+_{(1+t)R}} \left(\frac{|Du|^2}{(tR)^2} + |Df|^2 + |g|^2 \right) dx \,. \tag{10.12}$$

Finally, covering $\partial\Omega$ with a finite number of neighborhoods, and repeating for each of them the preceding argument, we arrive at the following:

Theorem 10.3 *Let u be a weak solution of the equation* (10.4) *in Ω, taking the value U on $\partial\Omega$. Assume that the coefficients $a^{ij}_{\alpha\beta}(x)$ are Lipschitz-continuous in $\bar{\Omega}$, and that $U \in W^2(\Omega)$, $f \in W^1(\Omega)$ and $g \in L^2(\Omega)$. Then, $u \in W^2(\Omega)$, and we have*

$$\int_\Omega |D^2 u|^2 \, dx \leq c \int_\Omega (|Du|^2 + |D^2 U|^2 + |Df|^2 + |g|^2) \, dx \,. \tag{10.13}$$

Note that, by virtue of Theorem 3.18, in the above estimate can replace the term $|Du|^2$ with $|u|^2$

We can now prove a general theorem of internal regularity.

Theorem 10.4 *Assume that the coefficients $a^{ij}_{\alpha\beta}(x)$ belong to $W^{k,\infty}(\Omega)$, that $f \in W^k(\Omega)$ and $g \in W^{k-1}(\Omega)$. Let $u \in W^1(\Omega)$ be a weak solution of* (10.4). *Then, u belongs to $W^{k+1}_{\text{loc}}(\Omega)$, and for every $\Sigma \subset\subset \Omega$ we have*

$$\|u\|_{k+1,\Sigma} \leq c(k,\Sigma)(\|Du\|_\Omega + \|f\|_{k,\Omega} + \|g\|_{k-1,\Omega}) \,. \tag{10.14}$$

Proof. The theorem has already been proved for $k = 1$. Let us assume that it holds for $k \geq 1$, and let us prove it for $k+1$. For that purpose, assume that $a^{ij}_{\alpha\beta}(x) \in W^{k+1,\infty}$, $f \in W^{k+1}$ and $g \in W^k$.

By assumption, the solution u belongs to W^{k+1}, and any of its derivatives $D_s u$ is a weak solution of the equation

$$\begin{aligned} D_i(a^{ij}_{\alpha\beta} D_j D_s u^\beta) &= D_i D_s f^i_\alpha - D_s g_\alpha - D_i(D_s a^{ij}_{\alpha\beta} D_j u^\beta) \\ &=: D_i F^i_\alpha \,, \end{aligned}$$

where

$$F^i_\alpha = D_s f^i_\alpha + \delta^i_s g_\alpha - D_s a^{ij}_{\alpha\beta} D_j u^\beta \,. \tag{10.15}$$

Let now Λ be an open set such that $\Sigma \subset\subset \Lambda \subset\subset \Omega$. By virtue of our assumptions, the functions F^i_α belong to $W^k(\Lambda)$; whence $D_s u \in W^{k+1}(\Sigma)$, and $u \in W^{k+2}(\Sigma)$. Finally, the inequality (10.14) for $k+2$ follows from the same estimates for $h < k+2$, taking into account that

$$\|F\|_k \leq \|f\|_{k+1} + \|g\|_k + c\|u\|_{k+1} \,. \qquad \square$$

If we want to extend the above result up to the boundary, we must consider again a solution u to (10.4) in the half-ball B^+, with $u = 0$ on P. We have in this case:

Theorem 10.5 *Let $u \in W^1(B^+)$ be solution of the equation (10.4) in B^+, with $u = 0$ on P. Assume that the coefficients $a^{ij}_{\alpha\beta}(x)$ belong to $W^{k,\infty}(\overline{B^+})$, and that $f \in W^k(B^+)$ and $g \in W^{k-1}(B^+)$. Then, $u \in W^{k+1}(B^+_r)$ for each $r < 1$, and for every cube $Q^+_R \subset\subset B^+$ of side sufficiently small, and every $\varrho < R$ we have the estimate*

$$\|u\|_{k+1,Q_\varrho} \leq c(k,\varrho,R)(\|u\|_{1,Q_R} + \|f\|_{k,Q_R} + \|g\|_{k-1,Q_R}) \,. \tag{10.16}$$

Proof. For $k = 1$, inequality (10.16) is nothing but (10.12). Assuming that it holds for $k \geq 1$, let us prove it for $k+1$. As above, the function $D_s u$ is a solution of the equation

$$\int a^{ij}_{\alpha\beta} D_j(D_s u^\beta) D_i \varphi^\alpha \, dx = \int F^i_\alpha D_i \varphi^\alpha \, dx$$

with F given in (10.15). If $s \neq n$, $D_s u$ is zero on P, and hence

$$\begin{aligned}
\|D_s u\|_{k+1,Q_\varrho} &\leq c(k,\varrho,R)(\|D_s u\|_{1,Q_{\frac{R+\varrho}{2}}} + \|F\|_{k,Q_{\frac{R+\varrho}{2}}}) \\
&\leq c(k,\varrho,R)(\|u\|_{k+1,Q_{\frac{R+\varrho}{2}}} + \|f\|_{k+1,Q_R} + \|g\|_{k,Q_R}) \\
&\leq c(k+1,\varrho,R)(\|u\|_{1,Q_R} + \|f\|_{k+1,Q_R} + \|g\|_{k,Q_R}) \,.
\end{aligned}$$

In this way we have estimated all the derivatives $D^\sigma u$, $|\sigma| = k+2$, except $D^{k+2}_n u$. The estimate for that derivative can be obtained as above from Eq. (10.10). $\square$

Coming back to our original DIRICHLET problem for the Eq. (10.4) with data U on the boundary, we have the following:

Theorem 10.6 *Let $u \in W^1(\Omega)$ be a solution of the equation (10.4) with $u = U$ on $\partial\Omega$. Assume that the coefficients are of class $W^{k,\infty}(\bar{\Omega})$, that the* *boundary of Ω is of class C^{k+1}, and that $U \in W^{k+1}(\Omega)$, $f \in W^k(\Omega)$ and $g \in W^{k-1}(\Omega)$. Then, $u \in W^{k+1}(\Omega)$ and we have the estimate*

$$\|u\|_{k+1} \leq c(\|u\|_1 + \|f\|_k + \|g\|_{k-1} + \|U\|_{k+1}) \tag{10.17}$$

with a constant c depending on Ω and on the coefficients.

10.2 Constant Coefficients

We continue our study by establishing some estimates for the solutions in the upper half-space $\mathbf{R}^n_+$ of the Eq. (10.4) with constant coefficients $a^{ij}_{\alpha\beta}$.

The estimates in question have the same form for the cubes $Q_R \subset \mathbf{R}^n_+$ and for their intersections $Q^+_R = Q_R \cap \mathbf{R}^n_+$, provided in the last case that $u = 0$ on the part of the boundary of Q_R lying on the hyperplane $P = \partial\mathbf{R}^n_+ = \{x \in \mathbf{R}^n : x_n = 0\}$. In the following, we shall treat only the latter situation; the proof in the other case is simpler, and we leave it to the reader, who can carry it through simply eliminating the unnecessary complications.

Let us begin from a solution v of the homogeneous equation:

$$a^{ij}_{\alpha\beta} D_{ij} v^\beta = 0\,, \tag{10.18}$$

with $v = 0$ on P.

Since the coefficients $a^{ij}_{\alpha\beta}$ are constant, the equation above can be written in the form

$$D_i a^{ij}_{\alpha\beta} D_j v^\beta = 0\,,$$

which after multiplication by an arbitrary test function $\varphi \in C^\infty_0(Q^+, \mathbf{R}^N)$ and integration by parts becomes

$$\int a^{ij}_{\alpha\beta} D_j v^\beta D_i \varphi^\alpha \, dx = 0\,. \tag{10.19}$$

We have already proved in the preceding section that $v \in W^k_{\mathrm{loc}}(Q^+)$[1] for every integer k, and hence by the SOBOLEV theorem (Theorem 3.12), it belongs to C^∞. Moreover, we have the CACCIOPPOLI estimate (6.68), with $U = g_1 = 0$:

$$\int_{Q^+_t} |Dv|^2 \, dx \leq \frac{c}{(s-t)^2} \int_{Q^+_s} |v|^2 \, dx\,. \tag{10.20}$$

---

[1]Since the coefficients are constant, it is not necessary to assume that the radii are small.

Since every derivative of v is itself a solution of the same Eq. (10.19), we easily have

$$\sum_{|\sigma|=k} \int_{Q_t^+} |D^\sigma v|^2 \, dx \le \frac{c(k)}{(s-t)^{2k}} \int_{Q_s^+} |v|^2 \, dx \,. \tag{10.21}$$

Taking now $s = 1$, $t = \frac{1}{2}$ and $k > 2 + \frac{n}{2}$, we can apply the SOBOLEV Theorem 3.12 and we can conclude that $v \in C^2(Q_{1/2}^+)$ and

$$\sup_{Q_{\frac{1}{2}}^+}(|Dv|^2 + |D^2 v|^2) \le c \int_{Q^+} |v|^2 \, dx \le c \int_{Q^+} |D_n v|^2 \, dx. \tag{10.22}$$

In particular we have for $t \le \frac{1}{2}$:

$$\int_{Q_t^+} |Dv|^2 \, dx \le (2t)^n \sup_{Q_{\frac{1}{2}}^+} |Dv|^2 \le ct^n \int_{Q^+} |D_n v|^2 \, dx \tag{10.23}$$

and moreover

$$\int_{Q_t^+} |Dv - (Dv)_t|^2 \, dx \le ct^2 \int_{Q_t^+} |D^2 v|^2 \, dx$$

$$\le ct^{n+2} \int_{Q^+} |D_n v|^2 \, dx \,. \tag{10.24}$$

If in the last inequality we replace v with $v - x_n \xi$, $\xi \in \mathbf{R}^N$ (which is a solution of (10.18) and is zero for $x_n = 0$), we get

$$\int_{Q_t^+} |Dv - (Dv)_t|^2 \, dx \le ct^{n+2} \int_{Q^+} |D_n v - \xi|^2 \, dx \,. \tag{10.25}$$

Writing $D_s v$, $s \ne n$, instead of v, we find an estimate for the derivatives $DD'v$. The remaining derivative $D_{nn} v$ can be estimated using the Eq. (10.18), so that we arrive to the estimate

$$\int_{Q_t^+} |D^2 v|^2 \, dx \le ct^n \int_{Q^+} |D^2 v|^2 \, dx \,. \tag{10.26}$$

In a similar way, we can replace v with $D_s v^\sigma - x_n \lambda_{sn}^\sigma$ in (10.24), and obtain

$$\int_{Q_t^+} |DD_s v - (DD_s v)_t|^2 \, dx \le ct^{n+2} \int_{Q^+} |D_{sn} v - \lambda_{sn}|^2 \, dx \,.$$

Once again, the derivative $D_{nn}v$ can be extracted from the equation, and hence

$$\int_{Q_t^+} |D^2v - (D^2v)_t|^2\,dx \le ct^{n+2}\int_{Q^+} |D^2v - \lambda|^2\,dx\,. \tag{10.27}$$

The estimates (10.26) and (10.27) continue to hold, possibly with a different constant, for every $t < 1$. The same estimates hold for a generic cube Q_R, and can be proved simply by reducing to the unit cube Q by means of a homotethy. We have therefore proved the following:

Theorem 10.7 *Let v be a solution of the homogeneous Eq.* (10.18) *in Q_R^+, with $v = 0$ on P. Then, for every $\varrho < R$ we have:*

$$\int_{Q_\varrho^+} |Dv|^2\,dx \le c\left(\frac{\varrho}{R}\right)^n \int_{Q_R^+} |Dv|^2\,dx\,, \tag{10.28}$$

$$\int_{Q_\varrho^+} |Dv - (Dv)_\varrho|^2\,dx \le c\left(\frac{\varrho}{R}\right)^{n+2} \int_{Q^+} |Dv - \xi|^2\,dx\,, \tag{10.29}$$

$$\int_{Q_\varrho^+} |D^2v|^2\,dx \le c\left(\frac{\varrho}{R}\right)^n \int_{Q_R^+} |D^2v|^2\,dx\,, \tag{10.30}$$

and

$$\int_{Q_\varrho^+} |D^2v - (D^2v)_\varrho|^2\,dx \le c\left(\frac{\varrho}{R}\right)^{n+2} \int_{Q_R^+} |D^2v - \lambda|^2\,dx \tag{10.31}$$

for every $\xi = \{\xi_i^\alpha\} \in \mathbf{R}^{nN}$ and every $\lambda = \{\lambda_{ij}^\sigma\} \in \mathbf{R}^{n^2N}$.

The same estimates hold if $Q_R \subset \Omega$.

Remark 10.1 The preceding estimates remain valid if we substitute everywhere the exponent 2 with $p \ge 1$. Actually, if we follow the proof of the preceding theorem, we will note that it is sufficient to prove (10.22) with 2 replaced by p. Now this is trivial if $p > 2$; for, calling U^2 the left-hand side of (10.22), we have

$$U^p \le c\left(\int_{Q^+} |D_nv|^2\,dx\right)^{\frac{p}{2}} \le c\int_{Q^+} |D_nv|^p\,dx\,.$$

If instead $1 \le p < 2$, we must remark that from (10.21) with $2k > n+4$ we can deduce that

$$\sup_{Q_t^+}(|Dv|^2 + |D^2v|^2) \le \frac{c}{(s-t)^{2k}}\int_{Q_s^+} |D_nv|^2\,dx$$

and hence, calling U_t^2 the quantity on the left-hand side,

$$U_t^2 \le \frac{cU_s^{2-p}}{(s-t)^{2k}} \int_{Q_s^+} |D_n v|^p \, dx$$

$$\le \frac{1}{2} U_s^2 + \left(\frac{c}{(s-t)^{2k}} \int_{Q_s^+} |D_n v|^p \, dx \right)^{\frac{2}{p}} .$$

At this point, a simple application of Lemma 6.1 gives the required estimate. □

Let us consider now the non-homogeneous Eq. (10.4), that we write in the form:

$$D_i a_{\alpha\beta}^{ij} D_j w^\beta = D_i f_\alpha^i + g_\alpha \tag{10.32}$$

with constant coefficients $a_{\alpha\beta}^{ij}$ verifying the condition of LEGENDRE–HADAMARD (10.3).

We can write w as sum of two functions: $w = v + z$, where v is a solution of the DIRICHLET problem

$$\begin{cases} a_{\alpha\beta}^{ij} D_{ij} v^\beta = 0 & \text{in } Q_R^+ , \\ v = w & \text{on } \partial Q_R^+ , \end{cases}$$

and $z = w - v$ is a solution of the DIRICHLET problem relative to the non-homogeneous equation with zero boundary value on ∂Q_R^+.

For what concerns z we have the following:

Proposition 10.2 *Let z be a weak solution of the* DIRICHLET *problem*

$$\begin{cases} D_i a_{\alpha\beta}^{ij} D_j z^\beta = D_i f_\alpha^i - g_\alpha & \text{in } Q_R^+ , \\ z = 0 & \text{on } \partial Q_R^+ . \end{cases}$$

For every $\pi = \{\pi_\alpha^i\}$ we have

$$\int_{Q_R^+} |Dz|^2 \, dx \le c \int_{Q_R^+} (|f - \pi|^2 + R^2 |g|^2) \, dx . \tag{10.33}$$

Proof. We have

$$\int_{Q_R^+} |Dz|^2 \, dx \le \int_{Q_R^+} a_{\alpha\beta}^{ij} D_i z^\alpha D_j z^\beta \, dx$$

$$= \int_{Q_R^+} (f_\alpha^i - \pi_\alpha^i) D_i z^\alpha \, dx + \int_{Q_R^+} g_\alpha z^\alpha \, dx$$

$$\leq c\left(\int_{Q_R^+}|f-\pi|^2\,dx\right)^{\frac{1}{2}}\left(\int_{Q_R^+}|Dz|^2\,dx\right)^{\frac{1}{2}}$$

$$+c\left(\int_{Q_R^+}|g|^2\,dx\right)^{\frac{1}{2}}\left(\int_{Q_R^+}|z|^2\,dx\right)^{\frac{1}{2}}$$

and (10.33) follows immediately by estimating the last integral by means of the inequality of POINCARÉ. □

It is now easy to get bounds for the second derivatives. We have the following:

Proposition 10.3 *Let z be a solution of the* DIRICHLET *problem*

$$\begin{cases} a_{\alpha\beta}^{ij}D_{ij}z^\beta = g_\alpha & \text{in } Q_R^+\,,\\ z=0 & \text{on } \partial Q_R^+\,. \end{cases}$$

Then,

$$\int_{Q_{R/2}^+}|D^2z|^2\,dx \leq c\int_{Q_R^+}|g|^2\,dx\,. \tag{10.34}$$

Proof. From (10.12) we get

$$\int_{Q_{R/2}^+}|D^2z|^2\,dx \leq c\int_{Q_R^+}(|g|^2+R^{-2}|Dz|^2)\,dx\,. \tag{10.35}$$

The last term can be estimated by means of (10.33) with $f=\pi=0$. We have

$$\int_{Q_R^+}|Dz|^2\,dx \leq cR^2\int_{Q_R^+}|g|^2\,dx\,,$$

from which (10.34) follows at once. □

We can now prove the following theorem:

Theorem 10.8 *Let $w(x)$ be a weak solution in Q_R^+ of the elliptic equation*

$$D_i a_{\alpha\beta}^{ij}D_j w^\beta = D_i f_\alpha^i - g_\alpha$$

with constant coefficients, and assume that $w=0$ on P. Then, for every $\varrho<R$ we have

$$\int_{Q_\varrho^+}|Dw|^2\,dx \leq c\left\{\left(\frac{\varrho}{R}\right)^n\int_{Q_R^+}|Dw|^2\,dx+\int_{Q_R^+}(|f|^2+R^2|g|^2)\,dx\right\}, \tag{10.36}$$

$$\int_{Q_\varrho^+} |Dw - (Dw)_\varrho|^2\,dx \le c\Big\{\Big(\frac{\varrho}{R}\Big)^{n+2}\int_{Q_R^+} |Dw-\xi|^2\,dx$$
$$+\int_{Q_R^+}(|f-\pi|^2+R^2|g|^2)\,dx\Big\}. \qquad (10.37)$$

Proof. Setting as above $w = v + z$, we have

$$\int_{Q_\varrho^+} |Dw - (Dv)_\varrho|^2\,dx \le c\int_{Q_\varrho^+}(|Dv-(Dv)_\varrho|^2+|Dz|^2)\,dx\,,$$

and therefore, using (10.29):

$$\int_{Q_\varrho^+} |Dw - (Dw)_\varrho|^2\,dx \le \int_{Q_\varrho^+} |Dw - (Dv)_\varrho|^2\,dx$$
$$\le c\left\{\Big(\frac{\varrho}{R}\Big)^{n+2}\int_{Q_R^+}|Dv-\xi|^2\,dx + \int_{Q_R^+}|Dz|^2\,dx\right\}$$
$$\le c\left\{\Big(\frac{\varrho}{R}\Big)^{n+2}\int_{Q_R^+}|Dw-\xi|^2\,dx + \int_{Q_R^+}|Dz|^2\,dx\right\}.$$

The first inequality (10.37) follows at once from (10.33). The proof of (10.36) is similar, and we leave it to the reader. □

We can prove now the basic estimates for the second derivatives.

Theorem 10.9 *Let $w(x)$ be a solution in Q_R^+ of the elliptic equation*

$$a^{ij}_{\alpha\beta}D_{ij}w^\beta = g_\alpha$$

with constant coefficients, and let $w = 0$ on P. Then, for every $\varrho < R$ we have

$$\int_{Q_\varrho^+}|D^2w|^2\,dx \le c\left\{\Big(\frac{\varrho}{R}\Big)^{n}\int_{Q_R^+}|D^2w|^2\,dx + \int_{Q_R^+}|g|^2\,dx\right\}, \qquad (10.38)$$

$$\int_{Q_\varrho^+}|D^2w-(D^2w)_\varrho|^2\,dx \le c\Big\{\Big(\frac{\varrho}{R}\Big)^{n+2}\int_{Q_R^+}|D^2w-\lambda|^2\,dx$$
$$+\int_{Q_R^+}|g-g_R|^2\,dx\Big\}. \qquad (10.39)$$

Proof. We shall prove (10.38) first. Writing as above $w = v + z$ we have

$$\int_{Q_\varrho^+}|D^2w|^2\,dx \le c\int_{Q_\varrho^+}(|D^2v|^2+|D^2z|^2)\,dx\,,$$

and hence, using (10.30) and assuming that $\varrho < R/2$:

$$\int_{Q_\varrho^+} |D^2 w|^2\, dx \le c\left\{ \left(\frac{\varrho}{R}\right)^n \int_{Q_{R/2}^+} |D^2 w|^2\, dx + \int_{Q_{R/2}^+} |D^2 z|^2\, dx \right\}.$$

The conclusion follows at once from Proposition 10.3, possibly changing the constant c if $\varrho \ge R/2$.

Let us consider now the second inequality. Denoting by A_{nn} the $N \times N$ matrix with components $a_{\alpha\beta}^{nn}$, and setting

$$y = w - \frac{1}{2} x_n^2 A_{nn}^{-1} g_R\,,$$

we get

$$a_{\alpha\beta}^{ij} D_{ij} y^\beta = g_\alpha - (g_\alpha)_R\,.$$

As above, we can split $y = v + z$, getting

$$\begin{aligned}\int_{Q_\varrho^+} |D^2 w - (D^2 w)_\varrho|^2\, dx &= \int_{Q_\varrho^+} |D^2 y - (D^2 y)_\varrho|^2\, dx \\ &\le 2\int_{Q_\varrho^+} |D^2 v - (D^2 v)_\varrho|^2\, dx + 2\int_{Q_\varrho^+} |D^2 z|^2\, dx\,.\end{aligned}$$

We can estimate v by means of inequality (10.31):

$$\begin{aligned}&\int_{Q_\varrho^+} |D^2 w - (D^2 w)_\varrho|^2\, dx \\ &\quad \le c\left\{ \left(\frac{\varrho}{R}\right)^{n+2} \int_{Q_{R/2}^+} |D^2 v - \lambda|^2\, dx + \int_{Q_{R/2}^+} |D^2 z|^2\, dx \right\} \\ &\quad \le c\left\{ \left(\frac{\varrho}{R}\right)^{n+2} \int_{Q_{R/2}^+} |D^2 w - \lambda|^2\, dx + \int_{Q_{R/2}^+} |D^2 z|^2\, dx \right\}.\end{aligned}$$

On the other hand, from Proposition 10.3 we deduce

$$\int_{Q_{R/2}^+} |D^2 z|^2\, dx \le c \int_{Q_R} |g - g_R|^2\, dx$$

and the conclusion follows. □

10.3 Continuous Coefficients

Once the proper estimates for the solutions of equations with constant coefficients have been obtained, we can deal with the regularity of solutions to equations with continuous coefficients. We shall consider both weak solutions of equations in divergence form:

$$\int_\Omega a^{ij}_{\alpha\beta}(x) D_j u^\beta D_i\varphi^\alpha \, dx = \int_\Omega [f^i_\alpha D_i\varphi^\alpha + g_\alpha\varphi^\alpha] \, dx \tag{10.40}$$

and pointwise solutions of non-divergence equations:

$$a^{ij}_{\alpha\beta}(x) D_{ij} u^\beta = g_\alpha(x) \quad \text{a.e. in } \Omega . \tag{10.41}$$

In both cases we shall assume that the coefficients $a^{ij}_{\alpha\beta}$ are continuous and satisfy the ellipticity condition (10.3). Moreover, since we are looking for local results, we can assume that the coefficients are uniformly continuous. We shall indicate by ω their modulus of continuity:

$$\omega(t) = \sup_{\substack{i,j=1,\dots,n \\ \alpha,\beta=1,\dots,N}} \sup_{|x-y|<t} |a^{ij}_{\alpha\beta}(x) - a^{ij}_{\alpha\beta}(y)| .$$

As above, we shall consider the case in which Ω is the upper half-space $\mathbf{R}^n_+$. Having proved the desired results for this situation, it will not be difficult to extend them, with a suitable change of variables, to the general case, provided the boundary of Ω is regular enough.

Lemma 10.1 *Let $u(x) \in W^{1,2}$ be a weak solution in $Q^+_R(x_0)$ of the elliptic Eq.* (10.40), *and assume that $u = 0$ on P.*

Then, for every $\varrho < R$ we have

$$\int_{Q^+_\varrho} |Du|^2 \, dx \le c\Big\{ \Big[\Big(\frac{\varrho}{R}\Big)^n + \omega(R)^2\Big] \int_{Q^+_R} |Du|^2 \, dx + \int_{Q^+_R} [|f|^2 + R^2|g|^2] \, dx\Big\}, \tag{10.42}$$

$$\int_{Q^+_\varrho} |Du - (Du)_\varrho|^2 \, dx \le c\Big\{ \Big(\frac{\varrho}{R}\Big)^{n+2} \int_{Q^+_R} |Du - \xi|^2 \, dx + \omega(R)^2 \int_{Q^+_R} |Du|^2 \, dx + \int_{Q^+_R} [|f - \pi|^2 + R^2|g|^2] \, dx\Big\}. \tag{10.43}$$

Proof. We have

$$\int_\Omega a^{ij}_{\alpha\beta}(x_0) D_j u^\beta D_i\varphi^\alpha\, dx = \int_\Omega [a^{ij}_{\alpha\beta}(x_0) - a^{ij}_{\alpha\beta}(x)] D_j u^\beta D_i \varphi^\alpha\, dx + \int_\Omega [f^\alpha D_i\varphi^\alpha + g_\alpha(x)\varphi^\alpha]\, dx$$

and the conclusion follows immediately from (10.36) and (10.37). □

In a similar way, writing Eq. (10.41) in the form

$$a^{ij}_{\alpha\beta}(x_0) D_{ij} u^\beta = [a^{ij}_{\alpha\beta}(x_0) - a^{ij}_{\alpha\beta}(x)] D_{ij} u^\beta + g_\alpha$$

we can obtain from (10.38) and (10.39) the following:

Lemma 10.2 *Let u be a solution of Eq. (10.41) in Q^+_R, and assume that $u = 0$ on P. Then, for every $\varrho < R$ we have*

$$\int_{Q^+_\varrho} |D^2 u|^2\, dx \le c\Big\{ \Big[\Big(\frac{\varrho}{R}\Big)^n + \omega(R)^2\Big] \int_{Q^+_R} |D^2 u|^2\, dx + \int_{Q^+_R} |g|^2\, dx \Big\} \tag{10.44}$$

$$\int_{Q^+_\varrho} |D^2 u - (D^2 u)_\varrho|^2\, dx \le c\Big\{ \Big(\frac{\varrho}{R}\Big)^{n+2} \int_{Q^+_R} |D^2 u - \lambda|^2\, dx + \int_{Q^+_R} |g - g_R|^2\, dx + \omega(R)^2 \int_{Q^+_R} |D^2 u|^2\, dx \Big\}. \tag{10.45}$$

The same inequalities hold at the interior, for cubes Q_ϱ and Q_R.

Remark 10.2 Estimates similar to (10.42) and (10.43) hold for solutions of the complete equation

$$\int_\Omega a^{ij}_{\alpha\beta}(x) D_j u^\beta D_i \varphi^\alpha\, dx = \int_\Omega [f^i_\alpha + b^i_{\alpha\beta}(x) u^\beta] D_i\varphi^\alpha\, dx + \int_\Omega [g_\alpha + c^j_{\alpha\beta}(x) D_j u^\beta + d_{\alpha\beta} u^\beta] \varphi^\alpha\, dx\,, \tag{10.46}$$

provided the functions b, c and d are bounded.

In fact, if we write (10.42) with the functions

$$F^i_\alpha = f^i_\alpha + b^i_{\alpha\beta} u^\beta ,$$
$$G_\alpha = g_\alpha + c_j^{\alpha\beta}(x) D_j u^\beta + d_{\alpha\beta} u^\beta ,$$

and we assume, as we are allowed to do, that $R \leq 1$, we easily get

$$\int_{Q^+_\varrho} |Du|^2\, dx \leq c\Big\{ \Big[\Big(\frac{\varrho}{R}\Big)^n + \chi(R)^2\Big] \int_{Q^+_R} |Du|^2\, dx + \int_{Q^+_R} [|f|^2 + |u|^2 + R^2 |g|^2]\, dx \Big\} \tag{10.47}$$

with $\chi(R)^2 = \omega(R)^2 + R^2$.

If instead we start from (10.43), we get, similarly,

$$\int_{Q^+_\varrho} |Du - (Du)_\varrho|^2\, dx \leq c\Big\{ \Big(\frac{\varrho}{R}\Big)^{n+2} \int_{Q^+_R} |Du - \xi|^2\, dx + \chi(R)^2 \int_{Q^+_R} |Du|^2\, dx + \int_{Q^+_R} [|f - \pi|^2 + |u|^2 + R^2 |g|^2]\, dx \Big\}. \tag{10.48}$$

In the same way, the solutions of the complete equation

$$a^{ij}_{\alpha\beta}(x) D_{ij} u^\beta + b^i_{\alpha\beta}(x) D_i u^\beta + c_{\alpha\beta}(x) u^\beta = g_\alpha(x) \quad q.o.\ in\ \Omega \tag{10.49}$$

satisfies estimates different from (10.44) and (10.45) only for the addition of the term

$$\int_{Q^+_R} (|Du|^2 + |u|^2)\, dx$$

on the right-hand side. □

Proceeding now as in Theorem 8.4, we get without particular difficulties the following regularity results for the first derivatives.

Theorem 10.10 *Let $u \in W^{1,2}(\Omega)$ be a weak solution of the equation*

$$\int_\Omega a^{ij}_{\alpha\beta}(x) D_j u^\beta D_i \varphi^\alpha\, dx = \int_\Omega [f^i_\alpha D_i \varphi^\alpha + g_\alpha \varphi^\alpha]\, dx \tag{10.50}$$

with the coefficients a_{ij} continuous and satisfying the ellipticity condition (10.3).

If the functions f belong to $L^{2,\lambda}(\Omega)$, with $\lambda < n$, and $g \in L^{2,\lambda-2}(\Omega)$, then the derivatives of u belong to $L^{2,\lambda}_{\text{loc}}(\Omega)$, and for every open set $\Sigma \subset\subset \Omega$ it holds that

$$\|Du\|_{2,\lambda,\Sigma} \leq c\{\|Du\|_{2,\Omega} + \|f\|_{2,\lambda,\Omega} + \|g\|_{2,\lambda-2,\Omega}\}. \tag{10.51}$$

If in addition $\partial\Omega$ is regular, and u is a solution of the DIRICHLET *problem with zero boundary data, the derivatives Du belong to $L^{2,\lambda}(\Omega)$, and the preceding estimate holds with Σ replaced by Ω.*

Similarly, for the solutions of (10.41), we have the following:

Theorem 10.11 *Let $u \in W^2(\Omega)$ be a solution of the equation*

$$a^{ij}_{\alpha\beta}(x)D_{ij}u^\beta = g_\alpha \quad \text{a.e. in } \Omega$$

with the coefficients a_{ij} continuous and satisfying the ellipticity condition (10.3).

If the function g belongs to $L^{2,\lambda}(\Omega)$, with $\lambda < n$, then the second derivatives of u belong to $L^{2,\lambda}_{\text{loc}}(\Omega)$, and for every open set $\Sigma \subset\subset \Omega$ we have

$$\|D^2u\|_{2,\lambda,\Sigma} \leq c\left\{\|D^2u\|_{2,\Omega} + \|g\|_{2,\lambda,\Omega}\right\}. \tag{10.52}$$

If in addition $\partial\Omega$ is regular, and u is a solution of the DIRICHLET *problem with zero boundary data, then the second derivatives D^2u belong to $L^{2,\lambda}(\Omega)$, and the preceding estimate holds for Ω.*

Proof. The proofs of the two Theorems 10.10 and 10.11 are practically identical, so that we can limit ourselves to one of them, for instance the second.

Let us begin from the interior regularity. Let $\Sigma \subset\subset \Omega$ and let $R < R_0 = \frac{1}{\sqrt{n}}\text{dist}(\Sigma, \partial\Omega)$. From (10.44), in which in our situation we have $Q^+_R = Q_R \subset\subset \Omega$, setting $\varphi(t) = \int_{Q_t} |D^2u|^2\,dx$, we deduce

$$\varphi(\tau R) \leq c(\tau^n + \omega(R)^2)\varphi(R) + cR^\lambda\|g\|_{2,\lambda}.$$

If we choose τ in such a way that $2c\tau^{\frac{n-\lambda}{2}} = 1$, and R_0 so small that $\omega(R_0)^2 \leq \tau^n$, we obtain

$$\varphi(\tau R) \leq \tau^{\frac{n+\lambda}{2}}\varphi(R) + cR^\lambda\|g\|_{2,\lambda},$$

and the conclusion follows immediately from Lemma 7.3.

The same argument leads to the regularity at the boundary. □

Finally, with a proof completely similar to that of Theorems 8.5 and 8.6, we get the following results:

Theorem 10.12 *Let $u \in W^{1,2}(\Omega)$ be a weak solution of Eq, (10.50), with coefficients $a_{\alpha\beta}^{ij}$ of class $C^{0,\sigma}$ and satisfying the ellipticity condition (10.3).*

If the functions f_α belong to $\mathcal{L}^{2,\lambda}(\Omega)$, with $\lambda \leq n + 2\sigma < n+2$, and $g \in L^{2,\lambda-2}$, then the derivatives of u belong to $\mathcal{L}^{2,\lambda}_{\mathrm{loc}}(\Omega)$, and for every open set $\Sigma \subset\subset \Omega$ we have

$$\|Du\|_{2,\lambda,\Sigma} \leq c\{\|Du\|_{2,\Omega} + \|f\|_{2,\lambda,\Omega} + \|g\|_{2,\lambda-2,\Omega}\}. \tag{10.53}$$

If in addition $\partial\Omega$ is regular, and u is a solution of the DIRICHLET *problem with zero boundary data, then the derivatives Du belong to $\mathcal{L}^{2,\lambda}(\Omega)$, and the preceding estimate holds with Σ replaced by Ω.*

Theorem 10.13 *Let $u \in W^2(\Omega)$ be a solution of the equation*

$$a_{\alpha\beta}^{ij}(x)D_{ij}u^\beta = g_\alpha \quad a.e.\ in\ \Omega$$

with coefficients $a_{\alpha\beta}^{ij}$ of class $C^{0,\sigma}$ and satisfying the ellipticity condition (10.3).

If the function g belongs to $\mathcal{L}^{2,\lambda}(\Omega)$, with $\lambda \leq n + 2\sigma < n+2$, then the second derivatives of u belong to $\mathcal{L}^{2,\lambda}_{\mathrm{loc}}(\Omega)$, and for every open set $\Sigma \subset\subset \Omega$ we have

$$\|D^2u\|_{2,\lambda,\Sigma} \leq c\left\{\|D^2u\|_{2,\Omega} + \|g\|_{2,\lambda,\Omega}\right\}. \tag{10.54}$$

If in addition $\partial\Omega$ is regular, and u is a solution of the DIRICHLET *problem with zero boundary data, then the second derivatives D^2u belong to $\mathcal{L}^{2,\lambda}(\Omega)$, and the preceding estimate holds with Σ replaced by Ω.*

In particular, a weak solution of the elliptic equation

$$\int_\Omega a_{\alpha\beta}^{ij}(x)D_ju^\beta D_i\varphi^\alpha\, dx = \int_\Omega f_\alpha^i D_i\varphi^\alpha\, dx$$

with Hölder-continuous coefficients and the right-hand side, has Hölder-continuous first derivatives, whereas a solution of the equation

$$a_{ij}(x)D_{ij}u = g(x)$$

always with Hölder-continuous coefficients and the right-hand side, has Hölder-continuous second derivatives.

Remark 10.3 Similar results hold for the solutions of the complete Eqs. (10.46) and (10.49). For simplicity, we shall only sketch the proof of the analogue of Theorem 10.10, restricted to the part relative to estimates on the whole Ω.

For that, we shall assume that Du belongs to $L^{2,\vartheta}$, $0 \leq \vartheta < \lambda$, with the estimate[2]

$$\|Du\|_{2,\vartheta} \leq c\{\|u\|_2 + \|Du\|_2 + \|f\|_{2,\lambda} + \|g\|_{2,\lambda-2}\}. \qquad (10.55)$$

By Proposition 3.7, we have $u \in \mathcal{L}^{2,\vartheta+2}$, with the estimate

$$\begin{aligned}\|u\|_{2,\vartheta+2} &\leq c(\|u\|_2 + \|Du\|_{2,\vartheta}) \\ &\leq c\{\|u\|_2 + \|Du\|_2 + \|f\|_{2,\lambda} + \|g\|_{2,\lambda-2}\}.\end{aligned}$$

Starting from (10.47), and arguing as in Theorem 10.10, we conclude that $Du \in L^{2,\mu}$, with $\mu = \min(\vartheta + 2, \lambda)$, and that estimate (10.55) holds with μ instead of ϑ. With a finite number of steps, starting from $\vartheta = 0$, we conclude that $Du \in L^{2,\lambda}$, with the estimate

$$\|Du\|_{2,\lambda} \leq c\{\|u\|_2 + \|Du\|_2 + \|f\|_{2,\lambda} + \|g\|_{2,\lambda-2}\}. \qquad (10.56)$$

□

At this point we could continue, as in the case of the Hilbert regularity, proving regularity theorems for the derivatives of higher order. We limit ourselves to a short statement of the results, leaving with the reader the task of completing the proofs.

Theorem 10.14 *Let u be a solution of Eq.* (10.50). *Assume that the coefficients are of class $C^k(\Omega)$* [resp. $C^{k,\alpha}(\Omega)$], *and that the derivatives of order k of the functions f^i_α and of order $(k-1)$ of g_α belong to $L^{2,\lambda}(\Omega)$* [resp. $\mathcal{L}^{2,\lambda}(\Omega)$]. *Then, the derivatives of order $(k+1)$ of u belong to $L^{2,\lambda}(\Sigma)$* [resp. $\mathcal{L}^{2,\lambda}(\Sigma)$] *for every $\Sigma \subset\subset \Omega$.*

If in addition $\partial\Omega$ is regular, and u is a solution of the DIRICHLET *problem with zero boundary data, the above estimates hold with $\Sigma = \Omega$.*

A similar result is valid for solutions of the equation

$$a^{ij}_{\alpha\beta} D_{ij} u^\beta = g_\alpha .$$

If the coefficients are in C^{k-1} [resp. $C^{k-1,\alpha}$] and the derivatives of order $(k-1)$ of g belong to $L^{2,\lambda}$ [resp. $\mathcal{L}^{2,\lambda}$], the derivatives of order $(k+1)$ of u are in $L^{2,\lambda}_{\rm loc}$ [resp. $\mathcal{L}^{2,\lambda}_{\rm loc}$], with global result if $u = 0$ on $\partial\Omega$.

[2]We have omitted the reference to the open set Ω.

10.4 L^p Estimates

The results of the preceding section, and the interpolation theorem of STAMPACCHIA (Theorem 2.14), can be used to obtain L^p estimates for the derivatives of the solutions of elliptic equations. The following theorem describes the simplest case, yet of some interest.

Theorem 10.15 *Let Ω be an open set of $\mathbf{R}^n$, with regular boundary and homeomorphic to a cube. Let $u \in W_0^1(\Omega)$ be a weak solution of the problem of* DIRICHLET *(with zero boundary data) for the equation*

$$D_i a_{\alpha\beta}^{ij} D_j u^\beta = D_i f_\alpha^i \tag{10.57}$$

with constant coefficients $a_{\alpha\beta}^{ij}$, satisfying the ellipticity condition (10.3).

If $f \in L^p(\Omega)$, $p > 2$, the derivatives Du belong to $L^p(\Omega)$, with the estimate

$$\|Du\|_p \leq c(\Omega)\|f\|_p\,. \tag{10.58}$$

Proof. Let us consider the operator T mapping every $f \in L^2(\Omega)$ into the gradient Du of the solution of problem (10.57). T is obviously linear, and by Lemma 5.1 we have

$$\|Tf\|_2 \leq c\|f\|_2\,.$$

On the other hand, by Theorem 10.12, T maps $\mathcal{L}^{2,n}$ into $\mathcal{L}^{2,n}$, with the estimate

$$\|Tf\|_{2,n} \leq c(\|Tf\|_2 + \|f\|_{2,n}) \leq c\|f\|_{2,n}\,.$$

We can therefore apply the theorem of STAMPACCHIA, and we conclude that T maps L^p into L^p, with the estimate (10.58). □

Remark 10.4 The constant c in (10.58) obviously depends on Ω. If however Ω is a ball of radius R, it can be taken independently of R.

Let u be a solution of problem (10.57) in B_R, and let $w(x) = u(Rx)$, $\eta(x) = \varphi(Rx)$ and $F(x) = f(Rx)$. We have $Dw(x) = RDu(Rx)$ and $D\eta(x) = RD\varphi(Rx)$, and hence the function $w \in W_0^1(B)$ is a solution of the equation

$$\int_B a_{\alpha\beta}^{ij} D_j w^\beta D_i \varphi^\alpha \, dx = R \int_B F_\alpha^i D_i \varphi^\alpha \, dx\,.$$

By the previous theorem, if $f \in L^p(B_R)$ we have $F \in L^p(B)$ and

$$\|Dw\|_{p,B} \leq c(B) R \|F\|_{p,B}\,,$$

which is equivalent to

$$\|Du\|_{p,B_R} \leq c(B)\|f\|_{p,B_R} . \qquad \square$$

If instead v is a solution of the homogeneous equation

$$\int_\Omega a^{ij}_{\alpha\beta} D_j v^\beta D_i \varphi^\alpha \, dx = 0$$

and takes the value U at the boundary, we can apply the preceding result to the function $u = v - U$, which is solution of the problem (10.57) with $f^i_\alpha = a^{ij}_{\alpha\beta} D_j U^\beta$. We have therefore

$$\|Du\|_p \leq c(\Omega)\|DU\|_p$$

and hence

$$\|Dv\|_p \leq c(\Omega)\|DU\|_p .$$

Remarking that the function $v - P = v - a - \langle \pi, x\rangle$ is a solution of the homogeneous equation and takes the value $U - P$ at the boundary, we have also

$$\|Dv - \pi\|_p \leq c(\Omega)\|DU - \pi\|_p . \tag{10.59}$$

In particular, the preceding estimate holds in a ball of radius R, with c independent of R. The last estimate is exactly the one we have used frequently in the previous chapters.

Under the same assumptions on Ω, a similar result holds for the second derivatives of the solutions of the DIRICHLET problem with zero boundary data for the equation

$$a^{ij}_{\alpha\beta} D_{ij} u^\beta = g_\alpha \text{ a.e. in } \Omega \tag{10.60}$$

with constant coefficients $a^{ij}_{\alpha\beta}$ satisfying the LEGENDRE–HADAMARD condition.

Since the coefficients are constant, we can write (10.60) in the weak form

$$\int_\Omega a^{ij}_{\alpha\beta} D_j u^\beta D_i \varphi^\alpha \, dx = -\int_\Omega g_\alpha \varphi^\alpha \, dx . \tag{10.61}$$

We can therefore apply Theorem 10.3, and we can conclude that

$$\int_\Omega |D^2 u|^2 \, dx \leq c \int_\Omega (|Du|^2 + |g|^2) \, dx .$$

On the other hand, taking $\varphi = u$ in (10.61), and using (10.5) and POINCARÉ's inequality, we easily get

$$\int_\Omega |Du|^2\,dx \le c\int_\Omega |g|^2\,dx \tag{10.62}$$

and hence

$$\int_\Omega |D^2u|^2\,dx \le c\int_\Omega |g|^2\,dx\,.$$

To the same function u we can apply Theorem 10.13, and in particular we can write (10.54) with $\Lambda = \Omega$. We have therefore

$$\|D^2u\|_{2,n} \le c(\|D^2u\|_2 + \|g\|_{2,n}) \le c\|g\|_{2,n}\,.$$

In conclusion, the linear operator T, which maps any $g \in L^2$ into the second derivatives of the solution of problem (10.60), maps L^2 into L^2, and $\mathcal{L}^{2,n}$ into $\mathcal{L}^{2,n}$. By the theorem of STAMPACCHIA, T maps L^p into L^p, with the estimate

$$\|D^2u\|_p \le c(\Omega)\|g\|_p\,. \tag{10.63}$$

We remark once more that if Ω is the ball of radius R, the constant c in the preceding estimate does not depend on R.

We shall continue now by showing on one hand that, always under assumptions of regularity of $\partial\Omega$, it is possible to avoid the assumption that Ω is homeomorphic to a cube; and on the other by extending the previous results to equations with continuous coefficients. Moreover, we will show that these results are local in character; in other words, if $A \subset \Omega$, and $f \in L^p(A)$, then $Du \in L^p(\Lambda)$ for every open set $\Lambda \subset\subset A$.

For that, let us consider a ball $B_R = B(x_0, R)$ and let $\eta \in C_0^\infty(B_R)$, with $0 \le \eta \le 1$, and $\eta = 1$ in $B_{R/2}$. Let u be a solution of the equation

$$\int_{B_R} a_{\alpha\beta}^{ij} D_j u^\beta D_i\varphi^\alpha\,dx = \int_{B_R} f_\alpha^i D_i\varphi^\alpha\,dx \tag{10.64}$$

for every φ with support in B_R. Writing $\eta\varphi$ instead of φ, we get easily

$$\int a_{\alpha\beta}^{ij} D_j(\eta u^\beta) D_i\varphi^\alpha\,dx = \int [\eta f_\alpha^i + a_{\alpha\beta}^{ij} u^\beta D_j\eta] D_i\varphi^\alpha\,dx$$

$$+ \int [f_\alpha^i D_i\eta - a_{\alpha\beta}^{ij} D_j u^\beta D_i\eta]\varphi^\alpha\,dx\,. \tag{10.65}$$

Let now w be the solution of the DIRICHLET problem

$$\begin{cases} \Delta w^\alpha = g^\alpha =: f^i_\alpha D_i\eta - a^{ij}_{\alpha\beta} D_j u^\beta D_i\eta & \text{in } B_R\,, \\ w = 0 & \text{on } \partial B_R\,. \end{cases}$$

The function w satisfies the inequality (10.63) with c independent of R; and hence if $g \in L^m$ for some $m \geq 2$, the second derivatives of w belong to L^m, and therefore $Dw \in L^{m^*}$. Introducing in (10.65) the preceding equation written in weak form, namely:

$$\int D_i w^\alpha D_i \varphi^\alpha \, dx = -\int [f^i_\alpha D_i\eta - a^{ij}_{\alpha\beta} D_j u^\beta D_i\eta]\varphi^\alpha \, dx\,,$$

we eventually get

$$\int a^{ij}_{\alpha\beta} D_j(\eta u^\beta) D_i\varphi^\alpha \, dx = \int F^i_\alpha D_i\varphi^\alpha \, dx\,, \tag{10.66}$$

where

$$F^i_\alpha = \eta f^i_\alpha + a^{ij}_{\alpha\beta} u^\beta D_j\eta - D_i w^\alpha\,. \tag{10.67}$$

We remark that if $f \in L^p$ and $u \in W^{1,m}$, then $F \in L^s$, with $s = \min(p, m^*)$.

Until now, no assumptions have been made on the coefficients. Let us assume now that they are continuous in Ω, and satisfy the condition of LEGENDRE–HADAMARD. The Eq. (10.66) can be written in the form

$$\int a^{ij}_{\alpha\beta}(x_0) D_j(\eta u^\beta) D_i\varphi^\alpha \, dx$$
$$= \int \{F^i_\alpha + [a^{ij}_{\alpha\beta}(x_0) - a^{ij}_{\alpha\beta}(x)] D_j(\eta u^\beta)\} D_i\varphi^\alpha \, dx\,. \tag{10.68}$$

Assume finally that $F \in L^s$, and for $v \in W^{1,s}_0(B_R)$ let w be the solution of the problem of DIRICHLET for the equation

$$\int a^{ij}_{\alpha\beta}(x_0) D_j w^\beta D_i\varphi^\alpha \, dx$$
$$= \int \{F^i_\alpha + [a^{ij}_{\alpha\beta}(x_0) - a^{ij}_{\alpha\beta}(x)] D_j v^\beta\} D_i\varphi^\alpha \, dx\,, \tag{10.69}$$

taking the value zero on ∂B_R

By Theorem 10.15 we have

$$\|Dw\|_s \leq c(\|F\|_s + \omega(R)\|Dv\|_s) \tag{10.70}$$

with c independent of R, where $\omega(R)$ is the oscillation of the coefficients in B_R. Moreover, if w_1 and w_2 are the solutions correspondent to v_1 and v_2, we have

$$\|D(w_1 - w_2)\|_s \le c\omega(R)\|D(v_1 - v_2)\|_s .$$

Choosing now R such that $c\omega(R) \le \frac{1}{2}$ (note that the value of R depends *only* on the modulus of continuity of the coefficients), the map $S : v \to w$ is contractive, and hence it has a unique fixed point, which cannot but coincide with ηu. By (10.70) we have

$$\|D(\eta u)\|_s \le 2c\|F\|_s .$$

In conclusion, we have proved that if the function F given by (10.67) belongs to $L^s(B_R)$, then

$$\|Du\|_{s,B_{R/2}} \le c\|F\|_{s,B_R} . \tag{10.71}$$

Let us assume now that, for some $r < R/2$, $u \in W^{1,m}(B_r)$ with $m < p$, and that

$$\|u\|_{1,m,B_r} \le c(\|f\|_{p,B_{2r}} + \|u\|_{1,2,B_{2r}}) .$$

By what we have seen, F will belong to $L^s(B_r)$, $s = \min(p, m^*)$, and therefore $u \in W^{1,s}(B_{r/2})$, with the estimate

$$\begin{aligned}\|u\|_{1,s,B_{r/2}} &\le c\|F\|_{s,B_r} \le c(\|f\|_{p,B_r} + \|u\|_{1,m,B_r}) \\ &\le c(\|f\|_{p,B_{2r}} + \|u\|_{1,2,B_{2r}}) .\end{aligned}$$

Starting then from $m = 2$, with a finite number of steps (dependent only on the dimension n) we reach the exponent p. We have therefore the following:

Theorem 10.16 *Let u be a solution of the equation*

$$\int_\Omega a^{ij}_{\alpha\beta}(x) D_j u^\beta D_i \varphi^\alpha \, dx = \int_\Omega f^i_\alpha D_i \varphi^\alpha \, dx \tag{10.72}$$

with continuous coefficients. If for $A \subset\subset \Omega$ the function f belongs to $L^p(A)$, $p \ge 2$, then u belongs to $W^{1,p}(\Lambda)$ for every $\Lambda \subset\subset A$, with the estimate

$$\|u\|_{1,p,\Lambda} \le c(\Lambda, A)(\|f\|_{p,A} + \|u\|_{1,2,A}) . \tag{10.73}$$

We remark that by CACCIOPPOLI's inequality, the norm W^1 on the right-hand side can be replaced by the L^2 norm.

A similar result holds at the boundary for the solutions of the DIRICHLET problem with boundary datum $U \in W^{1,p}$. As usual, we can assume $U = 0$, and by means of a diffeomorphism we can reduce to the case of flat boundary. In order that the coefficients of the transformed equation remain continuous, it is necessary that $\partial\Omega$ is of class C^1. Assume therefore that u is a weak solution of the equation (10.72) in the half-ball $B^+(0)$, taking the value zero on the flat part P of ∂B^+. Let us consider as above a function η with support in the ball $B_R = B(0, R)$; the function ηu is zero on ∂B_R^+ and satisfies the Eq. (10.65) in B_R^+.

At this point we cannot continue as above, since the regularity theorems in L^p require that the boundary of Ω is regular, whereas ∂B_R^+ is only Lipschitz-continuous. Luckily, the function η appearing in (10.65) has support which stays away from the singular part of the boundary, and therefore we can replace B_R^+ with a regular open set $\Lambda_R \subset B_R^+$ containing the support of η.

In order not to introduce unwanted dependence on R, we consider a function $\psi \in C_0^\infty(B)$, with $0 \le \psi \le 1$ and $\psi = 1$ in $B_{1/2}$, and a set $\Lambda \subset B^+$ with regular boundary containing $\operatorname{supp}\psi \cap B^+$. For $R < 1$, we set $\eta(x) = \psi(Rx)$ and

$$\Lambda_R = R\Lambda =: \{Rx : x \in \Lambda\}.$$

We can now continue our proof, writing Λ_R instead of B_R. Arguing as at the end of Theorem 10.15, we can conclude that the constants appearing in the estimates depend on Λ, but not on R.

In this way we can repeat without essential changes the proof of interior regularity, getting an L^p estimate up to the boundary. This result is essentially of local character: if $A \subset \Omega$ is an open set, and if $f \in L^p(A)$, then $u \in L^p(\Lambda)$ for any open set Λ whose closure is contained in $A \cup \partial\Omega$. If $A = \Omega$ one gets

Theorem 10.17 *Let u be a solution of the* DIRICHLET *problem* (10.57) *with zero boundary data, in an open set Ω with C^1 boundary. Assume that the coefficients are continuous in $\bar{\Omega}$ and that the right-hand side f belongs to $L^p(\Omega)$. Then, $u \in W^{1,p}(\Omega)$, and we have the estimate*

$$\|u\|_{1,p,\Omega} \le c(\|f\|_{p,\Omega} + \|u\|_{1,2,\Omega}). \tag{10.74}$$

The term $\|u\|_{1,2,\Omega}$ on the right-hand side of the preceding estimate can be replaced with $\|u\|_{2,\Omega}$ by the GÅRDING inequality (10.6), as is easily seen taking $\varphi = u$ in (10.64) and making the usual estimates.

10.5 Minima of Functionals

We shall now apply the results of the preceding section in order to prove the regularity of the minima of functionals

$$\mathcal{F}(u,\Omega) = \int_\Omega F(x,u,Du)\,dx\,,$$

where $F(x,u,z)$ is a regular function, satisfying the assumption of the previous chapter.

We have already remarked that every minimum $u(x)$ of $\mathcal{F}$ has second derivatives in L^2 and Hölder-continuous first derivatives in an open set $\Omega_0 \subset \Omega$,[3] and satisfies for almost every $x \in \Omega_0$ the equation

$$A^{ij}_{\alpha\beta}(x,u(x),Du(x))D_{ij}u^\beta = B_\alpha(x,u(x),Du(x))\,, \tag{10.75}$$

where

$$A^{ij}_{\alpha\beta}(x,u,z) = \frac{\partial^2 F}{\partial z_i^\alpha \partial z_j^\beta}(x,u,z)\,,$$

$$B_\alpha(x,u,z) = \frac{\partial F}{\partial u^\alpha} - \frac{\partial^2 F}{\partial z_i^\alpha \partial u^\beta} z_i^\beta - \frac{\partial^2 F}{\partial z_i^\alpha \partial x_i}\,.$$

Moreover the derivatives $D_s u$ are weak solutions of the equation

$$\int_\Omega \left[A^{ij}_{\alpha\beta}(x,u,Du)D_j D_s u^\beta + B^i_{\alpha,s}(x,u,Du)\right] D_i\varphi^\alpha\,dx = 0 \tag{10.76}$$

for every $\varphi \in C_0^\infty(\Omega_0,\mathbf{R}^N)$, with

$$B^i_{\alpha,s}(x,u,z) = \frac{\partial^2 F}{\partial z_i^\alpha \partial u^\beta} z_s^\beta + \frac{\partial^2 F}{\partial z_i^\alpha \partial x_s} - \frac{\partial F}{\partial u^\alpha}\delta_{is}\,.$$

We have the following:

Theorem 10.18 *Let $u \in W^2(\Omega_0) \cap C^{1,\sigma}(\Omega_0)$ be a minimum of the functional $\mathcal{F}$, and let the function $F(x,u,z)$ be of class $C^{k+2,\delta}$, $\delta < 1$ in its arguments. Then, u belongs to $C^{k+2,\delta}(\Omega_0)$.*

Proof. We consider first the case $k = 0$. The functions

$$a^{ij}_{\alpha\beta}(x) = A^{ij}_{\alpha\beta}(x,u(x),Du(x))$$

[3] In the scalar case we have $\Omega_0 = \Omega$; in the vector case the closed set $\Omega - \Omega_0$ has zero measure in general, whereas for quadratic functionals it has zero $(n-2)$-dimensional measure.

and

$$g_\alpha(x) = B_\alpha(x, u(x), Du(x))$$

are Hölder-continuous with some exponent $\sigma > 0$. By Theorem 10.13, the second derivatives of u belong to $C^{0,\sigma}$. In particular, the first derivatives are Lipschitz-continuous, and hence $a^{ij}_{\alpha\beta}$ and g_α belong to $C^{0,\delta}$. But then $D^2u \in C^{0,\delta}$ and the theorem is proved if $k = 0$.

Let us now discuss the case $k = 1$. The functions $a^{ij}_{\alpha\beta}(x)$ and $b^i_{\alpha,s}(x) = B^i_{\alpha,s}(x, u(x), Du(x))$ are of class $C^{1,\sigma}$, and therefore we can use the difference quotients method starting from (10.76), and we conclude that the second derivatives of u belong to $W^{1,2}_{\text{loc}}$. It follows that we can differentiate (10.75), getting

$$\begin{aligned} a^{ij}_{\alpha\beta}(x) D_{ij} D_s u^\beta &= D_s g_\alpha(x) + D_s a^{ij}_{\alpha\beta} D_{ij} u^\beta \\ &=: G_\alpha(x) \end{aligned} \tag{10.77}$$

with $G \in C^{0,\sigma}$. By Theorem 10.13 the third derivatives of u belong to $C^{0,\sigma}$, and hence, arguing as above, to $C^{0,\delta}$.

Finally, let us assume that the theorem holds for $k \geq 1$, and that $F \in C^{k+3,\delta}$. By the inductive hypothesis we can assume that $u \in C^{k+2,\delta}$, and hence the function $G(x)$ in (10.77) is of class $C^{k,\sigma}$ for some $\sigma > 0$. But then the derivatives of u belong to $C^{k+2,\sigma}$, and consequently $G \in C^{k,\delta}$. Applying once again the inductive assumption, we conclude that $Du \in C^{k+2,\delta}$, and consequently $u \in C^{k+3,\delta}$. □

In the same way, but with a little more effort, one can prove the Hölder-continuity up to the boundary of the derivatives, in the case $\Omega_0 = \Omega$.

10.6 Notes and Comments

The results of this chapter, in particular those relative to linear equations and systems, can be defined as "classical," and it is difficult nowadays to establish their origin with some precision. Theorems 10.12 and 10.13 are known as SCHAUDER *estimates* [1], and were obtained originally by means of potential theory. Here, in contrast we have followed the method of CAMPANATO [3, 4], which is based on integral estimates, and can be extended without difficulty to linear elliptic systems. In particular, Theorems 10.10 and 10.11, as well as the proofs of Theorems 10.12 and 10.13 are due to CAMPANATO.

The L^p estimates, originally obtained by means of potential theory,[4] are proved here by a method of STAMPACCHIA [3], later simplified by CAMPANATO [5]. For the extension to continuous coefficients, we have followed an idea of TRUDINGER.

[4] The most general results can be found in AGMON, DOUGLIS and NIRENBERG [1].

References

ACERBI E. and G. BUTTAZZO

[1] Semicontinuous envelopes of polyconvex integrals. *Proc. Roy. Soc. Edimburgh*, **96A** (1984), 51–54.

ACERBI E., G. BUTTAZZO and N. FUSCO

[1] Semicontinuity and relaxation for integrals depending on vector-valued functions. *J. Math. Pures Appl.*, **62** (1983), 371–387.

ACERBI E. and N. FUSCO

[1] Semicontinuity problems in the calculus of variations. *Arch. Rat. Mech. and Anal.*, **86** (1984), 125–145.
[2] A regularity theorem for minimizers of quasiconvex integrals. *Arch. Rat. Mech. and Anal.*, **99** (1987), 261–281.
[3] Local regularity for minimizers of non-convex integrals. *Ann. Sc. Norm. Sup. Pisa*, **(IV) 16** (1989), 603–636.
[4] Regularity of minimizers of non-quadratic functionals: the case $1 < p < 2$. *J. Math. Anal. Appl.*, **140** (1989), 115–135.

ADAMS R. A.

[1] *Sobolev spaces.* Academic Press, New York, 1975.

AGMON S., A. DOUGLIS and L. NIRENBERG

[1] Estimates near the boundary for solutions of elliptic partial differential equations satisfying general boundary conditions I, II. *Comm. Pure Appl. Math.*, **12** (1959), 623–727 and **17** (1964), 35–92.

ALIBERT J. J. and B. DACOROGNA

[1] An example of a quasiconvex function that is not polyconvex in two dimensions. *Arch. Rat. Mech. and Anal.*, **117** (1992), 155–166.

ALMGREN F.

[1] Existence and regularity almost everywhere of solutions of elliptic variational problems among surfaces of varying topological type and singularity structure. *Ann. of Math.*, **87** (1968), 321–391.

AMBROSIO L.

[1] New lower semicontinuity results for integral functionals. *Rend. Accad. Naz. Sci. XL*, **11** (1987), 1–42.

ANZELLOTTI G.

[1] On the $C^{1,\alpha}$-regularity of $\omega(R)$-minima of quadratic functionals. *Boll. U. M. I.*, **(VI) 2** (1983), 195–212.

ANZELLOTTI G., G. BUTTAZZO and G. DAL MASO.

[1] Dirichlet problems for demicoercive functionals. *Nonlinear Anal.*, **10** (1986), 603–613.

ANZELLOTTI G. and M. GIAQUINTA.

[1] Funzioni *BV* and tracce. *Rend. Sem. Mat. Univ. Padova*, **60** (1978), 1–21.

[2] Convex functionals and partial regularity. *Arch. Rat. Mech. and Anal.*, **102** (1988), 243–272.

ARZELÀ C.

[1] Sul principio di Dirichlet. *Rend. Acad. Sci. Ist. Bologna*, **1** (1896–97), 71–84. *Opere*, **2**, Cremonese, Roma (1992), 377–390.

ATTOUCH H. C.

[1] *Variational convergence of functionals and operators*. Pitman, London, 1984.

ATTOUCH H. C. and C. SBORDONE.

[1] Asymptotic limits for perturbed functionals of calculus of variations. *Ricerche di Mat.*, **29** (1980), 85–124.

AUBERT G.

[1] A counterexample of a rank one convex function which is not polyconvex in the case $n = 2$. *Proc. Roy. Soc. Edinburgh*, **106A** (1987), 237–240.

BALL J. M.

[1] Convexity conditions and existence theorems in nonlinear elasticity. *Arch. Rat. Mech. and Anal.*, **63** (1977), 337–403.

[2] Strict convexity, strong ellipticity, and regularity in the calculus of variations. *Math. Proc. Cambridge Phil. Soc.*, **87** (1980), 501–513.

BALL J. M., J. C. CURRIE and P. J. OLVER.

[1] Null lagrangians, weak continuity, and variational problems of arbitrary order. *J. Funct. Anal.*, **41** (1981), 135–174.

BALL J. M. and F. MURAT.

[1] $W^{1,p}$-quasiconvexity and variational problems for multiple integrals. *J. Funct. Anal.*, **58** (1984), 225–253.

BALL J. M. and K.-W. ZHANG.

[1] Lower semicontinuity of multiple integrals and the biting lemma. *Proc. Roy. Soc. Edimburgh*, **114A** (1990), 367–379.

BAROZZI G. C.

[1] Su una generalizzazione degli spazi $L^{(q,\lambda)}$ di Morrey. *Ann. Sc. Norm. Sup. Pisa*, **(III) 19** (1965), 609–626.

BHATTACHARYA T. and F. LEONETTI.

[1] A new Poincaré inequality and its applications to the regularity of minimizers of integral functionals with nonstandard growth. *Nonlinear Anal.*, **17** (1991), 833–839.

BERKOWITZ L. D.

[1] Lower semicontinuity of integral functionals. *Trans. Amer. Math. Soc.*, **192** (1974), 51–57.

BERNSTEIN S.

[1] Méthode générale pour la résolution du problème de Dirichlet. *C. R. Acad. Sci. Paris*, **144** (1907), 1025–1027.
[2] Conditions nécessaires et suffisantes pour la possibilité du problème de Dirichlet. *C. R. Acad. Sci. Paris*, **150** (1910), 514–515.
[3] Sur la géneralisation du problème de Dirichlet. II. *Math. Ann.*, **69** (1910), 82–136.
[4] Sur les surfaces définies au moyen de leur courbure moyenne et totale. *Ann. Sci. École Norm. Sup.*, **27** (1910), 233–256.
[5] Sur les équations du calcul des variations. *Ann. Sci. École Norm. Sup.*, **29** (1912), 431–485.

BETHUEL F., H. BREZIS and J. M. CORON.

[1] Relaxed energies for harmonic maps. *Variational Problems*, H. Berestycki, J. M. Coron and I. Ekeland ed., Birkhauser, Basel, 1990.

BOCCARDO L., P. MARCELLINI and C. SBORDONE.

[1] L^∞-regularity for variational problems with sharp nonstandard growth conditions. *Boll. U. M. I.*, **4-A** (1990), 219–225.

BOCCARDO L. and R. SCHIANCHI.

[1] A remark on the L^s-regularity of the minima of functionals of the calculus of variations. *Rev. Mat. Univ. Compl. Madrid*, **2** (1989), 113–118.

BOMBIERI E.

[1] Variational problems and elliptic equations. *Proc. Intern. Congress of Mathematicians*, Vancouver 1974, Vol. 1, 53–63.
[2] Variational problems and elliptic equations (Hilbert's problem 20). *Mathematical developments arising from Hilbert problems*, F. Browder ed., *Proc. Symp. Pure Math.*, **XXVIII** (1976), 525–535.

BOMBIERI E., E. DE GIORGI and M. MIRANDA.

[1] Una maggiorazione *a priori* relativa alle ipersuperfici minimali non-parametriche. *Arch. Rat. Mech. and Anal.*, **32** (1969), 255–267.

BOMBIERI E. and E. GIUSTI.

[1] A Harnack's type inequality for elliptic equations on minimal surfaces. *Inv. Math.*, **15** (1972), 24–46.

BOJARSKI B.

[1] Generalized solutions of a system of differential equations of the first order of elliptic type with discontinuous coefficients. *Mat. Sb.*, **43** (1957), 451–503.
[2] Remarks on the stability of reverse Hölder inequalities and quasiconformal mappings. *Ann. Ac. Sci. Fenn. (Math.)*, **10** (1985), 291–296.

BOJARSKI B., C. SBORDONE and I. WIK.

[1] The Muckenhoupt class $A_1(\mathbf{R})$. *Studia Math.*, **101** (1992), 155–163.

BROWDER F. E.

[1] Remarks on the direct methods of the calculus of variations. *Arch. Rat. Mech. and Anal.*, **20** (1965), 251–258.

BUTTAZZO G.

[1] *Semicontinuity, relaxation, and integral representation in the calculus of variations*. Pitman Research Notes in Math. **207**. Longman, Harlow, 1989.

CACCIOPPOLI R.

[1] Sulle equazioni ellittiche a derivate parziali con n variabili indipendenti. *Rend. Accad. Naz. Lincei*, **18** (1933), 103–106.
[2] Sulle equazioni ellittiche a derivate parziali con due variabili indipendenti and sui problemi regolari del calcolo delle variazioni. *Rend. Accad. Naz. Lincei*, **19** (1934), 83–89.
[3] Limitazioni integrali per le soluzioni di un'equazione lineare ellittica a derivate parziali. *Giorn. Mat. Battaglini*, **80** (1950–51), 186–212.

CAFFARELLI L., R. KOHN and L. NIRENBERG.

[1] Partial regularity of suitable weak solutions of Navier-Stokes equations. *Comm. Pure Appl. Math.*, **35** (1982), 771–831.

CALDERON A. P. and A. ZYGMUND.

[1] On the existence of certain singular integrals. *Acta Math.*, **88** (1952), 85–139.

CALKIN J. W.

[1] Functions of several variables and absolute continuity (I). *Duke Math. J.*, **6** (1940), 170–185.

CAMPANATO S.

[1] Proprietà di hölderianità di alcune classi di funzioni. *Ann. Sc. Norm. Sup. Pisa*, **(III) 17** (1963), 175–188.
[2] Proprietà di una famiglia di spazi funzionali. *Ann. Sc. Norm. Sup. Pisa*, **(III) 18** (1964), 137–160.
[3] Equazioni ellittiche del secondo ordine and spazi $\mathcal{L}^{2,\lambda}$. *Ann. Mat. Pura Appl.*, **69** (1965), 321–380.
[4] Equazioni paraboliche del secondo ordine and spazi $\mathcal{L}^{2,\vartheta}(\Omega,\delta)$. *Ann. Mat. Pura Appl.*, **73** (1966), 55–102.
[5] Su un teorema di interpolazione di G. Stampacchia. *Ann. Sc. Norm. Sup. Pisa*, **(III) 20** (1966), 649–652.
[6] Partial Hölder continuity of the gradient of the solutions of some nonlinear elliptic systems. *Rend. Sem. Mat. Univ. Padova*, **59** (1978), 147–165.
[7] *Sistemi ellittici in forma divergenza. Regolarità all'interno.* Quaderni. Sc. Norm. Sup. Pisa, 1980.

CAMPANATO S. and G. STAMPACCHIA.

[1] Sulle maggiorazioni in L^p nella teoria delle equazioni ellittiche. *Boll. U. M. I.*, **20** (1965), 393–399.

CARRIERO M., A. LEACI and E. PASCALI.

[1] On the lower semicontinuity of a class of integral functionals. *Proc. Int. Workshop Trieste 1985*; *Rend. Circ. Mat. Palermo*, **(2)**, Suppl. (1987), 155–161.

CESARI L.

[1] Lower semicontinuity and lower closure theorems without seminormality conditions. *Ann. Mat. Pura Appl.*, **98** (1974), 381–397.

CHOE H. J.

[1] Regularity of minimizers of certain degenerate functionals with nonstandard growth conditions. *Comm. Part. Diff. Eq.*, **16** (1991), 363–372.
[2] Interior behaviour of minimizers for certain functionals with nonstandard growth. *Nonlinear Anal.*, **19** (1992), 933–946.

DACOROGNA B.

[1] *Direct methods in the calculus of variations.* Appl. Math. Sci. **78**, Springer Verlag, Heidelberg and New York, 1989.

DACOROGNA B. and P. MARCELLINI.

[1] A counterexample in the vectorial calculus of variations. *Material instabilities in continuum mechanics*, J. M. Ball ed., Oxford Univ. Press 1988, 77–83.

[2] Semicontinuité pour des intégrandes polyconvexes sans continuité des déterminants. *C. R. Acad. Sci. Paris*, **311** (1990), 393–396.

DAL MASO G.

[1] *An introduction to* Γ*-convergence*. S.I.S.S.A., Trieste, 1992.

DAL MASO G. and L. MODICA.

[1] *A general theory of variational functionals*. Quad. Sc. Norm. Sup. Pisa, 1982.

DA PRATO G.

[1] Spazi $\mathcal{L}^{p,\vartheta}(\Omega,\delta)$ e loro proprietà. *Ann. Mat. Pura Appl.*, **69** (1965), 383–392.

D'APUZZO L. and C. SBORDONE.

[1] Reverse Hölder inequalities. A sharp result. *Rend. di Mat.*, **(VIII) 10** (1990), 357–366.

DE FIGUEIREDO D. G.

[1] *Lectures on the Ekeland variational principle with applications and detours.* Tata Inst. of Fund. Research, Bombay 1989.

DE GIORGI E.

[1] Sulla differenziabilità and l'analiticità delle estremali degli integrali multipli regolari. *Mem. Accad. Sci. Torino* (*Classe di Sci. mat., fis. and nat.*), **(3) 3** (1957), 25–43.

[2] Sulla proprietà isoperimetrica dell'ipersfera, nella classe degli insiemi aventi frontiera orientata di misura finita. *Mem. Accad. Naz. Lincei*, **(VIII) 5** (1958), 33–44.

[3] Frontiere orientate di misura minima. *Sem. Mat. Scuola Norm. Sup.*, Editrice Tecnico Scientifica, Pisa, 1961.

[4] *Teoremi di semicontinuità nel calcolo delle variazioni.* INdAM, Roma, 1968–69.

[5] Un esempio di estremali discontinue per un problema variazionale di tipo ellittico. *Boll. U.M.I*, **4** (1968), 135–137.

[6] Sulla convergenza di alcune successioni di integrali del tipo dell'area. *Rend. di Mat.*, **8** (1975), 277–294.

[7] Operatori elementari di limite ed applicazioni al calcolo delle variazioni. *Atti del convegno "Studio di problemi-limite dell'Analisi funzionale"*, Bressanone 1981. Pitagora, Bologna, 1982.

[8] Some continuity and relaxation problems. *Proc. of E. De Giorgi Colloquium.* Pitman, London, 1985.

De Giorgi E., G. Buttazzo and G. Dal Maso.

[1] On the lower semicontinuity of certain integral functionals. *Atti Acc. Naz. Lincei*, **74** (1983), 274–282.

Di Benedetto E.

[1] $C^{1,\alpha}$ local regularity of weak solutions of degenerate elliptic equations. *Nonlinear Anal.*, **7** (1983), 827–850.

[2] Harnack estimates in certain function classes. *Atti Sem. Mat. Fis. Univ. Modena*, **XXXVII** (1989), 173–182.

Di Benedetto E. and N. Trudinger.

[1] Harnack inequalities for quasi-minima of variational integrals. *Ann. Inst. H. Poincaré (Analyse non-lineaire)*, **1** (1984), 295–308.

Dolcini A., L. Esposito and N. Fusco.

[1] $C^{0,\alpha}$ regularity of ω-minima. *Boll. U. M. I.*, **(VII), 10-A** (1996), 113–125.

Douglas J.

[1] Solution of the problem of Plateau. *Trans. Am. Math. Soc.*, **33** (1931), 263–321.

Dunford N. and J. T. Schwartz.

[1] *Linear operators*, Part I. Pure and Appl. Math., J. Wiley, New York, 1958.

Edwards R. E.

[1] *Functional analysis. Theory and applications.* Holt, Rinehart & Wilson, New York, 1965.

Eisen G.

[1] A counterexample for some lower semicontinuity results. *Math. Z.*, **162** (1978), 241–243.

Ekeland I.

[1] Sur les problèmes variationnels. *C. R. Acad. Sci. Paris*, **275** (1972), 1057–1059.

[2] On the variational principle. *J. Math. Anal. Appl.*, **47** (1974), 324–353.

[3] Nonconvex minimization problems. *Bull. Am. Math. Soc.*, **(3) 1** (1979), 443–474.

Esposito L. and G. Mingione

[1] A regularity theorem for the ω-minimizers of integral functionals. *Rend. di Mat.*, **(VII) 19** (1999), 17–44.

EULER L.

[1] *Methodus inveniendi lineas curvas maximi minimive proprietate gaudentes, sive solutio problematis isoperimetrici latissimo sensu accepti.* Leonhardi Euleri Opera Omnia **24**. Füssli, Bern, 1952.

EVANS L. C.

[1] Quasiconvexity and partial regularity in the calculus of variations. *Arch. Rat. Mech. and Anal.*, **95** (1986), 227–252.

EVANS L. C. and R. F. GARIEPY.

[1] Blow-up, compactness and partial regularity in the calculus of variations. *Ind. Univ. Math. J.*, **36** (1987), 361–371.
[2] Some remarks concerning quasiconvexity and strong convergence. *Proc. Roy. Soc. Edimburgh*, **106A** (1987), 53–61.
[3] Partial regularity for constrained minimizers of convex and quasiconvex functionals. *Rend. Sem. Mat. Univ. Polit. Torino*, Fasc. speciale (1989) 75–93.

FEDERER H.

[1] Some properties of distributions whose partial derivatives are representable by integration. *Bull. Am. Math. Soc.*, **74** (1968), 183–186.
[2] *Geometric measure theory.* Die Grund. der Math. Wiss. **153**, Springer Verlag, Heidelberg and New York, 1969.

FEDERER H. and W. H. FLEMING.

[1] Normal and integral currents. *Ann. of Math.*, **72** (1960), 458–520.

FEDERER H. and W. P. ZIEMER.

[1] The Lebesgue set of a function whose distribution derivatives are pth power summable. *Ind. Univ. Math. J.*, **22** (1972), 139–158.

FLEMING W. H.

[1] On the oriented Plateau problem. *Rend. Circ. Mat. Palermo*, **(2) 11** (1962), 69–90.

FLEMING W. H. and R. RISHEL.

[1] An integral formula for the total gradient variation. *Arch. Math.*, **11** (1960), 218–222.

FINN R.

[1] Remarks relevant to minimal surfaces and to surfaces of constant mean curvature. *J. d'Anal. Math.*, **14** (1965), 139–160.

FONSECA I.

[1] Quasi-convex integrands and lower semicontinuity in L^1. *SIAM J. Math. Anal.*, **23** (1992), 1081–1098.

[2] Lower semicontinuity of surface energies. *Proc. Roy. Soc. Edinburgh*, **120A** (1992), 99–115.

FRANCIOSI M. and G. MOSCARIELLO.

[1] Higher integrability results. *Manus. Math.*, **52** (1985), 151–170.

FREHSE J.

[1] Una generalizzazione di un controesempio di De Giorgi nella teoria delle equazioni ellittiche. *Boll. U. M. I.*, **6** (1970), 992–1002.

[2] A discontinuous solution of a mildly nonlinear system. *Math. Z.*, **134** (1973), 229–230.

[3] A note on the Hölder continuity of solutions of variational problems. *Abh. Math. Sem. Hamburg*, **43** (1975), 59–63.

[4] On two-dimensional quasi-linear elliptic systems. *Manus. Math.*, **28** (1979), 21–50.

FUBINI G.

[1] Sul principio di Dirichlet. *Rend. Circ. Mat. Palermo*, **22** (1906), 383–386.

FUCHS M. and M. WIEGNER.

[1] The regularity of minima of variational problems with graph obstacles. *Ark. der Math.*, **53** (1989), 75–81.

FUSCO N.

[1] Quasi convessità and semicontinuità per integrali multipli di ordine superiore. *Ricerche di Mat.*, **29** (1980), 307–323.

FUSCO N. and J. HUTCHINSON.

[1] $C^{1,\alpha}$ partial regularity of functions minimizing quasiconvex integrals. *Manus. Math.*, **54** (1985), 121–143.

[2] Partial regularity for minimisers of certain functionals having non-quadratic growth. *Ann. Mat. Pura Appl.*, **155** (1989), 1–24.

FUSCO N. and C. SBORDONE.

[1] Higher integrability of the gradient of minima of certain functionals. *Rend. Accad. Sci. Fis. Mat. Napoli*, **56** (1989), 23–27.

[2] Local boundedness of minimizers in a limit case. *Manus. Math.*, **69** (1990), 19–25.

[3] Some remarks on the regularity of minima of anisotropic integrals. *Comm. Part. Diff. Eq.*, **18** (1993), 153–167.

GAGLIARDO E.

[1] Caratterizzazione delle tracce sulla frontiera relative ad alcune classi di funzioni in più variabili. *Rend. Sem. Mat. Univ. Padova*, **27** (1957), 284–305.

[2] Proprietà di alcune classi di funzioni in più variabili. *Ricerche di Mat.*, **7** (1958), 102–137.

[3] Ulteriori proprietà di alcune classi di funzioni in più variabili. *Ricerche di Mat.*, **8** (1959), 25–51.

GANGBO W.

[1] On the continuity of the polyconvex, quasiconvex and rank-one-convex envelopes with respect to growth condition. *Proc. Roy. Soc. Edinburgh*, **123A** (1993), 707–729.

GAVIOLI A.

[1] Necessary and sufficient conditions for the lower semicontinuity of certain integral functionals. *Ann. Univ. Ferrara*, **34** (1988), 219–236.

GEHRING F. W.

[1] The L^p-integrability of the partial derivatives of a quasi conformal mapping. *Acta Math.*, **130** (1973), 265–277.

GIAQUINTA M.

[1] *Multiple integrals in the calculus of variations and nonlinear elliptic systems.* Annals of Math. Studies **105**, Princeton Univ. Press, 1983.
[2] Growth conditions and regularity, a counterexample. *Manus. Math.*, **59** (1987), 245–248.
[3] The problem of the regularity of minimizers. *Proc. Int. Congr. of Math.*, Berkeley 1986, Am. Math. Soc. 1987, 1072–1083.

GIAQUINTA M. and E. GIUSTI.

[1] Nonlinear elliptic systems with quadratic growth. *Manus. Math.*, **24** (1978), 323–349.
[2] On the regularity of the minima of variational integrals. *Acta Math.*, **148** (1982), 31–46.
[3] Differentiability of the minima of non-differentiable functionals. *Inv. Math.*, **72** (1983), 285–298.
[4] Quasi-minima. *Ann. Inst. H. Poincaré (Analyse non-lineaire)*, **1** (1984), 79–107.
[5] Sharp estimates for the derivatives of local minima of variational integrals. *Boll. U.M.I.*, **(6) 3-A** (1984), 239–248.
[6] The singular set of the minima of certain quadratic functionals. *Ann. Sc. Norm. Sup. Pisa*, **(IV) 11** (1984), 45–55.

GIAQUINTA M. and P. A. IVERT.

[1] Partial regularity for minima of variational integrals. *Ark. für Math.*, **25** (1987), 221–229.

GIAQUINTA M. and G. MODICA.

[1] Regularity results for some classes of higher order nonlinear elliptic systems. *J. Reine Ang. Math.*, **311/312** (1979), 145–169.

[2] Partial regularity of minimizers of quasiconvex integrals. *Ann. Inst. H. Poincaré (Analyse non-lineaire)*, **3** (1986), 185–208.
[3] Remarks on the regularity of the minimizers of certain degenerate functionals. *Manus. Math.*, **57** (1986), 55–99.

GIAQUINTA M., G. MODICA and J. SOUČEK.

[1] Functionals with linear growth in the calculus of variations I, II. *Comm. Math. Univ. Carolinae*, **20** (1979), 143–156, 157–172.
[2] Cartesian currents, weak diffeomorphisms and existence theorems in nonlinear elasticity. *Arch. Rat. Mech. and Anal.*, **106** (1989), 97–159. Erratum and addenda, **109** (1990), 385–392.

GIAQUINTA M. and J. SOUČEK.

[1] Harmonic maps into a hemisphere. *Ann. Sc. Norm. Sup. Pisa*, **(IV) 12** (1985), 81–90.

GILBARG D. and N. S. TRUDINGER.

[1] *Elliptic partial differential equations of second order*. Die Grund. der Math. Wiss. **224**, Springer Verlag, Heidelberg and New York, 1977 (second edition 1983).

GIUSTI E.

[1] Equazioni quasi ellittiche e spazi $\mathcal{L}^{p,\vartheta}(\Omega, \delta)$. (I) *Ann. Mat. Pura Appl.*, **75** (1967), 313–354, and (II) *Ann. Sc. Norm. Sup. Pisa*, **(III) 21** (1967), 353–372.
[2] Regolarità parziale delle soluzioni di sistemi ellittici quasi lineari di ordine arbitrario. *Ann. Sc. Norm. Sup. Pisa*, **(III) 23** (1969), 115–142 and **27** (1973), 161-166.
[3] Precisazione delle funzioni $H^{1,p}$ and singolarità delle soluzioni deboli di sistemi ellittici non-lineari. *Boll. U.M.I.*, **2** (1969), 71–76.
[4] Boundary behavior of non-parametric minimal surfaces. *Ind. Univ. Math. J.*, **22** (1972), 435–444.
[5] *Equazioni ellittiche del secondo ordine*. Quaderni dell'U. M. I., **6**. Pitagora, Bologna 1978 (second edition 2001).
[6] *Minimal surfaces and functions of bounded variation*. Monographs in Mathematics, **80**. Birkhäuser, Basel 1984.

GIUSTI E. and M. MIRANDA.

[1] Un esempio di soluzioni discontinue per un problema di minimo relativo ad un integrale regolare del calcolo delle variazioni. *Boll. U.M.I.*, **2** (1968), 1–8.
[2] Sulla regolarità delle soluzioni di una classe di sistemi ellittici quasi lineari. *Arch. Rat. Mech. and Anal.*, **31** (1968), 173–184.

GOLDSTINE H.

[1] *A history of the calculus of variations from the 17th century through the 19th century*. Studies in Hist. of Math. and Phys. Sci. **5**. Springer Verlag, Heidelberg and New York, 1980.

GORI M. and F. MAGGI

[1] The common root of the geometric conditions in Serrin's lower semicontinuity theorem, preprint.

GORI M. and P. MARCELLINI

[1] An extension of the Serrin's lower semicontinuity theorem. *J. Convex Analysis,* **9** (2002), 1–28.

GRANLUND S.

[1] An L^p-estimate for the gradient of extremals. *Math. Scand.*, **50** (1982), 66–72.

HAAR A.

[1] Über das Plateausche Problem. *Math. Ann.* **97** (1927), 124–258.

HADAMARD J.

[1] Sur le principe de Dirichlet. *Bull. Soc. Math. France*, **34** (1906), 135–138.

HAMBURGER C.

[1] Regularity of differential forms minimizing degenerate elliptic functionals. *Math. Inst. Univ. Bonn*, 1988.

HARNACK A.

[1] *Die Grundlagen der Theorie des logarithmiscen Potentiales und der eindeutigen Potentialfunktion in der Ebene.* Teubner, Leipzig, 1887.

HARTMANN P. and G. STAMPACCHIA.

[1] On some nonlinear elliptic differential functional equations. *Acta Math.*, **115** (1966), 271–310.

HILBERT D.

[1] Über das Dirichletsche Prinzip. *Jber. Deut. Math. Ver.*, **8** (1900), 184–188.
[2] Über das Dirichletsche Prinzip. *Math. Ann.*, **59** (1904), 161–168.

HILDEBRANDT S.

[1] *Variationsrechnung Mehrdimensionaler Integrale.* Teil 2. Univ. Bonn 1975.

HONG M. C.

[1] Existence and partial regularity in the calculus of variations. *Ann. Mat. Pura Appl.*, **149** (1987), 311–328.
[2] Some remarks on the minimizers of variational integrals with nonstandard growth conditions. *Boll. U. M. I.*, **6-A** (1992), 91–101.
[3] Partial regularity of minimizers of certain quadratic functionals with unbounded obstacles. *Ann. Mat. Pura Appl.*, **161** (1992), 113–138.

HOPF E.

[1] Zum analytischen Charakter der Lösungen regulärer zweidimensionaler Variationsprobleme. *Math. Z.*, **30** (1929), 404–413.

IOFFE A. D.

[1] On lower semicontinuity of integral functional. I. II. *SIAM J. Control Opt.*, **15** (1977), 521–538, 991–1000.

IVERT P. A.

[1] Continuity of quasiminima under the presence of irregular obstacles, *Partial Differential Equations*, Banach Center Publ., Warsaw 1984, 155–167.

JENKINS H. and J. SERRIN.

[1] Variational problems of minimal surface type. II. Boundary value problems for the minimal surface equation. *Arch. Rat. Mech. and Anal.*, **21** (1966), 321–342.

[2] The Dirichlet problem for the minimal surface equation in higher dimensions. *J. Reine Ang. Math.*, **229** (1968), 170–187.

JOHN F. and L. NIRENBERG.

[1] On functions of bounded mean oscillation. *Comm. Pure Appl. Math.*, **14** (1961), 415–426.

JOHN O., J. MALÝ and J. STARÀ.

[1] Nowhere continuous solutions to elliptic systems. *Comm. Math. Univ. Carolinae*, **30** (1989), 33–43.

JOST J. and M. MEIER.

[1] Boundary regularity for minima of certain quadratic functionals. *Math. Ann.*, **262** (1983), 549–561.

KNOPS R. J. and C. A. STUART.

[1] Quasiconvexity and uniqueness of equilibrium solutions in nonlinear elasticity. *Arch. Rat. Mech. and Anal.*, **86** (1984), 233–249.

KOŠELEV A. I. and S. I. ČELKAK.

[1] *Regularity of solutions of quasilinear elliptic systems.* Teubner-Texte zur Math. **77**, Leipzig, 1985.

KRUŽALOV O. A.

[1] Primer kvazivypukloj, no ne polivypukloj funkcii. (An example of a quasiconvex function which is not polyconvex). *Vestn. Mosk. Univ.*, (1987), (6), 3–6.

KRYLOV N. V. and M. V. SAFONOV.

[1] Certain properties of solutions of parabolic equations with measurable coefficients (English transtation). *Math. USSR, Izvestia*, **16** (1981), 151–164. Russian original in *Izv. Akad. Nauk SSSR*, **40** (1980), 161–175.

KUFNER A., O. JOHN and S. FUČIK.

[1] *Function spaces.* Academia, Praha, 1977.

LADYŽENSKAYA O. and N. URAL'CEVA.

[1] *Linear and quasilinear elliptic equations.* Academic Press, New York, 1968 (second Russian edition, Nauka 1973).

[2] Local estimates for the gradient of solutions of non-uniformly elliptic and parabolic equations. *Comm. Pure Appl. Math.*, **23** (1970), 677–703.

LANDES R.

[1] Quasilinear elliptic operators and weak solutions of the Euler equations. *Manus. Math.*, **27** (1979), 47–72.

LEBESGUE H.

[1] Sur le problème de Dirichlet. *Rend. Circ. Mat. Palermo*, **24** (1907), 371–402.

[2] Intégrale, longueur, aire. *Ann. Mat. Pura Appl.*, **7** (1902), 231–359.

LEONETTI F.

[1] Regolarità L^p dei quasi-minimi. *Boll. U. M. I.*, **(VI), C-5** (1986), 341–355.

[2] On the Hölder-continuity of ω-minima. *Ann. Univ. Ferrara*, **34** (1988), 49–58.

[3] Two-dimensional regularity of minima of variational functionals without standard growth conditions. *Ricerche di Mat.*, **38** (1989), 41–50.

[4] On the regularity of ω-minima. *Boll. U.M.I*, **B-5** (1991), 21–38.

LEVI B.

[1] Sul principio di Dirichlet. *Rend. Circ. Mat. Palermo*, **22** (1906), 293–360.

LIEBERMAN G. M.

[1] On the regularity of the minimizer of a functional with exponential growth. *Comm. Math. Univ. Carolinae*, **33** (1992), 45–49.

LIONS J. L.

[1] Théorèmes de trace et d'interpolation. *Ann. Sc. Norm. Sup. Pisa*, **(III) 13** (1959), 389–403.

[2] *Contrôle optimal de systèmes gouvernés par des équations aux dérivées partielles.* Gauthier-Villars, Paris, 1968.

LIONS J. L. and E. MAGENES.

[1] *Non-homogeneous boundary value problems and applications*. Die Grund. der Math. Wiss. **181**, Springer Verlag, Heidelberg and New York, 1972.

LEWIS J. L.

[1] Regularity of the derivatives of solutions to certain degenerate elliptic equations. *Ind. Univ. Math. J.*, **32** (1983), 849–858.

LIU X. G.

[1] Quasiconvexity and partial regularity in variational methods. *Chinese Ann. of Math.*, **(A) 11** (1990), 420–428.

MAGENES E.

[1] Spazi di interpolazione ed equazioni alle derivate parziali. *Atti del VII Congresso U. M. I.* (Genova 1963). Cremonese, Roma, 1964, 134–197.

MALÝ J.

[1] Positive quasiminima. *Comm. Math. Univ. Carolinae*, **24** (1983), 681–691.

[2] Weak lower semicontinuity of polyconvex integrals. *Proc. Roy. Soc. Edimburgh*, **123A** (1993), 681–691.

MANFREDI J. J.

[1] Regularity of the gradient for a class of nonlinear possibly degenerate elliptic equations. Ph. D. Thesis, Washington Univ., Saint Louis, 1986.

[2] Regularity for minima of functionals with p-growth. *J. Diff. Equat.*, **76** (1988), 203–212.

MARCELLINI P.

[1] Approximation of quasiconvex functions and lower semicontinuity of multiple integrals. *Manus. Math.*, **51** (1985), 1–28.

[2] On the definition and the lower semicontinuity of certain quasiconvex integrals. *Ann. Inst. H. Poincaré (Analyse non-lineaire)*, **3** (1986), 391–409.

[3] Un exemple de solution discontinue d'un problème variationel dans le cas scalaire. *Ist. Mat. "U. Dini"*, Firenze 1987–88.

[4] Regularity of minimizers of integral of the calculus of variations with non-standard growth conditions. *Arch. Rat. Mech. and Anal.*, **105** (1989), 267–284.

[5] Non-convex integrals in the calculus of variations. *Methods of non-convex analysis*, A. Cellina ed. *Lecture notes in Math.*, **1446**, Springer Verlag, Heidelberg and New York, 1990.

[6] Regularity and existence of solutions of elliptic equations with p,q-growth conditions. *J. Diff. Equat.*, **90** (1991), 1–30.

MARCELLINI P. and C. SBORDONE.

[1] Semicontinuity problems in the calculus of variations. *Nonlinear Anal.*, **4** (1980), 241–257.

[2] On the existence of minima of multiple integrals in the calculus of variations. *J. Math. Pures Appl.*, **62** (1983), 1-9.

MARCHI S.

[1] Boundary regularity for weighted quasi-minima. (I) *Riv. Mat. Univ. Parma*, **15*** (1989), 63–74. (II) *Nonlinear Anal.*, **20** (1993), 461–467.
[2] A Wiener type criterion for weighted quasiminima. *Atti Acc. Naz. Lincei*, **(IX) 2** (1991), 25–28.
[3] Hölder continuity and Harnack inequality for De Giorgi classes related to Hörmander vector fields. *Ann. Mat. Pura Appl.*, **168** (1995), 171–188.

MASCOLO E.

[1] Local minima of non-convex problems. *Nonlinear Anal.*, **14** (1990), 593–605.
[2] Existence results for a class of non-coercive polyconvex integrals. *Boll. U. M. I.*, **5-A** (1991), 97–107.

MASCOLO E. and R. SCHIANCHI.

[1] Existence theorems for non-convex problems. *J. Math. Pures Appl.*, **62** (1982), 349–359.

MASCOLO E. and G. PAPI.

[1] Local boudedness of minimizers of integrals of the calculus of variations. *Ann. Mat. Pura Appl.*, **167** (1994), 323-339.

MASSARI U. and M. MIRANDA.

[1] *Minimal surfaces of codimension one*. North-Holland, Amsterdam, 1984.

MAZ'JA V. G.

[1] Primery nereguliarnych rešenij kvazilinejnych elliptičeskich uravnenij s analitičeskimi koefficentami (Examples of non-regular solutions for quasi-linear elliptic equations with analytic coefficients). *Funk. Anal. i Ego Pril.*, **2** (1968), 53–57.

MEYERS N.

[1] An L^p-estimate for the gradient of solutions of second order elliptic divergence equations. *Ann. Sc. Norm. Sup. Pisa*, **(III) 17** (1963), 189–206.
[2] Mean oscillation over cubes and Hölder continuity. *Proc. Am. Math. Soc.*, **15** (1964), 717–721.
[3] Quasiconvexity and lower semicontinuity of multiple variational integrals of any order. *Trans. Amer. Math. Soc.*, **119** (1965), 125–149.

MEYERS N. and J. SERRIN.

[1] H = W. *Proc. Nat. Acad. Sci.*, **51** (1964), 1055–1056.

MICHAEL J. H. and W. P. ZIEMER.

[1] Interior regularity for solutions to obstacle problems. *Nonlinear Anal.*, **10** (1986), 1427–1448.

[2] Existence of solutions to obstacle problems. *Nonlinear Anal.*, **17** (1991), 45–71.

MIRANDA M.

[1] Un teorema di esistenza and unicità per il problema dell'area minima in n variabili. *Ann. Sc. Norm. Sup. Pisa*, **(III) 19** (1965), 233–249.
[2] Un principio di massimo forte per le frontiere minimali and sua applicazione alle superfici di area minima. *Rend. Sem. Mat. Padova*, **45** (1971), 355–366.

MODICA G.

[1] Quasiminima of some degenerate functionals. *Ann. Mat. Pura Appl.*, **142** (1985), 121–143.

MODICA L. and S. MORTOLA.

[1] Un esempio di Γ^--convergenza. *Boll. U. M. I.*, **14-B** (1977), 285–299.

MORREY C. B. JR.

[1] Functions of several variables and absolute continuity (II). *Duke Math. J.*, **6** (1940), 187–213.
[2] Quasiconvexity and the lower semicontinuity of multiple integrals. *Pacific J. of Math.*, **2** (1952), 25–53.
[3] *Multiple integrals in the calculus of variations*. Die Grund. der Math. Wiss. **130**, Springer Verlag, Heidelberg and New York, 1966.
[4] Partial regularity results for nonlinear elliptic systems. *J. Math and Mech.*, **17** (1968), 649–670.

MOSCARIELLO G.

[1] Regularity results for quasi-minima of functionals with non-polynomial growth. *J. Math. Anal. Appl.*, **168** (1992), 500–510.

MOSCARIELLO G. and L. NANIA.

[1] Hölder continuity of minimizers of functionals with nonstandard growth conditions. *Ricerche di Mat.*, **40** (1991), 259–273.

MOSER J.

[1] A new proof of De Giorgi's theorem concerning the regularity problem for elliptic differential equations. *Comm. Pure Appl. Math.*, **13** (1960), 457–468.
[2] On Harnack's theorem for elliptic differential equations. *Comm. Pure Appl. Math.*, **14** (1961), 577–591.

MU L. B. and LI C. Z.

[1] A general notion of quasi-convexity and weak lower semicontinuity of nonlinear functionals. *Acta Math. Sci.*, **8** (1988), 55–59.

MUCKENHOUPT B.

[1] Weighted norm inequalities for the Hardy maximal function. *Trans. Am. Math. Soc.*, **165** (1972), 207–226.

MÜLLER S.

[1] Weak continuity of determinants and nonlinear elasticity. *C. R. Acad. Sci. Paris*, **307** (1988), 501–506.

[2] On quasiconvex functions which are homogeneous of degree 1. *Ind. Univ. Math. J.*, **41** (1992), 295–301.

MURAT F.

[1] Compacité par compensation II. *Atti del Conv. "Metodi recenti in analisi non-lineare"*, De Giorgi, Magenes and Mosco ed., Pitagora, Bologna, 1979, 245-256

NANIA L.

[1] On some reverse integral inequalities. *J. Austral. Math. Soc.*, **A-49** (1990), 327–346.

NASH J.

[1] Continuity of solutions of parabolic and elliptic equations. *Amer. J. of Math.*, **80** (1958), 931–954.

NEČAS J.

[1] *Les mèthodes directes en théorie des équations élliptiques*. Akademia, Praha 1967.

[2] Example of an irregular solution to a nonlinear elliptic system with analytic coefficients and conditions for regularity. *Theory of nonlinear operators*, Abhand. Akad. der Wiss. der DDR, 1977.

[3] *Introduction to the theory of nonlinear elliptic systems.* Teubner-Texte zur Math. **52**. Leipzig 1983.

NEČAS J., O. JOHN and J. STARÁ.

[1] Counterexample to the regularity of weak solutions of elliptic systems. *Comm. Math. Univ. Carolinae*, **21** (1980), 145–154.

OLECH C.

[1] A characterization of L^1-weak lower semicontinuity of integral functionals. *Bull. Acad. Pol. Sci.*, **25** (1977), 135–142.

PALLARA D.

[1] On the lower semicontinuity of certain integral functionals. *Rend. Accad. Naz. Sci. XL*, **15** (1991), 57–70.

PEETRE J.

[1] On convolution operators leaving $\mathcal{L}^{p,\lambda}$ spaces invariant. *Ann. Mat. Pura Appl.*, **72** (1966), 295–304.

PICCININI L. C.

[1] *Proprietà di inclusione e interpolazione tra spazi di Morrey and loro generalizzazioni*. Quad. Scuola Norm. Sup. Pisa, 1969.

RADÒ T.

[1] The problem of least area and the problem of Plateau. *Math. Z.*, **32** (1930), 763–796.
[2] *On the problem of Plateau*. Springer Verlag, Berlin, 1933.

RAYMOND J. P.

[1] Existence of monimizers for vector problems without quasiconvexity conditions. *Nonlinear Anal.*, **18** (1992), 815–828.

RIEMANN B.

[1] Grundlagen für eine allgemeine Theorie der Functionen einer veränderlichen complexen Grösse. *Bernhard Riemann's gesammelte mathematische Werke*, Leipzig, B. G. Teubner, 1892.

RIESZ F.

[1] Untersuchungen über Systeme integrierbarer Funktionen. *Math. Ann.*, **69** (1910), 449–497.

ROYDEN H. L.

[1] *Real Analysis*. Mac Millan, Toronto, 1968.

SAKS S.

[1] *Theory of the integral*. Dover, New York, 1964.

SBORDONE C.

[1] Lower semicontinuity and regularity of minima of variational integrals. *Nonlinear P. D. E. and applications*, H. Brezis and J. L. Lions ed. Pitman, London, 1983.
[2] Rearrangement of functions and reverse Hölder inequalities. *Proc. of E. De Giorgi Colloquium*. Pitman, london, 1985.
[3] Quasi-minima of degenerate functionals with non-polynomial growth. *Rend. Sem. Mat. Fis. Milano*, **59** (1989), 173–184.

SCHAUDER J.

[1] Über lineare elliptische Differentialgleichungen zweiter Ordnung. *Math. Z.*, **38** (1934), 257–282.

SCHIANCHI R.

[1] Teoremi di esistenza del minimo per funzionali non-convessi del calcolo delle variazioni. *Rend. Sem. Mat. Fis. Milano*, **54** (1984), 67–77.

SCHOEN R. and K. UHLENBECK.

[1] A regularity theory for harmonic maps. *J. Diff. Geom.*, **17** (1982), 307–335.

[2] Boundary regularity and the Dirichlet problem for harmonic maps. *J. Diff. Geom.*, **18** (1983), 253–268.
[3] Regularity of minimizing harmonic maps into the sphere. *Inv. Math.*, **78** (1984), 89–100.

SCORZA DRAGONI G.

[1] Un teorema sulle funzioni continue rispetto ad una and misurabili rispetto ad un'altra variabile. *Rend. Sem. Mat. Padova*, **XVII** (1948), 102–106.

SERRIN J.

[1] On a fundamental theorem of the calculus of variations. *Acta Math.*, **102** (1959), 1–32.
[2] On the definition and properties of certain variational integrals. *Trans. Amer. Math. Soc.*, **101** (1961), 139–167.
[3] The problem of Dirichlet for quasilinear elliptic differential equations with many independent variables. *Phil. Trans. Royal Soc. London*, **264** (1969), 413–496.

SOBOLEV S. L.

[1] Ob odnoj teoreme funkcional'nogo analiza (On a theorem in functional analysis). *Mat. Sb.*, **4** (1938), 471–497.

SOUČEK J.

[1] Singular solutions to nonlinear elliptic systems. *Comm. Math. Univ. Carolinae*, **25** (1984), 273–281.

SPANNE S.

[1] Sur l'interpolation entre les éspaces $\mathcal{L}_k^{p\phi}$. *Ann. Sc. Norm. Sup. Pisa*, **(III) 20** (1966), 625–648.

STAMPACCHIA G.

[1] Contributi alla regolarizzazione delle soluzioni dei problemi al contorno per equazioni del secondo ordine ellittiche. *Ann. Sc. Norm. Sup. Pisa*, **(III) 12** (1958), 223–245.
[2] Problemi al contorno ellittici, con dati discontinui, dotati di soluzioni Hölderiane. *Ann. Mat. Pura Appl.*, **51** (1960), 1–37.
[3] The spaces $L^{p,\lambda}$, $N^{p,\lambda}$ and interpolation. *Ann. Sc. Norm. Sup. Pisa*, **(III) 25** (1965), 443–462.
[4] Le problème de Dirichlet pour les équations elliptiques du second ordre à coefficients discontinus. *Ann. Inst. Fourier*, **15** (1965), 189–258.
[5] *Équations elliptiques du second ordre à coefficients discontinus*. Les Presses de l'Université de Montreal, 1966.

STARÁ J., O. JOHN and J. MALY.

[1] Counterexample to the regularity of weak solutions of quasilinear parabolic systems. *Comm. Math. Univ. Carolinae*, **27** (1986), 123–136.

STREDULINSKY E.

[1] Higher integrability from reverse Hölder inequalities. *Ind. Univ. Math. J.*, **29** (1980), 408–417.

STROFFOLINI B.

[1] Global boundedness of solutions of anisotropic variational problems. *Boll. U. M. I.*, **5-A** (1991), 345–352.

ŠVERÁK V.

[1] Quasiconvex functions with subquadratic growth. *Proc. Roy. Soc. London*, **(A) 433** (1991), 723–725.

[2] Rank-one convexity does not imply quasiconvexity. *Proc. Roy. Soc. Edimburgh*, **120A** (1992), 185–189.

[3] New examples of quasiconvex functions. *Arch. Rat. Mech. and Anal.*, **119** (1992), 293–300.

TANG Q.

[1] Regularity of minimizers of nonisotropic integrals in the calculus of variations. *Ann. Mat. Pura Appl.*, **164** (1993), 77–87.

TAUSCH E.

[1] A class of variational problems with linear growth. *Math. Z.*, **164** (1975), 159–191.

TEMAM R.

[1] A characterization of quasiconvex functions. *Appl. Math. and Opt.*, **8** (1982), 287–291.

TERPSTRA F. J.

[1] Die Darstellung biquadratischer Formen als Summen von Quadraten mit Anwendung auf die Variations-rechnung. *Math. Ann.*, **116** (1938), 116–180.

TOLKSDORF P.

[1] Regularity for a more general class of quasilinear elliptic equations. *J. Diff. Equat.* **51** (1984), 126–150.

[2] Remarks on quasi(sub)minima. *Non-linear Anal.*, **10** (1986), 115–120.

TONELLI L.

[1] La semicontinuità nel calcolo delle variazioni. *Rend. Circ. Mat. Palermo*, **44** (1920), 167–249.

[2] *Fondamenti del calcolo delle variazioni*. Vol. 1-2. Zanichelli, Bologna, 1921–23.

TRUDINGER N. S.

[1] On Harnack type inequalities and their application to quasilinear elliptic equations. *Comm. Pure Appl. Math*, **20** (1967), 721–747.

[2] A new proof of the interior gradient bound for the minimal surface equation in n dimensions. *Proc. Nat. Acad. Sci.*, **69** (1972), 821–823.

UHLENBECK K.

[1] Harmonic maps: a direct method in the calculus of variations. *Bull. Amer. Math. Soc.*, **76** (1970), 1082–1087.
[2] Regularity for a class of nonlinear elliptic systems. *Acta Math.*, **138** (1977), 219–240.

URAL'CEVA N.

[1] Vyroždajuščiesja kvazilinejnye elliptičeskie sistemy (Degenerate quasi-linear elliptic systems). *Leningrad Otdel. Mat. Inst. Steklov*, **7** (1968), 184–222.

WIDMAN K.-O.

[1] Hölder continuity of solutions of elliptic systems. *Manus. Math.*, **5** (1971), 299–308.

WIESER W.

[1] Parabolic Q-minima and solutions to variational flow. *Manus. Math.*, **59** (1987), 63–107.

WIK I.

[1] On John and Nirenberg's theorem. *Ark. for Math.*, **28** (1990), 193–200.
[2] Reverse Hölder inequalities with constant close to 1. *Ricerche di Mat.*, **39** (1990), 151–157.

XU HAI-XIANG

[1] Regularity of minima of nondifferentiable functionals of higher order. *Acta Math. Sci.*, **10** (1990), 14–24.

ZIEMER W. P.

[1] Boundary regularity of quasi-minima. *Arch. Rat. Mech. and Anal.*, **92** (1986), 371–382.
[2] Regularity of quasi-minima and obstacle problems, *Geometric measure theory and Calculus of variations*, Symp. in Pure Math. **44** (1986), 429–439.

Index

www.ingramcontent.com/pod-product-compliance
Ingram Content Group UK Ltd.
Pitfield, Milton Keynes, MK11 3LW, UK
UKHW020401150926
8624IPUK00004B/289

9 789812 380432